Ethics, Law and Military

C000162156

Also by David Whetham

JUST WARS AND MORAL VICTORIES: Surprise, Deception and the Normative Framework of European War in the Later Middle Ages

Ethics, Law and Military Operations

Edited by
David Whetham

First published 2011 by
PALGRAVE MACMILLAN

Palgrave Macmillan in the UK is an imprint of Macmillan Publishers Limited, registered in England, company number 785998, of Houndmills, Basingstoke, Hampshire RG21 6XS.

Palgrave Macmillan in the US is a division of St Martin's Press LLC, 175 Fifth Avenue, New York, NY 10010.

Palgrave Macmillan is the global academic imprint of the above companies and has companies and representatives throughout the world.

Palgrave® and Macmillan® are registered trademarks in the United States, the United Kingdom, Europe and other countries

ISBN 978–0–230–22170–3 hardback
ISBN 978–0–230–22171–0 paperback

This book is printed on paper suitable for recycling and made from fully managed and sustained forest sources. Logging, pulping and manufacturing processes are expected to conform to the environmental regulations of the country of origin.

A catalogue record for this book is available from the British Library.

A catalog record for this book is available from the Library of Congress.

10 9 8 7 6 5 4 3 2 1
20 19 18 17 16 15 14 13 12 11

Printed and bound in Great Britain by the MPG Books Group,
Bodmin and King's Lynn

This volume is dedicated to the students from all around the world who have passed through the Joint Services Command and Staff College at Shrivenham, UK. Their dedication, professionalism, enthusiasm and humour have made it a pleasure to work with them and I have learnt at least as much from them as I have been able to teach them over the years.

Twenty-five percent of the author royalties from this volume will be donated to the International Committee of the Red Cross to help support the vital work that they do in protecting the lives and dignity of the victims of conflict around the world.

Contents

List of Abbreviations

AP	additional protocol
APC	armoured personnel carrier
Bn	Battalion
CBRN	chemical, biological, radiological or nuclear
COIN	counter-insurgency
Comd Med	medical commander
EBAO	effects-based approach to operations
ECOWAS	Economic Community of West African States
GC	Geneva Convention
GMC	General Medical Council
ICC	International Criminal Court
ICRC	International Committee of the Red Cross
IEDs	improvised explosive devices
IHL	international humanitarian law
IO	international organization
ISAF	International Security Assistance Force
ISTAR	intelligence, surveillance, target acquisition and reconnaissance
JSCSC	Joint Services Command and Staff College
MOD	Ministry of Defence
MOU	memorandum of understanding
NATO	North Atlantic Treaty Organization
NGO	non-governmental organization
PGM	precision guided munition
PLAGOs	Personnel, Legal and General Orders
POW	prisoner of war
PTSD	post-traumatic stress disorder
R2P	responsibility to protect
RCGP	Royal College of General Practitioners
ROE	rules of engagement
SACEUR	Supreme Allied Commander Europe
UNAMIR	United Nations Assistance Mission for Rwanda
USMC	US Marine Corps
WMDs	weapons of mass destruction

Notes on Contributors

Dr Ted van Baarda is Associate Professor of Military Ethics at the Netherlands Defence College, Faculty of Military Science, in The Hague, The Netherlands. He is also a guest lecturer on military ethics at the Rwandan Military Academy. He was chief editor of *Military Ethics: The Dutch Approach* (2006) and *The Moral Dimension of Asymmetrical Warfare* (2009)

Surgeon Commander Duncan Blair is a General Practitioner in the Royal Navy with an interest in Pre-Hospital Emergency Care. He is currently working within the UK Developments, Concepts and Doctrine Centre (DCDC) as a concepts and doctrine author on behalf of the Headquarters of Surgeon General's Department.

Dr David J. Lonsdale is Lecturer in Politics and International Studies at the University of Hull, UK. His main areas of research are strategic studies and military history, specializing in strategic theory and its application to historical and contemporary strategic settings. His recent publications include *Alexander the Great: Lessons in Strategy* (2007) and a contribution to a new collection of studies on Carl von Clausewitz.

Brigadier Philip McEvoy OBE was commissioned into the Army Legal Corps in 1982. In 1984 he attended the US Army Judge Advocate General's School and was appointed honorary member of the US Court of Military Review. As well as numerous staff appointments, he has served in Northern Ireland and Cyprus. He was appointed Colonel and then Brigadier Operational Law at the UK's Land Warfare Centre and in 2008 he was posted to the Army Prosecuting Service as Brigadier Prosecutions. With the formation of the Service Prosecuting Authority in January 2009 he became Deputy Director Service Prosecutions.

Dr Alastair McIntosh is Visiting Professor in the Department of Geography and Sociology, University of Strathclyde, Scotland, UK, and a Fellow of Scotland's Centre for Human Ecology. His books include *Soil and Soul* (2001) and *Hell and High Water: Climate Change, Hope and the Human Condition* (2008).

Dr Srinath Raghavan joined the Defence Studies Department of King's College London based at the Joint Services Command and Staff College in the UK Defence Academy, in April 2007. He completed his MA and PhD at the Department of War Studies, King's College London, UK. Prior to that, he spent six years as an officer in the Indian Army. He is currently a Senior Fellow for the Centre for Policy Research.

Professor Henry Shue is Senior Research Fellow at the Centre for International Studies of the Department of Politics and International Affairs, University of Oxford, UK, as well as being Professor Emeritus of International Relations at Oxford. He edited *Nuclear Deterrence and Moral Restraint* (1989) and co-edited *Preemption* (2007) and *Just and Unjust Warriors* (2008).

Dr Paolo Tripodi is Professor of Military Ethics and Director of the Ethics Branch at the US Marine Corps University Lejeune Leadership Institute, Quantico. An international scholar, Dr Tripodi has held academic positions in Italy, the UK, Chile and Sweden. In addition to numerous articles, he is the author of *The Colonial Legacy in Somalia* (1999). He trained as an infantry officer and served as a First Lieutenant with the Carabinieri, the Italian Military Police.

General Sir Peter Wall is Commander-in-Chief of UK Land Forces. He has directed operations in Iraq and Afghanistan from the MOD and PJHQ and has commanded up to divisional level.

Dr Christopher P.M. Waters is Associate Dean of the Faculty of Law, University of Windsor, Canada. An international law specialist, he has extensive field experience in the Balkan and Caucasus regions and has frequently engaged with military audiences on law of armed conflict issues.

Dr David Whetham is Senior Lecturer in Defence Studies, King's College London, based at the Joint Services Command and Staff College in the UK Defence Academy. His main research interests are focused on the ethical dimensions of warfare and the development of the laws of war. He is the author of *Just Wars and Moral Victories*: *Surprise, Deception and the Normative Framework of European War in the Later Middle Ages* (2009), and is also a practising magistrate in Swindon, Wiltshire, UK.

Introduction: Purpose, Content and Structure

DAVID WHETHAM

Purpose

Public interest in military ethics and the laws of war has been heightened in recent years, beyond that of the traditional military practitioner community. Contemporary 'wars of choice' or 'discretionary conflicts', particularly where vital national interests are not obviously at stake, appear to pose different ethical and legal challenges for democracies when compared to wars of national survival. This has led to an increasing fascination with the subject around the world. Perceived ethical and legal failures, at both the *ad bellum* level (decision to go to war) and *in bello* level (the actual conduct of war), have caused some real problems in a number of different conflicts. At the same time, the potential range of problems is widened by the varied types of activity the military can become involved with – peacekeeping and peace enforcement or humanitarian relief operations can pose different types of dilemma to 'traditional' high-intensity, state-on-state warfare, while counter-insurgency and stabilization operations introduce their own ethical and legal issues and dilemmas that need to be engaged with and resolved if campaigns are to be successful in achieving their ultimate political goals.

While there are many legal textbooks on the subject of the International Law of Armed Conflict (or International Humanitarian Law, as it is also widely referred to) and many works exploring the normative aspects of war such as the Just War Tradition and its applicability to contemporary conflict, there is nothing that combines these in a way that would be of specific help to military personnel who work at, or are expected to understand, the operational level of war. This is the level at which campaigns and major military operations are planned and sustained: 'Operational art – the skilful employment of military forces to attain strategic goals through the design, organisation, integration and conduct of campaigns or major operations – links

1

military strategy to tactics.'[1] The operational level is therefore defined by activity rather than scale and one of the few certainties of the constantly changing strategic environment is the complexity of contemporary campaigns. This complexity means that the qualities that are most valuable for those working at this level are 'intellect, independent-mindedness, scepticism and creativity' and it is to these areas, along with others, that effective staff officer education is aimed.[2]

Around the world, different military staff colleges and other senior professional military education institutions attempt to provide the appropriate tools for their incoming students – officers with seniority and, often, many years of practical field experience behind them. Staff level education should enhance the student's familiarity with the joint environment (the way the land, sea and air services must work together to achieve desired effects), but also with the multinational and multi-agency one.

The student officers will be expected to develop an advanced appreciation of history as well as engage analytically with issues of contemporary and future defence and international security, from political, economic and international relations theory, through to strategy, security structures and equipment procurement policies.[3] Invariably, part of this education will include a legal dimension and, due to the way it impacts on every area of the military profession, increasingly it is becoming recognized that the ethical dimension also needs to be addressed as an explicit subject alongside the legal studies.

At the staff level of military organization, practitioners need to understand and apply concepts in the place where the political direction and the planning of practical military action meet. The law provides a framework within which decisions should be made; however, it can rarely provide the actual answers. Judgement is always required in order to answer questions relating not only to such issues as 'Would this action be lawful in this situation?' but also 'Is it actually the *right* thing to do?' For example, balancing proportionality and discrimination issues against the imperatives of military necessity when planning a major operation requires practical decision-making that is informed by both ethical and legal factors if the appropriate and best outcome is to be achieved (and, of course, while one will invariably have the advice of a military lawyer when planning military operations, there is unlikely to be an individual there specifically charged with providing ethical guidance – it is generally up to the commander and possibly his or her staff to settle on this). For this purpose, philosophical or legal treatises that are not very solidly grounded in military reality are often meaningless to the people who actually need them – abstract metaphysics are unlikely to impress. On the other hand, works on military ethics that deal with only tactical-level considerations are also unlikely to be of much practical use to someone who must make decisions at a fundamentally different, higher level of responsibility.

Addressing issues that are common to all of the international programmes of study that can be found at this stage of military career progression, the ethics and international law stages of the UK's Advanced Command and Staff Course are designed to develop an understanding of the factors which affect the way military power is applied and the implications of this for operational-level decision-making. The intention of this work is to capture the essence of this stage of the course and provide a clearly written, accessible work that covers both the practical implications of the law and the ethical foundations that underpin it. All of the contributors are either in the military, work alongside the military in various educational establishments, or have long-standing professional engagement with this environment. Obviously, this is important as the work has been written with military practitioners in mind. It is intended that the work should cover issues that can arise across the entire spectrum of military operations, from high-intensity war fighting, through to peace support operations, peacekeeping and the provision of humanitarian aid. As such, it should be of use to countries that partake in the full spread of military operations, and those who have a more restricted or refined mandate for the deployment of their military forces. The work should also be of interest to anyone studying or engaged with military, defence or strategic issues in a broader sense.

The key questions that this work addresses are:

- What is the relationship between values, morality, ethics and law?
- In what ways do international law and ethics influence whether force will be used and the way in which it is employed?
- What are the ethical and legal challenges facing the commander at the operational level?

While no single volume that aims to be accessible can possibly hope to be comprehensive across the range of subjects involved, it is hoped that as many of the key issues as possible can be brought out and explored. If the reader requires more depth or breadth than each chapter can provide, the material here can serve as a starting point for further investigation.

Content and structure

There is a need to understand the moral issues inherent in contemporary warfare, particularly the constraining effects of law and ethics on the use of force, but also how this restraint relates to the prospect of long-term success. In the opening chapter, the editor David Whetham introduces the normative aspects of strategy, starting with the distinctions between morality, ethics and law and the way they are related. It outlines the reasons why this area of warfare should be understood and appreciated. Challenges to applying ethical considerations will be touched upon, such as ethical relativism and the moral asymmetry found in many contemporary operations where only one side appears constrained by the rules in their conduct. This chapter also introduces deontological and consequentialist reasoning and the cosmopolitan concept of essential core values, common to all societies. The chapter provides the foundation on which the subsequent chapters will build, and introduces the Just War Tradition as a set of key ideas that allow a common language for normative debates to be conducted and examines the way that ethics and law can come together to affect strategic choices at all levels, across the spectrum of conflict.

Chapters 2 and 3 (David J. Lonsdale and Alistair McIntosh) should be considered together rather than as stand-alone offerings. They introduce the reader to a dialectic between value systems by providing two very contrasting accounts of ethical reasoning regarding conflict. Each represents an aspect of (and in this sense they can only be regarded as samples rather than surveys) much broader traditions of thought, with many variations and 'flavours'. The purpose is to demonstrate how the Just War Tradition emerges from a dialogue between these two broad bodies of thought. By looking at the areas they agree on – for example, the importance of power – this allows a profitable exploration of 'left and right of arc' and provides the context within which the Just War Tradition develops. Lonsdale presents a realist perspective on the role of ethics in war, beginning with the core concepts that underpin this view. From this perspective, the chapter offers three overriding factors that should influence the conduct of war: strategy, the nature of war, and victory. In doing so, the chapter seeks to take a rational, amoral approach to the subject. However, while recognizing the rational basis for war, the chapter still acknowledges that ethical considerations cannot entirely be ignored, even for the dedicated realist.

Chapter 3 provides a direct challenge and a very different approach to the Lonsdale chapter, arguing that such realist rationality misses the deeper meaning of what human life is. While not claiming to be representing anything more than one aspect of this broad canon of thought, the chapter starts by outlining some of the types of practical nonviolence that can be practised and moves on to arguing that the vicious spiral of meeting violence with violence leads to the atrophy of the soul. This leads to a discussion on power dynamics, psychology and spiritual foundations of deep nonviolence. The chapter concludes with three short case studies of its application as a tool of security. While there are many nonviolent or pacifist approaches that could have been presented here, the direct spiritual challenge that McIntosh poses has provided a provocative perspective for students in the UK for a number of years – even if the conclusions are ultimately rejected by many, the points raised still have resonance for many serving officers.

The mainstream Just War Tradition returns properly in Chapter 4 as David Whetham explores it as a 'compromise' or synthesis – accepting, on the one hand, the view that war is indeed a terrible thing, but also recognizing that it is still sometimes to be preferred to some of the alternatives. At the same time it also acknowledges that even if it can be justified, it must be conducted within strict limits – the end goal being a better peace. The chapter charts the development of the Just War Tradition as it was confronted with new challenges and considerations, and the way that it responded to these. Moral and ethical challenges to the Just War Tradition and contemporary issues are introduced and looked at from the different ethical standpoints. Demonstrating how the tradition continues to evolve and respond to contemporary events, the chapter will look at some of the current debates within the Tradition.

Once the ethical foundations have been soundly laid, the legal stage of the book commences. In Chapter 5, Christopher P. M. Waters looks first at the nature of international law and how it relates to the ethical foundations on which it sits, but also how it relates to strategy and the operational environment. The chapter demonstrates how legal questions have pervaded questions surrounding the use of military force. Using terms familiar to the language of the Just War Tradition, the chapter explores the legal articulation of both *jus ad bellum* and *jus in bello*, covering the inherent right of self-defence, military force as authorized by the UN Security Council, and then the International Law of Armed Conflict. Recent cases of military action, from Kosovo and Iraq through to Gaza in 2008, are referred to in order to illustrate the subject matter and demonstrate its very real applicability in the practice of international relations. The question of enforcing international law and the mechanisms present to do this are also covered.

In Chapter 6, a senior military lawyer, Brigadier Philip McEvoy, explores the implications and practice of the law at the military practitioner level. The chapter demonstrates how decisions taken, and the general conduct of troops at the tactical level, can have considerable consequences at the operational, strategic and political levels. The chapter examines the impact of law at all four levels but addresses the issues from the point of view of the commander and the units, subunits and soldiers he or she commands. The chapter recognizes that the law is often regarded as a constraint on the prosecution of effective military operations and examines whether, in reality, it is primarily the law that constrains action and, if it is not, what actually does. The points raised are illustrated with a number of contemporary examples from military operations, demonstrating the impact of international law on planning at the joint operational level, the commander's responsibilities and, crucially, focusing on the International Law of Armed Conflict as an enabler rather than simply as a restriction of actions.

Chapters 7 to 12 explore a number of specific issues in the contemporary operating environment through the use of case studies and specific examples. These are written by academics and/or serving military personal who have drawn upon their own operational and professional experiences.

In Chapter 7, Henry Shue examines the relationship between protecting civilians and protecting military forces. It examines the considerations pertaining to what is and what is not a legitimate target in contemporary military operations. In the light of the Kosovo campaign and more recent military operations in Iraq, Afghanistan, the Lebanon and Gaza, what is now considered to be an effective and/or an appropriate object of attack? The whole concept of military necessity is explored and explained with reference to the way the idea has developed over time and how it is interpreted (and often misapplied) today. After exploring the key ideas of proportionality and discrimination, the chapter concludes by looking at the kinds of reasons that can legitimately inform a judgement about whether civilian losses are excessive compared to a concrete and direct military advantage.

In Chapter 8, Ted van Baarda explores the moral and ethical challenges found in the contemporary complex security environment. It highlights the types of moral confusion and dilemmas that military operations can pose for those taking part in them, particularly today. It brings out the various tensions and potential conflicts within the codes of ethical reasoning and argues that moral integrity is an essential tool for navigating the complex moral environment in which Western societies choose to place their armed forces. In the most extreme of circumstances, there is a strong chance that individuals will only uphold moral values if those values are more than a thin, politically correct layer with no real deep roots in personal conviction. What is required is that those values genuinely become, as a result of moral (self) education, a part of one's personal identity.

Paolo Tripodi offers, in Chapter 9, an analysis of atrocities that can be perpetrated by even well-trained, regular troops, and what commanders can do to prevent them happening. The chapter seeks to explain how they can happen and the role that can be played by situational forces. It analyzes how the military system interacts in a positive or negative way with situational forces and uses various cases to argue that even in those situations in which there is some kind of dispositional explanation for wrong doing, the situation itself still plays an important role in shaping behaviour. The objective of the chapter is to raise awareness among commanders about the 'hidden' risks that exist in any deployment. It makes a powerful argument that nothing could be more dangerous than for commanders to convince themselves that they and their troops are somehow immune or invincible to situational forces.

In Chapter 10, Surgeon Commander Duncan Blair asks: to whom does a military medical commander owe a moral duty? This chapter looks at the inherent conflict engendered in the dual roles of being both a doctor and officer within a country's armed forces and some of the moral and ethical dilemmas that will inevitably present themselves in the course of carrying out those duties as a direct result of this duality. It is essential that commanders understand this relationship so that they understand both what their medical staff can and cannot do, even on military deployments. Having set out the two competing concepts of the duties of the doctor and those of the officer, the chapter draws out specific moral dilemmas that may come to light within the role of the medical officer performing the task of Medical Commander on a deployed operation. Examples regarding the treatment of combatants, noncombatants and medical confidentiality are used to refine and explore the competing demands placed upon the medical officer. Blair argues that the medical commander owes a prime moral duty to the patient but that this may be abrogated, in certain circumstances, to the wider needs of the military community served. The medical officer must be ever mindful of this duality, reflect upon it and act in accordance with the imperatives of law, tempered by one's personal ethical construct.

In Chapter 11, Srinath Raghavan explores the ethics of nuclear deterrence. With the end of the Cold War, many people imagined that the nightmare of nuclear war had now passed and, with it, the need to engage with the ethical debates that surround these weapons. In light of the proliferation of states in possession of these weapons since the end of the Cold War, some would argue that the issues are actually more relevant than ever today. Given the complex alliances and coalitions that characterize many contemporary operations, one cannot remain isolated from their presence, whether or not one's own state is in possession of them. This chapter considers the broader ethical questions surrounding the strategy and policy of nuclear deterrence. Nuclear weapons are often held to have constituted a revolution in warfare, challenging and altering the traditional notions of strategy and war-fighting, but equally they have called into question our settled views about ethics and the use of armed force. The development of ethical thinking about nuclear weapons has run parallel to, and has been influenced by, strategic thinking about these weapons. The chapter therefore begins by examining the strategic idea of deterrence, particularly in the nuclear context. It then goes on to discuss the three main ethical approaches to nuclear deterrence: consequentialism, the Just War Tradition approach and deontology. The chapter suggests that nuclear deterrence presents a real moral quandary because our commonly held approaches fail to give us clear-cut ethical guidance in this realm of conflict.

To conclude the volume, General Sir Peter Wall provides an experienced senior commander's perspective, drawing together the different themes and illustrating this in practice by highlighting the real-world implications for a senior operational-level commander.

There are, of course, limitations to what can be achieved in a single volume of this kind. There are many challenges that do not have simple answers and there are also ones that do not seem to have any (palatable) answers at all. The type of reader that this book is aimed at, those who exhibit the type of analytical skills required for success at the operational level, will no doubt pick up on a number of contradictions, or at least inconsistencies, between some of the contributors in certain areas. Beyond the quite deliberate, polemical, structure of the realist/pacifist debate, this reflects the fact that no amount of editing can disguise that there are areas within the otherwise broad ethical and legal consensus that governs the use of force by the majority of states where friction and disagreement still exists. For example, McEvoy and Waters reach slightly different conclusions as to the extent and potential problems caused by the idea of 'legal encirclement' of Western military forces and also the degree to which the 2003 invasion of Iraq was legally justifiable. In part this is simply a reflection of their backgrounds – while they are both lawyers, one is a serving military officer, while the other is a professor in a civilian university's law faculty. This then could be seen as a healthy tension and here, as well as elsewhere, it would be wrong to impose a 'party line' that might create a false impression that consensus is universal in every respect. In another area, trying to balance national policy requirements with broader universal ethical principles is unlikely to result in straightforward answers, nor are questions about the limits of what one's state can expect from its servants. These are the type of questions that simply do not have straightforward answers, but they are also the type of questions that need to be engaged with, thought about and discussed by the people for whom they are most pertinent. The hope is that this work can provide part of the preparation for *jus ante bellum* – the moral, ethical and legal preparation and thought that should be engaged with *before* conflict is considered or entered into.[4]

Notes

1. *British Military Doctrine: Joint Warfare Publication 0-01* (Ministry of Defence, 2001), pp. 1–2.
2. Lt General Sir John Kiszely, 'Thinking about the Operational Level', *Journal of the Royal United Services Institute* (December 2005), p. 42.
3. David Whetham, 'The Moral, Legal and Ethical Dimensions of War at the Joint Services Command and Staff College', in Don Carrick, James Connelly and Paul Robinson (eds) *Ethics Education for Irregular Warfare* (Farnham: Ashgate, 2009).
4. George R. Lucas Jr, 'Foreword: This is Not Your Father's War', in Carrick, Connelly and Robinson, *Ethics Education for Irregular Warfare*, p. xvi.

Chapter 1

Ethics, Law and Conflict

DAVID WHETHAM

Introduction

For many people perhaps unfamiliar with the subject, the idea of 'military ethics' or 'the law of war' appears somewhat of an oxymoron. However, despite the fact that warfare takes place when the normal rules of political interaction no longer apply, it has consistently been one of the most rule-bound activities that mankind conducts. This apparent contradiction presents us with something of a conundrum; one that deserves further exploration and will therefore be the focus of this chapter.

During conflict, the law provides a framework and a context within which commanders must make their decisions. Therefore, why bother with ethics or moral concerns when the law has developed in this area and already provides us with guidance? Unfortunately, the law itself can rarely provide the actual answers. Judgement is always required in order to answer questions relating to not just whether a particular action (or even inaction) is lawful, but also if it is actually the right thing to do in that particular situation. Furthermore, for law to be legitimate, it must be based on ethical principles. Observing the letter of the law is no guarantee of doing 'the right thing', just as breaking it is not *always* the 'wrong' thing to do (although, of course, in most circumstances, it will be). For example, when Rosa Parks refused to give up her seat on the bus to a white man in Montgomery, Alabama, in 1955, she was acting contrary to the racial segregation rules in place there at the time, and was therefore arrested. However, most people today (and indeed many people then) would not wish to claim that her action was illegitimate.[1] The NATO intervention in Kosovo in 1999 provides a case study of an international action that also appeared to violate international law but was still, arguably, legitimate.[2]

There is clearly a difference between knowing the detail of the law and understanding what it means, where it comes from and why we have it. For example, one can have the personal moral idea that stealing is wrong. This may lead to the broader ethical principle (ethics being an accepted system or collection of moral principles) 'thou shalt not steal'. The law, however, cannot operate on the basis of a rule that says 'stealing is wrong' as it needs to try and cover the specifics of all of the possible circumstances that might arise, from enforcing fair contracts, buying a house or an insurance premium, through to shoplifting, major fraud or embezzlement. Clearly the three levels are and must be related, but they are also different.

Rules of engagement (ROE) are laid down by governments and command authorities as an additional factor to take into account in military operations. They cover when and where force can be used, against whom and how that force should be used to achieve the desired ends. As such, they are informed by the political intent of the military operation and act as a set of additional considerations on top of the legal ones – for example, seeking to prevent a conflict from inadvertently escalating by providing clear parameters. They often consist of a large set of standard orders pertaining to a particular environment, setting out, for example, what actions can be taken by a commander or an individual soldier on their own authority, and what type of activity may require permission from higher up the chain of command. ROE can also be supplemented or issued for specific missions or operations or summarized to set out in clear language the particular circumstances when, for example, lethal force can be employed. Good ROE assist in the synchronization of political-diplomatic and military components of strategy. ROE are not laws themselves, but should be constructed within the law – ROE cannot make an otherwise illegal action legal, and they do not in themselves carry legal weight. They are simply an additional level of prudential consideration to ensure that military force remains a political instrument (see Chapter 6).

The contemporary legal framework of conflict is heavily informed by the Just War Tradition. Chapter 4 specifically explores the elements of the Tradition and how they have developed. In the same way that the laws of war apply to armed conflicts rather than only formally declared wars, so the principles within the Just War Tradition can be applicable even when a formal war has not been declared. It reflects over two millennia of attempts to look at conflict through an ethical framework. Its huge and enduring presence means that it has influenced the very language we use to debate the rights and wrongs of conflict. For example, while a pacifist and a realist may disagree over the rights and wrongs of the use of force, they can do so using a frame of reference in which the actual concepts such as proportionality or just cause themselves are understood by both parties. The same influence can be seen in the legal sphere with terms such as *jus ad bellum* (what is required to justify going to war) and *jus in bello* (the limits on the actual conduct of war) permeating the very structure of international legal discourse. The Tradition provides us with a framework for thinking about rights and also responsibilities and duties in time of conflict. It can help determine when military actions may be justified and legitimate, and when they are not, and in this sense, performs a very important function for those who have chosen arms as their profession:

> Without ethical and legal constraints on both the decision to wage it (*jus ad bellum*) and its conduct (*jus in bello*), war is nothing more than the application of brute force, logically indistinguishable from mass murder.[3]

There are few, if any, professional militaries around the world where you would not find such a sentiment, even if it is not necessarily present in formal doctrine. Such sentiments are certainly not lost on the British military:

> Only on the basis of absolute confidence in the justice and morality of the cause, can British soldiers be expected to be prepared to give their lives for others.[4]

All military operations pose a whole host of ethical questions and dilemmas. As Michael Ignatieff put it to the cadets at the US Naval Academy, Annapolis: 'your life is one continuous set of ethical challenges'.[5] What this chapter and the rest of this book will seek to demonstrate is that a genuine understanding of normative issues lies at the heart of sound strategic thinking and operational practice.

What tools are available?

Later chapters will show how different ethical and legal tools combine to provide a framework within which complex military decisions can be taken. The Just War Tradition will be discussed in depth in Chapter 4, the international legal framework is set out in Chapter 5, while the specifics of operational law are discussed in Chapters 6 and 7. However, all of the chapters will also refer directly or indirectly to two ethical schools of thought that influence and shape ethical and legal considerations: consequentialism and deontology.

Consequentialism is often associated with its most famous version, utilitarianism. The founding fathers of this particular consequentialist approach were Jeremy Bentham and John Stuart Mill. Within this approach, an act is considered right or wrong in the context of what it achieves rather than the act itself being judged in isolation. An act is a good one if it produces the greatest balance of happiness over unhappiness. Each and every person is considered equally capable of feeling pleasure and pain, and happiness and unhappiness are the same all over the world so the approach appears to offer a universal standard by which to measure actions. This approach appears quite intuitive. If a commander can save five crew on a small submarine threatened with sinking by ordering one crew member to close a hatch on a flooding compartment from the inside, then, assuming there was no other way to prevent the loss of the vessel, this action can be justified because it maximizes overall good: one life to save five, or to put it another way, one life lost instead of all six. Few people would argue that this was not the right thing to do in the circumstances, difficult though it may be to condemn an individual to drown. However, in many cases, this level of epistemic certainty is far from present; in many cases it is almost impossible to know what action will cause more good than harm except in retrospect. It is suggested that President Truman's decision to use atomic terror attacks against Hiroshima and Nagasaki in 1945 was justified in consequentialist terms, as fewer people died as a result of these two bombs than would have in the expected conventional invasion of Japan, which was the alternative course of action required to end the war. However, historians remain divided as to whether this would actually have been the case.[6] It is often extremely hard to work out in advance exactly what the balance of good versus harm will be of the events actually happening. A second profound objection to this line of reasoning is that, in this case, the consequentialist position also asks us to effectively accept 'the murder of non-combatants in order (possibly) to save some other combatants and non-combatants'.[7] Why should those particular civilians forgo their rights and how is it possible to say that this was ethically the right thing to do?

To use an analogy, a 5 member fireteam come into contact with incoming rounds and are consequently wounded, requiring immediate blood transfusions and organ donations if they are to survive. The medic arriving on the scene realises this and, noticing an unconscious but otherwise unharmed noncombatant nearby, decides to take the necessary tissues from the innocent individual in a bid to save 5 lives at the expense of 1 (as the noncombatant will unfortunately not survive this procedure). The moral logic within the consequentialist approach is clear – the overall good will have been maximized after all. Therefore, the conscience of the medic would be clear, wouldn't it? The ends still justify the means? This example does not feel as intuitively correct, particularly if you happen to be the unconscious civilian, who doesn't even know the other people and gets no say in the matter either way.

A deontological approach to ethical thinking provides a different outcome to this type of problem. The word is derived from *deon* – the Greek word for duty – and while it has several variations, the position is often associated with its most influential advocate, Immanuel Kant.[8] Deontological ethical reasoning stresses that ends cannot be used to justify means: one must do the right thing for the right reason, regardless of what the consequences may be. It also requires a kind of impartiality – we are not allowed to make exceptions for ourselves, nor to do that which we would not permit others to do.[9] Some acts are simply wrong no matter what their consequences are and they cannot be made right just because of a favourable result further down the line or because of who we happen to be. For example, the intentional targeting of noncombatants, torture or the use of rape as an instrument of war cannot be made acceptable just because they might help you to achieve your short-term military objectives, as might well be the case with relying on a consequentialist calculation. While the consequentialist, fundamentally, is not concerned with the act itself, only with what it is likely to achieve, for the deontologist, just because good consequences might come from bad acts, that is not enough to make those bad acts anything other than bad. In the same way, an act that was motivated by a right intention but has an unfortunate and unexpected result is judged on the intention of the act rather than the results of that act.

This deontological school of thought also appears quite honest and straightforward but, like the consequentialist position, can have far-reaching and sometimes undesirable results. For example, providing medical assistance to some injured children in a village within a contested area of military operations is obviously the right thing to do: vulnerable people are in need and you are in a position to help them and alleviate their suffering. However, if you know that such acts of kindness will result in the whole village including the children being killed by insurgents as a warning to other villages in the area against fraternizing with Westerners, the right thing to do suddenly becomes far more complicated. Even if one can be assured that one is doing the right thing within this type of ethical reasoning, can the horrific consequences really be ignored? Kant himself, took the deontological idea to be an absolute principle, for example, famously arguing that it is always wrong to lie, even if a murderer was asking you for the location of their next victim.[10] However, in such cases, many people would argue that doing the 'right thing' and telling the murderer where their next victim is hiding will actually result in far more harm than doing the 'wrong thing' and lying on this occasion.[11]

Clearly, both traditions of ethical reasoning can pose some big challenges for the person on the ground who has to try and apply them to determine the correct course of action. How does one make the decision? Unfortunately, despite many attempts over the centuries, there is still no single unified system of ethics to help the individual make such decisions. Instead, one must engage in the analytical process of reaching a 'reflective equilibrium' by balancing the competing demands, values and considerations one is being faced with.[12] One must reconcile the two types of reasoning by considering the context of the situation: the deontological or absolutist position 'operates as a limitation on utilitarian reasoning, not as a substitute for it'.[13] Effectively, this decision-making process involves reviewing one's judgement in light of theory and reviewing the theory in light of the particular case. As many of the pertinent factors as possible need to be weighed and balanced against each other to determine what is the right course of action given the circumstances (this process, and how best to prepare for it, is explored in more depth in Chapter 8).

It is, of course, accepted that when faced with an immediate life or death situation, a detailed examination of different ethical or legal principles and theories may be impossible. At the operational planning level, staff officers may still need to make very fast decisions, but they are also likely to have the slight luxury of deliberation (although given some time-sensitive decisions, this cannot be taken as a given). Therefore, it is essential that those who are expected to make such decisions are familiar with the types of choices they are likely to be faced with and the type of considerations that will need to be born in mind *before being placed in such a situation*. Through reflecting on one's own experience, drawing on that of others and analyzing case studies in more benign surroundings, one does not need to start from scratch when faced with an ethical or legal challenge that demands an immediate response. To quote from Vegetius, 1,600 years ago: 'no-one fears to do that which is learnt well'.[14] The role of training and education at all levels is essential to ensure that the person on the ground is prepared and supported in the appropriate way.[15]

The Just War Tradition provides a useful framework or tool for balancing ethical considerations precisely because it represents a set of principles that have emerged over time, in part, through a dialogue between deontological and consequentialist reasoning. It has survived as a 'fund of practical moral wisdom' by evolving to reflect the changing character of war and because it contains prudential calculations that acknowledge that context must be taken into account when determining a correct course of action.[16] This is why it should be regarded as a tradition of thought rather than a discrete theory. However, while its principles can be of assistance, rules and different ethical decision-making frameworks cannot make an actual decision for you. This still requires that the commander balance the relevant factors and principles to reach a reasoned and justifiable decision.

The relativist challenge

This chapter has already suggested that the Just War Tradition has provided a common language for use in this type of discussion, no matter what one's actual views on the use of force might be. However, once the idea that there is no such thing as 'military ethics' has been put to rest, the next challenge usually comes from those who insist that it is simply a Western or Christian preoccupation and has no relevance in much of the rest of the world.

It appears self-evident that different cultures can look at the same things in different ways. Herodotus gives a well-known example of ethical relativism in book III of his *History* when he explains an anthropological experiment carried out by Darius, the King of Persia. He noted that the Greeks practised cremation, regarding the funeral pyre as a natural way to honour their dead, while the Callatians believed that for true respect to be shown, the bodies of their dead fathers had to be eaten. Darius brought the Greeks and Callatians together to discuss their mutual practices. Predictably, both parties were shocked and appalled by the other's lack of respect for their dead. This example appears to demonstrate that ideas concerning right and wrong can vary substantially from culture to culture. Because there is no independent or objective standard by which to judge an action or custom, one can only refer to one's own culture and say 'in my culture, this is right or that is wrong'. The argument is that it would be nothing short of ethical imperialism to expect other cultures to conform to our own ideas and beliefs, wouldn't it?[17] In practical military terms, in a complex security environment, troops have 'neither the time nor the resources to impose Western standards of law and order by responding to the injustices and violence they see around them'.[18] Ethical relativism states that even if it were possible, it would also be wrong to try and impose those standards where they do not necessarily belong.

However, it is easy to overstate the extent of ethical differences. For example, in some societies, attempted suicide is an act that is prosecuted and punished by the law. In others, a suicide attempt is met with sympathy and emotional or psychological support rather than condemnation and prosecution. Therefore, does this tell us that suicide is ethically wrong in one culture but not another? Actually, of course, the disagreement is not about whether suicide is right or wrong – both societies accept that it is something that should be prevented. The difference is actually in the way one goes about doing this – through a legal approach to reinforce the idea that it is wrong by enshrining it in statute, or through the provision of support in a completely different way, attempting to address why someone would feel the need to take their own life. The fundamental value of that human life is absolutely 'shared in both cultures; they just have different ways of dealing with the outcome of that shared value'.[19] The same can be said of the Greeks and the Callatians – fundamentally, the same value is shared – respect for their dead. They just have different ways of expressing it. What at first appears to be a huge difference, quickly comes to look like something rather similar after all. This should not be that surprising because it is obvious that some values *have* to be universal rather than relative for societies to survive and flourish. At its most basic, all societies value human life and, in particular, the lives of their children. Because babies and small children simply cannot look after themselves, they are incredibly vulnerable and obviously need looking after. Any society that does not accept that it was necessary to look after and protect its infants in this way will simply die out.

Clearly, some ethical standards must be shared for a culture or society to exist in the first place. There *are* certain common ethical values that cannot vary substantially from culture to culture. Even if ethical relativism at first appears intuitive, it can only be taken a very short way before it runs into certain things that are accepted as wrong wherever you are. These core values fit within John Rawl's idea of an 'overlapping consensus' – essentially areas of agreement reached by all reasonable doctrines.[20] This argument does not go as far as claiming that these values are actually universal, but simply that they consistently appear again and again, presumably for very good reasons.[21] Many things can and will be different as one goes from place to place, but all societies will have certain areas of inter-subjective agreement: 'One can presumably find in all cultures condemnations of genocide, murder, torture and slavery.'[22] That is not to say, of course, that such things do not happen. But it does mean that attempts to justify them on ethical grounds are not going to be convincing.

On the plus side, even while rejecting ethical relativism as a comprehensive position, it can still act as a useful reminder that many ethical assumptions that we take for granted may not actually be based on anything more concrete than our own society's customs and traditions. Where ethical relativism goes too far is to conclude that because *some* of our practices are simply based on local tradition, *all* of our practices must be like this. It can still prompt us though, to keep an open mind when passing judgement on other people's actions and standards of behaviour.[23] However, even while accepting that there can be legitimate differences on some levels, this understanding does not, and should not be allowed to tolerate standing by while someone commits actions that outrage the common decency of all mankind. The overlapping consensus *may* not extend far beyond certain accepted core values, but, as it is precisely those core values that we are concerned with in most military operations, this actually provides a broad and accepted foundation from which to build.

The asymmetric challenge

There is a substantial and qualitative moral difference between accidentally causing loss of innocent life and deliberately targeting noncombatants in the way that terrorists did in the East African embassy bombing of 1998, Washington and New York in 2001, Bali in 2002, Madrid in 2004 or London in 2005.[24] As this book is being written, frequent car bombs are still going off in Iraq, deliberately timed to maximize the civilian death and destruction caused as a result. As Chapter 4 notes, there has been a long-running idea within the Just War Tradition that there should be an assumption of the 'moral equality of combatants'. However, there are profound conceptual and moral difficulties in treating a 'genocidal militia ... in Rwanda, a lawless, murderous mob ... in Haiti, or shadowy, non-state actors in the hills of Tora Bora' as morally equal to internationally sanctioned peacekeepers trying to uphold laws, moral values and human decency.[25] Whatever these reservations, rules still (and must) apply even when one's opponent is certainly not one's equal. To deny this is to open the door to wars akin to the medieval crusades where the value and worth of the other side is so diminished, while one's own moral certainty is so enhanced, that anything and everything can be excused as justified by the nature of the war that is being fought. Killing such enemies comes to be seen as a positive thing, perhaps even 'understood as a means of reconciling oneself with God and thereby obtaining eternal salvation', rather than as a regrettable necessity.[26] The consequences of such attitudes are unfortunately all too easy to see in the history of war, ancient, medieval and more recent. For example, chapters 8 and 9 both refer to events at My Lai and how they came to happen, in part, due to a process of dehumanization.

However, while not going as far as to say that none of the rules should apply, there are some who have argued that the situation that we find ourselves in today is unprecedented and the rules that we have to govern our behaviour in response to such threats are unsuited to the foe that we are now faced with. In addition to deliberately targeting noncombatants, many opponents are also deliberately creating a situation of moral asymmetry: they appear unaffected by public criticism and deliberately seek to manipulate our own moral scruples and the public opinion of the wider world. In basketball this would be called 'drawing the foul', where one deliberately places oneself in a position so that the opposing player cannot avoid initiating an illegal contact, thus conceding a penalty.[27] When this is practised in war:

this tactic involves a conventionally weaker side's attempt to neutralize the technological and/or numerical advantage of its enemy by shaping the conditions of combat so that the enemy cannot act without violating the rules of *jus in bello*.[28]

In such circumstances, it seems unfair to restrain your own side when the enemy is deliberately seeking to provoke a reaction in this way. Responding to this challenge, in February 2006 the then UK Defence Secretary John Reid argued that:

> There is no curtailment of systematic violence against civilians by Al Qaeda; quite the opposite. But they and their apologists will be the first to complain and exploit isolated unlawful acts by those ranged against them. In this life and death struggle they want both of their hands free and both of our hands tied behind our back ... one could argue that our forces are fighting at a disadvantage. Yet this so-called disadvantage is often what we are fighting for. It is the rule of law and the virtue of freedom of expression versus barbarism.[29]

As Dr Reid states, it is precisely those rules that we are fighting to uphold and it is therefore essential to avoid descending to the same moral level as the insurgents and terrorists as this simply undermines what we are fighting for in the first place. While not ruling out a return to the type of large-scale industrial warfare that dominated Western military thinking in the twentieth century, it appears obvious that counter-insurgency operations of various types will be a major feature of many future military operations. Legitimacy itself is the battleground of this environment and winning the narrative of the situation is just as significant as winning any tactical engagement. The principle goal of an insurgency is 'not to defeat the armed forces, but to subvert or destroy the government's legitimacy, its ability and moral right to govern'.[30] The ubiquitous nature of the contemporary media and the speed with which information is transmitted (although not necessarily interpreted or analyzed) makes this a powerful tool to shape and influence public opinion on any number of different levels. The news media 'constitute the crux of the legitimization process, framing discourse about the just character and rightfulness of both cause and conduct of whatever action is taken'.[31] As Chapter 6 notes, if something controversial happens in the contemporary operating environment, it is only going to be a matter of time before it comes out in the global news arena. That is one of the very practical reasons why it so essential *not only to act correctly, but to be seen to be acting correctly* and in support of the wellbeing of the people that Western military forces are supposedly there to protect.[32] If it really is impossible to engage in the fight without abandoning the values that we are supposedly fighting for, then it has to be considered that it is time to leave: 'most likely, if there is no way to fight a war justly, the war itself is [or becomes] unjust'.[33]

When considering the subject of asymmetric warfare we are most familiar with referring to terrorism or guerrilla warfare employed *against* the West. However, it is also worth bearing in mind that virtually all warfare conducted *by* the West against non-Western forces is also inherently asymmetric. This huge imbalance in military capabilities and military technologies is itself providing a challenge to the existing norms of warfare possibly far more significant than the insurgent or terrorist threat outlined above. The widespread use of standoff weaponry provides a good illustration of why this is so. For example, in 2003 the Iraq military had no effective counter to the US-led coalition's cruise missiles, and beyond the rather limited 'Scud' missile, nor did they possess this type of capability themselves. This represented a significant asymmetric advantage and this has certain moral implications, both benign and more problematic. Many of the advantages of such precision standoff weaponry are obvious: more precision means fewer weapons are required to achieve the same result; a fewer number of warheads means a reduction in 'collateral damage'; because the weapons are generally more accurate, a smaller warhead can be used to achieve the desired effect, resulting in fewer civilian deaths and less damage to the infrastructure. When compared to the aerial 'carpet bombing' tactics employed during the Second World War, the advantages from a moral perspective are very clear: a single missile can replace whole squadrons of heavy bombers. Because fewer weapons are needed, where platforms are still required to deliver them, fewer of these are needed. Therefore, either fewer pilots need to put their lives at risk, or, increasingly, pilots can be removed from the loop altogether, with unmanned vehicles being remotely piloted far from the area of danger. Such weapons can therefore help to preserve life from both perspectives (and are more financially efficient into the bargain). These considerations combine to reduce the fear of the 'CNN effect', where it is believed that bad press may undermine one's cause with domestic public opinion – an essential element of wars fought by democratic states.[34] Because the threat to ordinary civilians is supposedly minimized through ever-improving precision technology, it is easier to maintain the moral high ground. At the same time, there are fewer body bags coming home to challenge public resolve. When combined, these effects can significantly lower the political cost of military action, by allowing it to be portrayed as controlled, precise, humane and efficient.[35]

Of course, in practice, any precision weapon system is only as good as the information that goes into it. For example, in 2007 a wedding party in Afghanistan was incorrectly identified as a legitimate target by a Predator's remote operator, leading to 30 civilian deaths, and such mistakes in contemporary conflicts are still proving common.[36] Even without operator mistakes being added to the equation, while the actual technology is constantly improving, the idea of 100 per cent efficacy is still far from reality and error rates can still be high. These factors, when combined, can lead to a 'paradox of precision'.[37] If and when such precision weapons do fail to hit their targets, the implications can be profound due to the raised public expectations. If the public is constantly led to believe that precise use of military force is achievable, when it fails, and innocent civilians die, people want to know why at the least, and such events can be used to demonstrate that civilians are being deliberately targeted – if the weapons are as good as we make out, obviously the things destroyed by them 'must' have been the intended target after all. This can lead to a very real issue about managing expectation and not misleading the people on whose behalf military operations are being carried out. Another potential ethical implication is that, at least in this context, 'can implies ought'. If a commander has two weapon systems available, both capable of destroying a legitimate target, one of which is likely to cause far less collateral damage than the other, the commander is morally obliged to use that weapon rather than the less focused one (this is an issue explored at some length in Henry Shue's chapter on military necessity). However, if one does not have the option of a precision guided munition (PGM), then one is not faced with the same concern; the calculation of military necessity balanced against unintended harm is more straightforward as a consequence as the choice is simply whether the attack be carried out or not rather than adding the additional question of which weapon to employ to do it. Therefore, proportionality is, to at least some extent, relative rather than absolute. This has the implication that in some ways the technologically advanced West *is* held to a higher standard, both legally and morally, than some of its potential or real opponents.[38]

This does raise a number of implications. For example, most commentators would now accept that pursuing a radical force protection policy (where a military presence seeks to remove the burden of risk as far as possible, withdrawing behind defensive positions, maximizing the use of firepower and taking as few risks as possible) can often be counterproductive in contemporary military operations. For example, refusing to patrol during the hours of darkness for fear of casualties leaves the local population effectively unprotected (as well as creating blind spots in information gathering about threats for the rest of the time). Such policies can therefore be counterproductive, cutting the soldiers off from the very people they need to engage with and protect. At the same time, by reinforcing the 'them and us' idea, the dehumanization process can work in both directions. An over-reliance on standoff weaponry risks taking this moral distance even further as it can create a perception that a 'normal' or conventional response by an opponent is simply futile as it has no chance of success. In effect, such weapons may be seen by some people to generate what Michael Walzer refers to as a 'supreme emergency' for the enemy they are employed against (see below).[39] This situation acknowledges that in such circumstances extreme measures *might* be justified. In this case, if you have a 'just cause' but cannot hurt an enemy apart from by taking the war directly to their society, terrorist attacks on civilians become easier to sell to a sympathetic audience as the weapon of the weak against the strong.

The lower political cost associated with using standoff PGMs, suggested above, clearly means that such weapons can be easier to justify and therefore employ. This can lead to the phenomenon of 'drive-by wars' where there is no real moral commitment to the struggle. The practical implications of very senior military officers or even their direct political masters being present at and directly making targeting decisions – the very long screwdriver – also provides a profound challenge to the traditional *ad bellum/in bello* distinction between the two levels of war (see Chapter 4). While some have argued that such a distinction is unsustainable anyway for other reasons, this type of technological development in war seems to provide a very practical objection to maintaining what arguably becomes an artificial separation between the two levels.[40]

Such weapons may also have implications for the very nature of victory itself. Specifically, why should the 'losing side' accept their defeat? New generations of technologically advanced weapons seem to be designed to demonstrate an ability to kill *but little or no willingness to die* for the West's causes. This appears to provide a fundamental or even existential shift in the context of the use of force. As David J. Lonsdale points out in Chapter 2, Clausewitz argues that war is a duel, a clash of wills, a moral struggle where both actors seek to overthrow their opponent. However, if one side refuses to engage on this moral level and fails to demonstrate their resolve, why should the other side accept that they have been beaten?[41] Clausewitz acknowledges that one can have a victory without fighting, but to achieve this, one must also be *willing* to shed blood to get there. Technologically advanced weapon systems that seek to isolate their operators from danger *may* actually demonstrate a fundamental lack of resolve rather than an effective coercive capability in many situations. For example, some argue that the use of unmanned drones along Pakistan's border with Afghanistan 'feed a perception that the US is a cowardly enemy, too frightened to shed blood in battle'.[42] Such accusations of cowardice can be highly damaging: 'We can't afford to be seen as people who fight from afar, who don't even dare to put a pilot in our planes.'[43] Fully autonomous military systems take this idea further by removing the person from the decision-making loop all together, but the otherwise laudable attempt to eliminate risk in one part of the equation may also actually make attaining victory harder as a result.

It may be that we have to concede that accepting an element of risk is necessary – relying too heavily upon technologies that remove our own people from the battlefield may actually prove counterproductive in this regard. While the technology can make the decision to employ force easier, at the same time it may actually be making it harder to achieve a successful outcome. Successful coercion of an opponent requires that one demonstrates resolve and 'by demonstrating a willingness to kill but not to die for a cause, that fundamental resolve can be questioned, making the use of force less credible even as it becomes increasingly sophisticated'.[44]

Can the rules be overridden?

As mentioned above, the idea of a 'supreme emergency' was developed by the hugely influential political philosopher Michael Walzer to describe 'an imminent catastrophe' to a people.[45] The phrase comes from one of Winston Churchill's wartime speeches and refers to a genuine existential threat faced by a society and suggests that the way one balances ends and means may change under certain circumstances. Thus, the threatened state or political entity may take whatever means is necessary to defend itself from such a threat, and one may even 'be required to override the rights of innocent people and shatter the war convention'. Walzer uses the example of RAF Bomber Command during the Second World War to illustrate this point, arguing that for some time it was the sole instrument available to the British people to take the war to Germany and relieve their hard-pressed Russian allies. However, to have a chance of getting through the anti-air defences, it was necessary for the bombers to fly at night. Technological limitations meant that only very large targets such as cities could be realistically targeted in the darkness, but that violated the absolute principle that one cannot deliberately target a civilian population. However, the terrible cost of defeat to the Nazi regime meant that the strategic bombing of German cities and the deliberate targeting of their residents could be justified as a necessity given the lack of alternatives available. As that extreme necessity passed, the policy became less and less justifiable: new allies joined the struggle and new theatres of potential operations opened up, long-range escort fighters were developed (making daylight flying more feasible), targeting technology improved (making it feasible to hit smaller targets) and, most importantly, the supreme emergency itself passed because Britain was no longer faced with imminent invasion, defeat and all the horrors that would have come with Nazi occupation.

On the face of it, few people are going to contest the idea that that most states will be prepared to push the ethical boundaries further in the face of a genuine existential threat, than they would be prepared to go in a discretionary 'war of choice'. However, whether or not the 'supreme emergency' idea even presents an ethically justifiable position is still debated at some considerable length and it divides ethicists.[46] Michael Ignatieff asks, as a warning: 'what lesser evils may a society commit when it believes it faces the greater evil of its own destruction?'[47] In the type of discretionary conflicts seen since the world wars (and certainly since the end of the Cold War), where continued national survival has not been an issue and perhaps even vital national interests themselves are not obviously being challenged, the contentious leeway offered by the 'supreme emergency' idea is clearly simply not applicable: 'dirty hands aren't permissible (or necessary) when anything less than the ongoingness of the community is at stake, or when the danger we face is anything less than communal death'.[48] Of course, that is not to say that the emergence of a new (or indeed the re-emergence of an old) existential threat is utterly inconceivable, but this test demonstrates a very high threshold and there are no current threats that come remotely close to it. Political rhetoric aside, it is the challenge of discretionary wars that we are routinely faced with today and in this type of conflict, there is no conceivable excuse for abandoning the rules and violating the ethical norms of war.[49]

Conclusion

General Krulak coined the phrase 'strategic corporal' to describe the potential strategic impact actions at the tactical level could now have, particularly given the nature of the contemporary media:

> The inescapable lesson of Somalia and of other recent operations, whether humanitarian assistance, peace-keeping, or traditional war fighting, is that their outcome may hinge on decisions made by small unit leaders, and by actions taken at the *lowest* level.[50]

In such environments, it is obviously not just the actions of senior commanders that can have profound affects on the success or failure of a mission. Private Lynndie England's 'smiling poses in photos of detainee abuse' at Abu Ghraib, and the political shockwaves these sent around the world, demonstrate just how important maintaining the moral high ground can be, as well as the damage that losing it can inflict on one's cause.[51] In a counter-insurgency operation, if the perception of legitimacy is lost, subsequent military action almost becomes an irrelevance as it can no longer bring victory at the level that counts – the political level.[52] If international support is lost, military action can become unsustainable for all but the most powerful of states, and even then, the political price can be enormous. If one's own domestic support drops below a certain point, it becomes impossible to justify foreign military endeavours as being in the interests of those people. In the type of military operations we have seen in recent decades, it is authority and legitimacy that are the prizes at stake, and success is likely to go to the side that best 'mobilises and energises its global, regional and local support bases'.[53]

The way that a conflict is conducted has a direct bearing on how successful its outcome is likely to be. Any military operation must contribute to the overall political and strategic end state or it cannot be considered successful in any meaningful way. It is essential to consider the second- and third-order effects before 'success' can be adequately assessed for any given action. The purpose of strategy is to match the means available to the political ends sought. If the means employed actually make achieving the desired political end state harder rather than easier, then it is clearly a bad strategy. That is why the rules of war – the legal principles and the ethical foundations on which they sit – are such an important part of military operations.

Notes

1. Rosa Parks with James Haskins, *Rosa Parks: My Story* (New York: Scholastic Inc., 1992).
2. The House of Commons Select Committee on Foreign Affairs concluded that the military operation was 'illegal but necessary'. See Patrick Wintour, *The Guardian*, 7 June 2000. http://www.guardian.co.uk/politics/2000/jun/07/balkans.politicalnews.
3. Alex J Bellamy, *Just Wars: From Cicero to Iraq* (Cambridge: Polity Press, 2006), p. 1.
4. Army Doctrine Publication, Vol. 5, *Soldiering, the Military Covenant* (2002), pp. 3–13.
5. M. Ignatieff, Address to Officer Cadets, US Naval Academy, Annapolis (21 March 2001).
6. For example, see R. Rhodes, *The Making of the Atomic Bomb* (New York: Simon and Schuster, 1995), and W. Walker, *Prompt and Utter*

Destruction: President Truman and the Use of Atomic Weapons Against Japan (Raleigh: University of North Carolina Press, 1997).

7.   Bellamy, *Just Wars*, p.142.
8.   See Immanuel Kant, *The Moral Law: Groundwork of the Metaphysics of Morals*, trans. H. J. Paton (London: Routledge, 2005).
9.   J. B. Schneewind, 'Autonomy, Obligation and Virtue: An Overview of Kant's Moral Philosophy', in Paul Guyer (ed.) *The Cambridge Companion to Kant* (Cambridge: CUP, 1993), p. 322.
10. Christine Korsgaard, 'Kant on Dealing with Evil', in James P. Sterba (ed.), *Ethics: The Big Questions* (Oxford: Blackwell, 1998).
11. Sissela Bok, *Lying: Moral Choice in Public and Private Life* (London: Vintage, 1989), p. 39.
12. See J. Rawls, *A Theory of Justice* (Cambridge, MA: Harvard University Press, 1971).
13. T. Nagel, 'War and Massacre', in *Philosophy and Public Affairs* (1972), 1, p. 128.
14. Vegetius, *Epitome of Military Science*, N. P. Milner (ed.) (Liverpool: Liverpool University Press, 2001), pp. 2 and 12.
15. See Don Carrick, James Connelly and Paul Robinson (eds) *Ethics Education for Irregular Warfare* (Farnham: Ashgate, 2009).
16. James T. Johnson, *Can Modern War be Just?* (New Haven, CT: Yale University Press, 1984), p. 15.
17. See David Whetham 'The Challenge of Ethical Relativism in a Coalition Environment', in *Journal of Military Ethics* (7[4], November 2008) where the issues of ethical relativism are discussed in more depth.
18. Bob Breen, *Struggling for Self Reliance: Four Case Studies of Australian Regional Force Projection in the Late 1980s and the 1990s* (Canberra: Australian National University, 2008), p. 13.
19. Whetham, *Ethical Relativism*, p. 306.
20. J. Rawls, *Political Liberalism*, 2nd revised edn (New York: Columbia University Press, 2005).
21. This later view does not go as far as Rawls' earlier *Theory of Justice* in this regard, but the result is that the concept is arguably more useful as a result.
22. J. Bauer and D. Bell, 'Conditions of an Unforced Consensus on Human Rights', in *The East Asian Challenge for Human Rights* (Cambridge University Press, 1999), p. 125.
23. Whetham, *Ethical Relativism*, p. 307.
24. Unfortunately, this is a sample rather than an exhaustive list.
25. George R. Lucas Jr, 'Foreword: This is Not Your Father's War', in Carrick, Connelly and Robinson, *Ethics Education for Irregular Warfare*, xiii.
26. D. S. Bachrach, *Religion and the Conduct of War c.300–1215* (Woodbridge: Boydell, 2003), p. 103.
27. David Whetham, 'Killing Within the Rules', *Small Wars and Insurgencies*, 18(4) December 2007, pp. 721–33.
28. Michael Skerker, 'Just War Criteria and the New Face of War: Human

seizure of a piece of land falls into this category. However, the relationship between the two is not always clear or direct. For example, when a state is engaged in the complex process of nation building, the role of military force is less clear.

When faced with difficult strategic cases, we must return to the notion of power in order to understand the relationship between policy and military force. Clausewitz defines war as 'an act of force to compel our enemy to do our will'.[16] This is remarkably similar to the aforementioned definition of power offered by Aron: 'the capacity of a political unit to impose its will upon other units'.[17] Either through coercion or physical control, a successfully executed war enables one actor to impose its will on others. An imbalance in power is established by the depletion, either physically and/or psychologically, of the enemy's capability and willingness to resist. Thus, the use of military force does not directly have to lead to the attainment of the policy objective, although it may. Rather, it can create the circumstances required to pursue the policy objective via the other instruments of grand strategy. So, for example, to return to the case of nation building, military force provides a secure physical environment within which development (economic, social and political) can occur. Ideally, military success presents the victor with the possibility for physical presence, and thereby control of the situation: 'The ultimate determinant in war is the man on the scene with the gun. This man is the final power in war. He is control.'[18]

Regardless of the circumstances, including those where war merely serves to shift the power balance (and therefore the relationship between war and policy is more abstract), warfare does not operate exclusively within its own logic. Factors external to immediate military concerns influence the conduct of war. The desired post-war state of affairs (the ultimate realization of the policy objective) must inform the use of force during a conflict. The costs and implications of war must be constantly assessed during a conflict, so that they do not undermine (economically, socially, or politically) the chosen strategy.

It has thus been established that military operations (even in the type of total war seen in the first half of the twentieth century) are never free from influences external to the military world. Clausewitz's abstract concept of 'absolute war' can rarely, if ever, be achieved in the real world. As a philosophical tool, Clausewitz distinguished between 'absolute' and 'real' war. While the former represents the theoretical realisation of war's unrestricted nature, Clausewitz realised that real war would be restricted by a range of factors, including friction and the influence of policy. Real war is, therefore, always restrained because it takes place in the real world, bounded by the frictions of real life such as the length of time it takes to get from A to B or for an order to be written, transmitted, understood and then followed etc. It is also restrained in part because it is a rational act: as an instrument of policy, military force must be used in a rational manner if it is to be sensical. This impingement of policy upon military activities brings with it various political concerns, including those of a moral character. War cannot operate in a moral vacuum. Human beings are moral creatures, thus at some level their actions must be influenced by ethical considerations. It is not just one's own ethical judgements that matter. The reaction of others (the international community, for example) may also be an important consideration. As Gray states:

> Although war has a grammar of its own which must be respected, insofar as is practicable, it should be waged militarily in such a manner as not to sabotage its political goals.[19]

The conduct of war: respecting the nature of war

The above discussion suggests that to some degree war may reflect the cultural values of the belligerents involved. In this sense, Anthony Coates argues that war does not have an independent nature; rather, its nature is moulded by those who practice it.[20] This is a central issue for our debate. If the nature of war is something we control, then theoretically we could instil it with a significant ethical dimension. However, it is argued here that war does have its own nature, which if not entirely independent from the belligerents, is not entirely under their control either. War may reflect our cultural preferences, but the relationship between the two may not be direct or especially significant. At the most basic level of analysis, the relationship between culture and war is complicated by the fact that there is always more than one belligerent involved. What if the different sides in a war exhibit very different cultural preferences? For example, what if they disagree on the prohibition of targeting civilians? Indeed, even if they agree on that point, they may disagree on what constitutes a civilian. In such circumstances, which cultural outlook will the nature of the war reflect? In the heated environment of war, will one cultural outlook influence the other? It may be the case that the likelihood for success will decide which approach comes to dominate. Clausewitz certainly believes that a less restrained approach will offer better prospects for victory: 'If one side uses force without compunction, undeterred by the bloodshed it involves, while the other side refrains, the first will gain the upper hand.'[21]

A telling historical example of Clausewitz's warning is the battle of Chaeronea in 338BC, as a result of which Macedonia gained hegemony over the Greek world. To use Plato's distinction (see Chapter 4), the Greek city-states facing Macedonia thought they were engaged in a conflict of 'discord' rather than 'war'. The former denotes a conflict in which the violence is limited because the belligerents are of a similar cultural mindset. Such an approach created a form of warfare in Greece that had fairly extensive rules of engagement, and was essentially quasi-ritualistic in nature. However, Macedonia was operating on the basis of 'war', which called for a much more total approach to conflict. The result of this cultural, and therefore operational, mismatch was a crushing victory for Macedonia. At Chaeronea, the Greek city-states could not impose their own culturally inspired approach to war. Indeed, Macedonia's success was so decisive precisely because it operated in tune with the true nature of war.[22]

Rather than closely reflecting our peacetime cultural values, war may serve to change certain ethical standards. After all, the taking of a human life is regarded as a legitimate action in war. This is clearly not the case in peacetime, except under the most extreme circumstances.

As noted by Clausewitz's above warning, and assuming that victory is sought, the nature of war operates largely against the impulse to regulate war for ethical reasons. By its very nature war has a rational goal, but it is also violent, uncertain and competitive. Ultimately, due to the play of friction, a war is never entirely under the strict control of any one of the belligerents involved. The role of violence in war all but guarantees that human suffering will result: 'War is a clash between major interests, which is resolved by bloodshed – that is the only way in which it differs from other conflicts.'[23] This is an important basic point to remember when attempting to mitigate suffering in war. Rather than engage in futile efforts to rid war of suffering, the latter must be anticipated and accepted. Indeed, naive attempts to humanize war may have negative consequences for one's proficiency in warfare. A strategic culture that does not accept suffering will be ill suited to deal with the harsh realities of war.

However, we must not make the mistake of regarding violence as merely incidental and regrettable in war. When we come to discuss coercion, it becomes apparent that suffering plays a crucial role in the process by which military force works towards policy ends. In this sense, rather than attempting to mitigate suffering, it must be recognized that suffering may actually be an intended outcome of violence in war: a route to the objectives sought. This implies that violence, and emotive responses to it, can be used and manipulated for strategic effect. Thus, overly enthusiastic ethical restraints to limit suffering may impede strategic performance.

Although strategy requires a controlled, rational approach to violence, the nonrational factors operating in war ensure that it is not necessarily an activity that can be tightly controlled and managed. Violence unleashes strong emotions, which can have a debilitating affect on those responsible for war. Under extreme emotional pressure, warriors will make mistakes. It is no surprise that practitioners turned theorists (such as Clausewitz and Sun Tzu) call for commanders with a cool temperament.[24] The depletion of their forces through violence may also limit a commander's available options. Thus, he may have to rely more heavily on less sophisticated expressions of violent force to succeed, perhaps even to survive. In such an environment it seems inappropriate to impose overly rigid laws and rules of engagement. A strict legal approach may jar against the nonrational aspects of war's nature. This argument is strengthened when one considers the influence of uncertainty and nonlinearity in war. Calls to impose stringent regulations on warfare are based on an unrealistic appreciation of the controllable nature of war.

At a practical level, beyond the rarified atmosphere of legal and moral debates, the complex, chaotic and uncertain nature of war has considerable implications for those responsible for command. In particular, commanders should heed Clausewitz's advice on adopting a simplified approach:

> It seems to us that this is proof enough of the superiority of the simple and direct over the complex ... rather than try to outbid the enemy with complicated schemes, one should, on the contrary, try to outdo him in simplicity.[25]

Overly complex and restrictive rules of engagement may simply add to the commander's woes. In addition, certain circumstances (depending upon the policy objectives and nature of the enemy) may demand direct, extreme and simple forms of violence. Certain strategies may not tolerate finely tuned, limited operational methods.

The last noteworthy aspect of war's nature is its competitive element. As Clausewitz recognizes, this produces an escalatory dynamic in war. Although one may wish to fight a very restrained form of warfare, the enemy may not. He may deploy increasing levels of force, demanding an increase in one's levels of violence to counter them. In addition, his will may be so strong that only significant levels of suffering will coerce him. Alternatively, perhaps he cannot be coerced at all. In which case, only his total destruction may achieve the policy objective.

Taken together, the above components of the nature of war produce significant levels of friction, which 'is the only concept that more or less corresponds to the factors that distinguish real war from war on paper'.[26] From an operational perspective, the United States Marine

Corps advises its marines to treat friction as inevitable, and therefore learn how to operate with it.[27] Lawyers and moralists would do well to heed this advice. We must all understand that war on paper, whether in a commander's plans, law books or moral treaties, is a very different beast to real war.

The conclusion from this discussion of the nature of war is that strategy has an inescapable logic. Strategy demands the application of violence in a complex, chaotic, competitive and escalatory environment. However, the violence must be used in a rational manner to serve the policy objective. Thus, war may be more or less limited, depending upon the nature of the objectives sought and the nature of the enemy. In addition, we must tolerate mistakes within such an environment, and not limit our choices or freedom of action too severely. A simple approach is preferable to one dominated by complex rules of engagement and moral philosophizing. In order to succeed, one must respect the logic of strategy.

The conduct of war: victory and coercion

Among the understandable anguish of dealing with war's inhumane nature, it must not be forgotten that victory (attainment of the policy objectives, or a military end state that promotes this) should be the aim. Too often, shortly after a conflict has begun calls for a ceasefire and a rapid exit strategy abound. There often appears to be more concern with the costs of the conflict than the achievement of victory. Victory should never be regarded as a secondary consideration. Indeed, it should be remembered that one of the criterion of Just War doctrine is that there should be reasonable prospects for success. It follows, that a war terminated prematurely on overly sensitive ethical grounds before success has been achieved, may be considered unjust. If the horrors of war are to be justified at all, one should endeavour to ensure that the effort bears some fruit.

It should also be borne in mind that victory and defeat can be poles apart. If one is engaged in a conflict with no obvious vital national interests involved (so-called 'wars of choice', such as Vietnam or Iraq) it is easy to underestimate the significance of victory. However, military defeat can serve as the prelude to enormous levels of suffering for the vanquished. The Jews of Europe, beyond those already trapped in Germany, were murdered in the holocaust precisely because their respective states were conquered by Nazi Germany. The holocaust is an extreme example that well illustrates Brian Bond's comment: 'terrible and destructive as war is, victory is usually sharply differentiated from defeat'.[28]

The ethics and war debate is not just complicated by the concept of victory. The route to victory may also be important. This is particularly the case with coercion. From an ethical standpoint, coercion is problematic because it relies upon the infliction of pain on others. For the strategist it is both an effective and efficient use of force: 'The power to hurt can be counted among the most impressive attributes of military force.'[29] If conducted well, coercion can produce significant results from relatively little effort. Nonetheless, the infliction of pain on others may trigger a significant ethical dilemma. This dilemma surrounding coercion is particularly pertinent in modern counter-insurgency (COIN) campaigns. As Colin Gray notes,

> The winning of 'hearts and minds' may be a superior approach to quelling irregulars, but ... military and police terror is swifter and can be effective. The proposition that repression never succeeds is, unfortunately, a myth. Half-hearted repression conducted by self-doubting persons of liberal conscience certainly does not work.[30]

Although legitimacy is normally an important aspect of COIN campaigns, it is a mistake to regard the latter as competitions in popularity. Rather, they can be more accurately described as competitions in authority. The enforcement of authority often requires the threat or actual infliction of punishment. During the successful British COIN campaign in Dhofar, military force was used 'to focus their [local population] minds. The population had to be convinced of the power of the government and the ability of the security forces to inflict punishment if support and assistance is extended to the rebels.'[31] One of the most highly regarded COIN campaigns was that conducted by the British in Malaya. Lessons from Malaya, such as operating within the rule of law, have been enshrined in Robert Thompson's highly regarded and influential book *Defeating Communist Insurgency*. However, it is also important to note that the British enacted some fairly harsh measures, including execution for ownership of explosives, curfews, detention without trial and the forced relocation of hundreds of thousands of locals.[32] It should also be noted how long the campaign took. Clearly, as well as being won over by a hearts and minds campaign, the population caught up in an insurgency arguably also need a degree of coercion to compel them to behave. However, it would be naive and unwise to assume that ethical concerns can be ignored. The aforementioned legitimacy can quickly be lost if the population and broader international community regard the COIN actor as unnecessarily brutal.

Strategic necessity

Thus, the modern strategist is left facing a dilemma. How does one wage war successfully, achieving one's policy objectives, while not undermining the whole project through moral outrage? The Just War Tradition provides a set of principles that seek to address this dilemma. However, there are problems when it comes to defining and applying some of the Tradition's key elements. For example, how does one adequately define proportionality or discrimination in the use of force? If military action just steps over the perceived line of proportionality does the war in question therefore become unjust and morally untenable? One could argue that the bombing campaigns against the Third Reich failed the Just War discrimination test. However, would it really be correct to thereby regard the Allied war against Nazi Germany as unjust? ROE that are vague, too complex, or inappropriate for the circumstances can add to this problem. One also has to ask the question, who should be responsible for deciding these complex moral and operational issues? Finally, within the context of a fast-moving, complex military environment, it is desirable to have lawyers and moralists peering over every operational detail?

A potential answer to the modern strategist's dilemma is the notion of 'strategic necessity', as opposed to the more limited idea of 'military necessity'. A purely rational, strategy-focused approach avoids many of the perceptual problems identified above. A strategist should not be concerned with reaching the ultimate moral truth on a matter. Rather, he should consider moral issues in a purely instrumentalist way. Rather than asking, 'Is this action right?', he should ask 'How will this action be perceived?', or 'What will the strategic consequences be of any moral judgements of our actions?' In addition to having a clearer position on moral issues, strategic necessity adopts an approach that incorporates contextual military requirements and political considerations. Strategic necessity has as its guiding principle the primacy of the policy objective. Thus, it deals with ethical and military issues on the basis of how they will affect the attainment of the policy objective.

Working on the basis of strategic necessity, violence is controlled, relative to the key needs. Unnecessary acts of violence, such as those perpetrated at Abu Ghraib, are an anathema to strategic necessity because they undermine support for the war, and thereby may negatively impact on attainment of the policy objective. Logically, the opposite must also be true, that acts of violence would be encouraged within the framework of strategic necessity, if they promote attainment of the policy objective. Clearly, the level and type of violence should be dictated by the context. Thus, within COIN or counter-terrorist campaigns the levels of violence are likely to be more limited. In contrast, within the context of a conflict akin to the Second World War (in terms of scale, objectives and nature of the belligerents), the levels of violence will be significantly higher. How much violence is required, whom or what it will be targeted against and what form it will take are all at the behest of the strategist's judgement.

Despite the inherent advantages of 'strategic necessity', it may not always provide the desired outcome (although it is more likely to achieve this than other approaches). The flawed nature of human beings suggests that the strategist's judgement will sometimes be wrong. In addition, all of the above assumes that violence can be controlled. As noted earlier, the chaos, uncertainty and passions involved in war may diminish the possible levels of control.

Realism and the refinement of war

The then Secretary of Defense, Donald Rumsfeld, and his colleagues believed they had found a technologically based route to rapid, low-cost victory when they invaded Iraq in March 2003.[33] As the ongoing conflict in Iraq demonstrates, overly optimistic, limited approaches to war (whatever the motive of such limitations) may actually lead to drawn-out, more costly wars. Those trying to operate against the nature of war, whether that be through the application of new technology or through adopting a misguided moral approach, will likely find victory much harder to achieve. Such perspectives create unrealistic expectations of what is possible in the conduct of war. In addition, unrealistic expectations of operational conduct can lead to overly harsh ROE and inappropriate legal restrictions on the commander and his forces, making victory harder and ultimately more costly to achieve.

Ironically, when considering the chances of restraining the brutality of war, the realist approach may offer a more positive outlook. It is plausible that within the modern framework of norms pertaining to human rights, strategists will have to be more cognizant of ethical issues, and therefore violence will be more controlled. Indeed, it is plausible to argue that those operating from an ideological (of whatever nature), rather than a pragmatic perspective, are more likely to use less controlled acts of violence. It is theoretically possible that a pragmatic realist approach to war will often result in less violence than some alternatives (aside from pacifism). Indeed, Christopher Coker argues that those who believe they have created a humane approach to warfare may be more likely to roll the iron dice.[34] However, it should be noted that while realism will normally seek to exert substantial levels of control (friction aside) over violence, it will also tolerate extreme levels of violence (even nuclear war) should the need arise.

Conclusion

As unpalatable as it may be, realism promotes an approach to war that is based upon how it understands war is, rather than how idealists would wish it to be. From this empirical basis, the strategist promotes a normative approach to war that is respectful of the logic of strategy and in tune with the nature of war. Of paramount importance within this logic is the idea of regarding war as a rational act, in the pursuit of the policy objective. Most often, this will result in limits on the use of force, including those that are based upon a respect for ethical issues. However, that respect is purely instrumentalist in nature. In order to work towards the policy objective, the notion of victory must be given due respect and prominence. The achievement of victory requires an approach to warfare that is motivated by the nature of war rather than ethical considerations. Indeed, it must be recognized that often that victory will be gained through the deliberate infliction (or threat) of pain on others. However, it would be naive to assume that the strategist can ignore ethical concerns. His chosen approach to war must not undermine the whole project through moral outrage. Hence, he is left in a quandary: how to achieve the policy objectives, respecting the nature of war, while being sensitive to ethical concerns. The answer may be 'strategic necessity', which is most competently described by Clausewitz:

> We can thus only say that the aims a belligerent adopts, and the resources he employs, must be governed by the particular characteristics of his own position; but they will also conform to the spirit of the age and its general character. Finally, they must always be governed by the general conclusions to be drawn from the nature of war itself.[35]

Notes

1. For a discussion of the relationship between realism and strategic studies, see John Baylis and James J. Wirtz, 'Introduction', in John Baylis *et al.* (eds) *Strategy in the Contemporary World* (Oxford: Oxford University Press, 2007), pp. 1–15.
2. Although clearly, proponents of both nonviolence and realism also argue that their perspectives are legitimate as stand-alone positions very much in opposition to the Just War Tradition as well as each other.
3. Raymond Aron, *Peace and War: A Theory of International Relations*, trans. Richard Howard and Annette Baker Fox (New York: Anchor Press/Doubleday, 1973), p. 44.
4. Morgenthau, quoted in Jack Donnelly, 'Realism', in S. Burchill *et al.*,

Theories of International Relations (Basingstoke: Palgrave Macmillan, 2005), p. 31.

5.   N. Machiavelli, *Discourses* (London: Penguin, 1998), pp. 111–12.
6.   Art and Waltz, quoted in Donnelly, *Realism*, p. 49.
7.   G. H. Snyder, 'Process Variables in Neorealist Theory', *Security Studies*, 5, 1996.
8.   Donnelly, *Realism*, p. 47.
9.   See, for example, Joel Rosenthal, *Righteous Realists: Political Realism, Responsible Power, and American Culture in the Nuclear Age* (Baton Rouge: Louisiana State University Press, 1991).
10. Quoted in B. H. Liddel Hart, *Sherman: Soldier, Realist, American* (New York: Praeger, 1958).
11. This quotation can be found in Sherman's memoirs, which can be accessed online at http://www.sonofthesouth.net/union-generals/sherman/memoirs/general-sherman-burning-atlanta.htm.
12. Sun Tzu, *The Art of War*, trans. Samuel B. Griffith (London: Oxford University Press, 1971), p. 63.
13. See Christopher Coker, *Humane Warfare* (London: Routledge, 2001).
14. Sun Tzu, *The Art of War*, p. 77.
15. Donnelly, *Realism*, pp. 36f.
16. Carl von Clausewitz, *On War*, ed. Michael Howard, Peter Paret and Bernard Brodie (Princeton, NJ: Princeton University Press, 1989), p. 75.
17. Aron, *Peace and War*, p. 44.
18. J.C. Wylie, *Military Strategy: A General Theory of Power Control* (Annapolis, MD: Naval Institute Press, 1967).
19. Colin S. Gray, *Strategy and History: Essays on Theory and Practice* (London: Routledge, 2006), p. 86. This point is supported by Machiavelli's writings, as noted in Paul R. Viotti and Mark V. Kauppi, *International Relations Theory: Realism, Pluralism, Globalism, and Beyond* (London: Allyn and Bacon, 1999), p. 60.
20. Anthony Coates, 'Culture, the Enemy and the Moral Restraint of War', in Richard Sorabji and David Rodin (eds) *The Ethics of War: Shared Problems in Different Traditions* (Farnham: Ashgate, 2006).
21. Clausewitz, *On War*, pp. 83f.
22. For a detailed description of the battle see David J. Lonsdale, *Alexander the Great: Lessons in Strategy* (Abingdon: Routledge, 2007).
23. Clausewitz, *On War*, p. 173.
24. Clausewitz, *On War*, pp. 100–12 and Sun Tzu, *The Art of War*, p. 136.
25. Clausewitz, *On War*, p. 271.
26. Clausewitz, *On War*, p. 119.
27. H. T. Hayden (ed), *Warfighting: Manoeuvre Warfare in the US Marine Corps* (London: Greenhill, 1995), p. 38.
28. Brian Bond, *The Pursuit of Victory* (Oxford: Clarendon Press, 1998), p. 1.
29. Thomas Schelling, *Arms and Influence* (New Haven: Yale University Press, 1966), p. 2.
30. Colin S. Gray, *Another Bloody Century: Future Warfare* (London: Weidenfeld & Nicolson, 2005), p. 223.

31. J. Newsinger, *British Counterinsurgency: From Palestine to Northern Ireland* (Basingstoke: Palgrave Macmillan, 2002), p. 144.
32. Newsinger, *British Counterinsurgency*, pp. 41–6.
33. Bernard Weinraub and Thom Shanker, 'War on the Cheap?', *International Herald Tribune*, 2 April 2003, pp. 1–2.
34. Coker, *Humane Warfare*, p. 150.
35. Clausewitz, *On War*, p. 718.

Chapter 3

A Nonviolent Challenge to Conflict

ALASTAIR McINTOSH

Introduction

So far, nobody has managed to rid the world of war using nonviolence. But neither have they done so using violence. Let us proceed from the basis of such mutual deficiency. We saw in the previous chapter how David Lonsdale laid out a 'realist' position on war – what he calls a 'rational, amoral approach'.[1] Whilst I would not wish to presume that this is necessarily his personal ethos, he has helpfully explained the value-free – one is tempted to say, 'valueless' – rationality of thinkers like Clausewitz who are driven by the singular premise that war is 'an act of force to compel our enemy to do our will'. It follows, says David, that military ethics should be 'purely instrumentalist in nature'. The commander's estimate therefore becomes an arid calculus to ensure that a 'chosen approach to war does not undermine the whole project through moral outrage'. Rather than asking whether an action is 'right', the realist commander should ask, 'How will this action be perceived?' Victory is all, and from that logic we are moved to the abject conclusion: 'while realism will normally seek to exert substantial levels of control ... over violence, it will also tolerate extreme levels of violence (even nuclear war) should the need arise'.

In this chapter I suggest that such rationality misses the deeper meaning of what human life is. I shall define violence as violation of the soul, including its extension into the body. Kinetic action will therefore be my main focus, but nonkinetic measures such as 'psy-ops' (psychological operations aimed at influencing a target audience's reasoning, beliefs or behaviour) must also be considered violent if they disrupt or distort fundamental human values and alignment with truth. I will start by outlining types of nonviolence and move on to arguing that the vicious spiral that violence sets up atrophies the soul. This will lead to exploring the power dynamics, psychology and spiritual foundations of nonviolence, concluding with three short case studies of its application as a tool of security.

Types of nonviolence

This chapter defines a 'pacifist' as one who recognizes that conflict is real and normal in human societies, but who seeks to process it nonviolently. This does not mean passivity. It means active, sustained nonviolence. Note the awkwardness of that word. 'Nonviolence' is more than just the opposite of violence, such as might be achieved with words like love or relationship. 'Nonviolence' implies an active challenge to the ethos of violence. And yet, there can be no self-righteous triumphalism in this. As Martin Luther King Jr said, 'I came to see the pacifist position not as sinless but as the lesser evil in the circumstances.'[2] To paraphrase Gandhi: 'All life entails violence. Our duty is to minimise it.'[3] To the philosopher, talk of 'duty' – *deon* in the Greek – implies a 'deontological' position. But Gandhi's position, and that of most committed pacifists, goes very much deeper than such willpower alone. Deontology is too arid a basis to be a singular motivating force for the committed and costly action that nonviolence can call for. I shall therefore not emphasize it here.

Another pacifist approach is the 'consequentialist' or utilitarian outlook. Here nonviolence is vindicated by its *consequences*. An example is the writing of the scholar, Gene Sharp, sometimes called 'the Clausewitz of nonviolent warfare' or 'the Machiavelli of nonviolence'. His meticulous expositions of nonviolent civilian-based defence are justified purely in secular terms as 'a pragmatic choice'. Only in passing is inner motivation hinted at. For example, in a case study of India's independence movement, just five words are accorded to Gandhi's 'philosophy or frequent religious explanations'.[4]

In my estimate, both deontology and consequentialism are vital parts of the picture, but neither on their own grip the deepest viscerals. For this we need to add a third category – spirituality. This sets our little lives into a much greater framework of meaning. Spirituality resonates with 'virtue ethics' – the philosopher's more usual third category of normative principles. But it goes deeper: for the modern philosopher, whose vision is usually limited to rationality, misses the point. The point is that virtue should not be considered an end in itself. Rather, it is the means by which a greater vision of spirituality is served.

To the secular mind spirituality is a delusion. That may be so; however, I would urge that it must be studied if we are to grasp the motivation of the world's greatest peace activists. Mairead Corrigan Maguire was the Northern Irish co-recipient of the Nobel Peace Prize.

Her family suffered intimately from sectarian violence. She explains the spiritual imperative as follows:

> Gandhi realized that the spirit of nonviolence begins within us and moves out from there. The life of active nonviolence is the fruit of an inner peace and spiritual unity already realized in us, and not the other way around ... As our hearts are disarmed by God of our inner violence, they become God's instruments for the disarmament of the world. Without this inner conversion, we run the risk of becoming embittered, disillusioned, despairing or simply burnt out, especially when our work for peace and justice appears to produce little or no result.[5]

Let me now outline four types of pacifism as it appears in practice; my emphasis hereafter is on the last of these.

- *Pacifism as cowardice.* We can dismiss further discussion of this with Gandhi's observation: 'It is better to be violent, if there is violence in our hearts, than to put on the cloak of non-violence to cover impotence ... There is hope for a violent man to become non-violent. There is no such hope for the impotent.'[6]
- *Pacifism as nuclear unilateralism.* Unilateralists often accept the necessity for conventional warfare but draw the line at weapons that are genocidal or threaten mutually assured destruction. This contingent pacifism is a variation of 'Just War' theory, explored elsewhere in this volume.
- *Pacifism as peacekeeping.* As the motto of the United States Air Force's erstwhile Strategic Air Command had it, 'Peace is our Profession'. In my experience, this principle can be a bridgehead between principled soldiers and nonpassive pacifists. Both are committed to confronting violence. Both refreshingly understand the need to engage with power. Peace as the end is the same. What differs is the means of achieving it. But the means can matter greatly. As Tacitus reported of the Roman conquest of Britain: 'they make a desolation and they call it peace'.[7]
- *Pacifism as Nonviolence.* If I might express this from a personal standpoint, since there are many variants: we all have the moral right to kill proportionately in self-defence. This is the right of 'just war'. The conscientious objector renounces that right. While both the soldier and the pacifist share in common a willingness to die for their values, *the pacifist refuses to kill for them.* If necessary the pacifist accepts the path of suffering and death. Superficially this may appear ineffectual. In practice, it sometimes yields a tremendous hidden power to transform conflict.

The spiral of violence

Violence starts with small and very normal beginnings. As a boy, I lost fights until my mid-teens. But one day a school bully positioned himself behind me in the music class. He set about making percussive jabs to my back. Half way through the teacher left the room and a berserker spirit set loose inside me. I turned round, calmly took my astonished adversary by his collar and tie, and laid in incessantly, pulverizing his face. It was exhilarating, and I became the talk of the school. The bully and his deflated chums henceforth restricted their ministrations to other kids.

If violence can be so effective, pleasurable and even heroic, why dampen its powder by raising the spectre of nonviolence? The problem is that my story is not so stitched up as might first appear. For sure, I had made my own little world safer. In a small way I had become drunk on the ecstasy of destruction.[8] But it did nothing to address the roots of violence, the continued bullying of other children, or what happens when retributive violence remains normalized and even spreads infectiously when we are grown up.[9] Violence in the adult world perpetuates itself through 'the myth of redemptive violence' – the belief that greater violence is a legitimate and effective way of resolving lesser violence.[10] Although it might be hoped that fire can extinguish fire like an explosive charge pitched by Red Adair at the base of a blazing oil wellhead, more commonly violence on violence simply fuels an escalating spiral. The expression 'spiral of violence' came from Dom Hélder Câmara, a Catholic archbishop who spent his life among the poor and downtrodden of Brazil. He observed that the *primary violence* of social injustice (or 'structural evil') leads to the *secondary violence* of revolt by the afflicted. That precipitates the *tertiary violence* of retaliation and repression by the powerful whose interests are threatened. The additional stress on a society's socio-economic framework perpetuates more primary violence.[11] Israel–Palestine is one of many cases in point.

The challenge for modern humankind is that war has advanced faster than our cultural evolutionary ability to fully absorb its moral implications. Consider Winston Churchill. As a young officer in 1898, amid much initial derring-do, he galloped into the Battle of Omdurman with pistol in hand and sabre by his side. It was one of the last cavalry charges of the British Army, aimed at putting down the upstart Mahdist Islamic state against a cultural backdrop where the Ottoman Arabs had been slavers at the expense of the indigenous Sudanese. As Churchill described it in the first paragraph of *The River Wars*, the Nile was the Sudan's 'only channel of progress' along which 'European civilisation can penetrate the inner darkness'. To British eyes, recolonization was noblesse oblige. The 50,000-strong enemy of 'Dervish skirmishers' defending Omdurman bore only light arms and flags inscribed with verses from the Qu'ran. To Churchill they were 'like the old representations of the Crusaders in the Bayeux tapestry'.[12] As Kitchener's forces turned on their Maxim guns and the cavalry charged, 'each man *saw the world* along his lance ... or through the back-sight of his pistol' (my emphasis). Meanwhile, out on the Nile, 'the terrible machine, floating gracefully on the waters – a beautiful white devil – wreathed itself in smoke'.

Caught between the shock and awe of fusillade and gunboat diplomacy, 'the darker side of war' took hold. 'Bullets were shearing through flesh, smashing and splintering bone; blood spouted from terrible wounds; valiant men were struggling on through a hell of whistling metal, exploding shells, and spurting dust – suffering, despairing, dying.' Churchill could not hide his empathy with the enemy. 'It seemed an unfair advantage to strike thus cruelly when they could not reply,' but defeat for 'these brave men' was now merely 'a matter of machinery'. With some 10,000 'Arabs' dead to just 48 on the British side – a ratio of 200:1 – Churchill concluded:

> Thus ended the battle of Omdurman – the most signal triumph ever gained by the arms of science over barbarians. Within the space of five hours the strongest and best-armed savage army yet arrayed against a modern European Power had been destroyed and dispersed, with hardly any difficulty, comparatively small risk, and insignificant loss to the victors.

But European civilisation wasn't done yet. Presented with film footage of the carpet bombing of the Ruhr in 1943, Churchill asked: 'Are we beasts? Are we taking this too far?'[13] Two years later nuclear weapons decimated civilian-packed Japanese cities. War's spiral had been to the rhythm of technoculture, not human culture. Seen down a lance or the back-sight of a pistol it always looked justified. But seen through a broader lens, Gandhi, when asked what he thought of 'European civilization', had to say, 'I think it would be a good idea.'

Today the spiral of violence has ratcheted further. Radical Muslims trace much of their ongoing angst to the secondary effects of primary colonial violence and what they see as the West's continued tertiary attempts to repress it.[14] In the West we forget that decolonization is less than a lifetime old. It suits us to be short-sighted to the fact that the Islamic world just happens to be the neighbour we most deeply colonized. And the stakes escalate. When addressing a summit on nuclear proliferation in November 2004, Mohamed ElBaradei, then head of the UN's International Atomic Energy Agency, said that there had been

630 confirmed incidents of trafficking nuclear and radioactive materials since 1993. He warned:

> We need to do all we can to work on the new phenomenon called nuclear terrorism, which was sprung on us after 9/11 when we realised terrorists had become more sophisticated and had shown an interest in nuclear and radioactive material ... We have a race against time because this was something we were not prepared for.[15]

It is this kind of development that makes violence a *spiral* and not just a circle. The ground qualitatively shifts so that today, as General Sir Rupert Smith puts it: 'Our opponents are formless, and their leaders and operatives are outside the structures in which we order the world and society.'[16] Like at Omdurman, asymmetrical warfare is at play, but then the boot was firmly on the West's foot. Now that is less sure. On the one hand, the West still maintains its 'beautiful white devil' or rather, her gunboat successors lurking nuclear depths. On the other hand, the field is no longer one of pitched battles tilted to the mechanical advantage of industrial warfare. Instead it has sublimated to what Smith calls 'war amongst the people'.

Violence can now exert a globalized leverage that exceeds confident military capacity to ensure deterrence and containment. Suicide bombers or lightly armed assassins can terrorize civilian life. A pleasure craft sailed up the Hudson with a primitive nuclear device, a civilian airliner targeted into a Trident submarine docked on the Clyde, or even a fertilizer bomb on a coastal dyke as rising sea levels from climate change kick in could pluck the heart from densely populated homelands. As the IRA used to say: 'We only have to get lucky once. You have to be lucky all the time.'

The Bomb is therefore our generation's basic call to consciousness. For the first time in history we have at our finger tips utter destructive power, but matched to it, all the possibilities for greater understanding opened up by globalized communications. Now is the time to press the reset button at many levels of depth. To borrow Churchill's expression, it is time collectively to address our 'inner darkness'. This is not terrain for the comfortably complacent, for as Conrad famously wrote within a year of Omdurman: 'We penetrated deeper and deeper into the heart of darkness. It was very quiet there.'[17]

Atrophy of the soul

Let us try and understand that uncanny quietness at Conrad's vortex;

that macabre sense of the familiar having been rendered foreign, of home becoming 'un-homed' in the German sense of *unheimlich* that characterizes 'trauma' – psychic injury – in all its 'mindlessness'; all its 'senselessness'.[18] In March 2009 Susan Tsvangirai, wife of Zimbabwe's Prime Minister, died in a road accident. There was no suspicion of foul play. But the BBC reported Morgan Tsvangirai's closest aide, Finance Minister Tendai Biti, saying at the funeral: 'We're so traumatised, brutalised, we couldn't feel the pain. Why, why, why?'[19]

A lack of reporting context left it unclear whether Biti was referring to the specific trauma of the accident, or to the wider brutalization of Zimbabwe under Mugabe. Whichever was in his mind, both merged to one in the world's media. The statement's depth needed no explanation. For violence unhealed destroys the capacity to feel. Psychic numbing whether from childhood or later traumatic stress disables empathy and with it, the capacity to love and be loved. As poet Alice Walker says: 'Tears left unshed/ turn to poison/ in the ducts/ Ask the next soldier you see/ enjoying a massacre/ if this is not so.'[20]

We need to realize that *violence is violation*. The French word-origin is clear: *le viol* means rape. Newton's third law of motion is similarly lucid: 'To every action there is an equal and opposite reaction.' And that's the trouble with violence. Yes, we can put on Tolkien's 'Ring of Power' and often get away with violence on a short temporal wavelength. It can appear effective in our archetypal battles against the Dark Lord. But on the wider horizon, violence ricochets around the echo chambers of the soul; like an addictive drug, it atrophies the soul.

In fourteen years of guest lecturing at British and overseas staff colleges I've been given the privilege of addressing more than 5,000 soldiers. The vast majority are people of undoubted courage, integrity, selflessness and remarkable depth of community spirit. Their ethos commands my admiration to an extent that sweeps aside sense of paradox. One makes friends in such circumstances, and I observe that some are not immune to the consequences of Newton's third law. Observers might call it borderline post-traumatic stress disorder (PTSD), but I should prefer to call it sentient humanity. Said one soldier, freshly home from Afghanistan: 'I feel, unclean.' And another, back from routing Saddam's conscripts in the Euphratean marshes: 'What did it feel like to have killed?' I impetuously asked. 'I'll tell you, Alastair,' he said, lowering his voice, this eminently decent man. 'I notice three things. I sleep less well than I used to. I get more irritable. And physically, I feel the cold more.' I could not help thinking that in Dante's *Inferno* Satan languishes in the ninth circle of Hell – not warmed, but frozen.[21] That is the still vortex of the spiral of violence. 'It was very quiet there.' Chillingly *unheimlich*.

In making such an observation there can be no room for finger pointing. We are all complicit, even if unconsciously so. Every time I press my foot on the car's accelerator, I too am complicit with oil that was fought for. And let us be careful how we judge others in whose moccasins we have not walked. Looking back on the twentieth century, we might say that we 'won' the First and Second World Wars, and that this proved the redemptive power of violence. Most certainly, nobody can deny the heroism of those darkest hours. But can 'we' really be so cocksure of our virtue if the lines of sight are widened beyond lance or pistol sights? Neither of the two world wars can be separated from the underlying cut and thrust of European imperialism. Germany's quest for *lebensraum* was unexceptional in the wider scheme of things. The root of its transgression was to foul the European nest by extrapolating from Maxim guns that had dispatched 'fuzzie wuzzies' at the rate of 600 rounds a minute.[22] And what might have happened if, from Versailles onwards, the international community had applied its efforts to take away, instead of aggravating, the causes that inflated Hitler's psychopathology into the psyche of his nation?

That is the trouble with the sword. Spiritually it truncates our enemy's possibility of redemption in this world and, politically, it lobotomizes other foreign policy options. We use violence with insufficient understanding of karmic retribution – the principle that 'what goes around, comes around'. Britain's willingness to throw its weight about in Iraq and Afghanistan therefore cannot be disaggregated from our triumphalistic history. Neither can it be so from domestic security consequences for our future.

The spectrum of socially expressed power

I am aware that the military response to what I am saying might be, 'Yes ... but ... These are thought-provoking points, but they don't address the moral imperative of peacekeeping in the "real world".' In this I am forced to concede much to my detractors. But if the court-martial might permit a short stay of execution, let me make this appeal. Power is socially expressed along a broad spectrum.[23] It progresses from:

- the hard sanction of military coercion, to
- the soft sanction of nonlethal policing, to
- the persuasive power of psychological convincement, to
- the spiritual power of 'metanoia' – which is inner transformation.

Only one foot belongs to the so-called 'real world'. The other belongs to the spiritually real world. Ought we not, then, walk with both feet, and play the full spectrum of violence or nonviolence as befits circumstances? The problem is the spectrum is asymmetric. Violence can always hope that nonviolence will tidy up its mess. But for nonviolence to sanction even 'surgical' violence would poison its inner integrity.

Let us recall our basic definition: 'violence is violation of the soul including its extension into the body'. Killing is a very ultimate action. It is not impossible to imagine how it could be justified within our definition; mercy killing would be a case in point. But generally, for the military peacekeeper or the pacifist alike the decision to kill or not to kill is the supreme conundrum. I can only answer as did the early Quaker, George Fox. When William Penn, the founder of Pennsylvania, asked whether he should continue to wear a sword, Fox replied: 'I advise thee to wear it as long as thou canst.' In other words, it is better to be prepared to fight than to renounce fighting before one is ready to live with the consequences of so doing. Later, Fox met Penn and saw that he was unarmed. 'William, where is thy sword?' Said he: 'Oh! I have taken thy advice; I wore it as long as I could.'[24] Of course, there could have been a cutting military riposte to the disarmed Penn. Namely: 'I advise thee to *unwear* it as long as thou canst'! At a 'real-world' level, none of us have got adequate answers. We must therefore press our enquiry deeper.

Spiritual dynamics of nonviolence

The asymmetry between violence and nonviolence derives from the observation that, far from being a passive lack of violence, nonviolence is active spiritual force. For Gandhi, *ahimsa*, or nonviolence, was driven by what he called *Satyagraha* – variously translated from the Sanskrit as 'truth force', 'love force' or 'soul force'.[25] He said, 'The badge of the violent is his weapon, spear, sword or rifle. God is the shield of the nonviolent.' This does not mean that the nonviolent will be physically shielded from dying. Many have died using nonviolence (though perhaps not so many as have died using violence). What it does mean is that with nonviolence we are shielded from spiritual death and even, perhaps, armed with spiritual power.[26] This is what gives nonviolence its oft-remarked out-of-the-blue dynamic that can transfigure conflict in unique and unpredictable ways.

Spiritual transformative capacity shows in the bearing or presence of a person. It is authored authority. It comes from a progressively deeper grounding in that level of being which includes, but utterly transcends, ego consciousness. The conscious 'I', the ego in our field of normal consciousness, may be considered as only the tip of who we are as people.[27] For the spiritually aware the deepest level of being – the soul – is always rooted in that of God within. This interconnects to our fellow humankind. As Hassidic Jewish mysticism teaches, *God is relationship*.[28] As biblical Christian mysticism teaches, we are all branches of the 'True Vine' of life, 'participants of the divine nature' and therefore, 'members one of another' – because 'God is love.'[29] Similar metaphysics unite the mystical traditions of all great faiths.

What obscures this from being self-evident is the narcissism of ego-centricity where we deny our own psychological 'shadow' – our ego's alter-ego. This comprises all that we repress, all that awaits resolution, all that has not yet flowered into maturity. As the late Adam Curle, a wartime army officer turned veteran mediator in Biafra and elsewhere has explained:

> we displace the guilt from which we all suffer to some degree, onto the enemy. In the case of leaders, the guilt we commonly feel for the inadequacy of our lives, the repressed conflicts of infancy and veiled fear that we are denying the truth of our being, is supplemented by a more rational guilt for the misery and slaughter they are causing. For them to accept all this as 'my fault' would be too much for the already sensitive 'I' to bear. But luckily it can be legitimately projected outwards onto the foe: it is he who is to blame. They only did, and reluctantly, what was necessary to defend *their* innocent people from *his* brutal and unjustified aggression.[30]

This process of compartmentalizing, splitting off and projecting the shadow psychologically drives the demonization of the enemy. The enemy may indeed be very real, but we must guard against also needing him to be so for our own virtuous self-definition. To do this dooms us to perpetual conflict, for example, psychologically transmogrifying IRA into IRA-q or IRA-n. On this basis it was unsurprising that the perceived threat of militant Islam rose in proportion to the Berlin Wall's fall. Similarly, the politics of 'good state; bad state' is always the projection of a conflicted, compartmentalized mind.

The only way out is to ground both our ego and our shadow selves more and more into that of 'God within'. Such spirituality means facing the shadow and wrestling, at least metaphorically, with both our own demons and those of others. The aim of nonviolence is to call back power that is 'fallen' to its higher, God-given vocation. Theologically, all power is God-given and should be redeemed, not destroyed.[31] Conflict resolution requires commitment to such a difficult but life-giving journey. To varying degrees the word's great faiths testify to this.

War, religion and nonviolence

Within Islam, the central Qu'ranic text pertaining to war is Surah 2:190: 'Fight in the cause of God/ Those who fight you,/ But do not transgress the limits;/ For God loveth not transgressors.'[32] This is pure Just War theory, and yet, the *Hadiths* – the authoritative oral traditions of the Prophet (peace be upon him) – go further and make explicit the limits. These include: not to kill women and children, POWs to be treated humanely, no one should be killed by burning and not to mutilate the dead.[33] As Philip Stewart points out:

> If the Islamic rules were followed today, much of modern warfare would be impossible, and terrorism would be unthinkable. There would be no attacks on civilians, no retaliation against innocent parties, no taking hostage of non-combatants, no incendiary devices.[34]

Strictly speaking, then, the problem with 'Islamic terrorism', like with its Christian equivalents, is not fundamentalism. The problem is that the terrorists are not fundamentalist enough. They appear unaware of Islam's considerable canon of witness and theology affirming nonviolence or highly proscribed violence.[35]

The Judeo-Christian tradition begins with the cultural context of Hebrew 'just war' teaching. The morality of this evolves historically. The rules of war laid down in Deuteronomy 20–1 are draconian, sanctioning slavery of the vanquished, the taking of women as booty and absolute genocide. But many commentators interpret Moses' 'an eye for an eye ...' as an injunction that aims to limit retribution.[36] Later Jewish prophets look to a world beyond war. For Isaiah and Micah: 'In days to come ... they shall beat their swords into ploughshares, and their spears into pruning hooks; nation shall not lift up sword against nation, neither shall they learn war any more.'[37] Jesus follows by totally repudiating violence.[38] Jesus does not teach Just War theory; he teaches nonviolence. It includes nonviolent direct action as when he turned over the moneychangers' tables that violated the temple – making a whip for use not against people (as is often misinterpreted), but to drive out 'both the sheep and the cattle' – thereby rescuing them from sacrifice.[39] Christian pacifists who break into military bases and hammer nuclear submarines and jet fighters tread this 'ploughshares' path. They do not run away like terrorists would, but await arrest to take further stands of witness during their trial.[40]

Jesus told his followers to love their enemies, to pray for (or do good towards) those who mistreat them and to turn the other cheek when struck.[41] He said: '... *until now* the kingdom of Heaven has suffered violence, and the violent take it by force'.[42] In other words, a new ful-filment of the law is to take ascendancy henceforth. When the brothers James and John – *Boanerges*, or the 'Sons of Thunder' as they were called – asked Jesus to draw down 'fire from Heaven' to burn up their enemies he refused, and rebuked them.[43] His dalliance with the sword was explicitly symbolic, serving only to fulfil prophesy.[44] Jesus there-fore told Peter: 'Put your sword away. For all they that take the sword shall perish with the sword ... No more of this!'[45] He also healed the severed ear of Malchus, the high priest's official, thereby symbolically restoring the enemy's capacity to listen – the prerequisite for peace.[46] He repudiated violence by absorbing suffering into his God-centred being, telling his would-be Master at Arms: 'Shall I refuse to drink the cup of sorrow which the Father has given me to drink?'[47] And later telling Pilate, 'My kingdom does not belong to this world; if my kingdom belonged to this world, my followers would fight to keep me from being handed over.'[48]

In these ways love transcends mere utility and conquers evil and death. As the Indian-Spanish Hindu-Christian theologian, Raimon Panikkar, reminds us: 'Peace is participation in the harmony of the rhythm of Being ... Only forgiveness, reconciliation, ongoing dialogue, leads to Peace, and breaks the law of karma.'[49] The Cross thereby stands as a cosmological symbol of nonviolence.[50]

Hinduism can deepen our understanding of this cosmology. The opening line of its most sacred text, the Bhagavad Gita, commences: 'On the field of Truth, on the battlefield of life, what came to pass, Sanjaya ... ?'[51] What comes to pass in everyday life is here portrayed as being situated on the wider battlefield of life, and that, in turn, is situ-ated on the field of cosmic truth – the *Dharma*. The Christian equiva-lent to *Dharma* would be the Grace of the Holy Spirit acting in Providence. As such, the Bhagavad Gita teaches us to step back from the daily tactical realities of what comes to pass, to know ourselves as standing upon the wider operational battlefield of life, and then to step back even further and see the whole shebang from the cosmological strategic perspective of a God's-eye view. And note who these charac-ters were! Sanjaya was the eagle-eyed charioteer to Dhritarashtra, the *blind* king. Political power on its own is always blind. To stop its chariot from sinking in the mire requires spiritual vision.

Case studies of nonviolence

The former US president, Jimmy Carter has said, 'Historically and cur-rently, we all realize that religious differences have often been a cause or a pretext for war. Less well known is the fact that ... religion can be a potent force in encouraging the peaceful resolution of conflict.'[52] It is precisely because spiritual development means processing our individual and collective shadows that war and religion so often cross each other's paths. Equally, the threads of violence and nonviolence invariably intertwine. Nonviolence gains traction precisely because the Damoclean alternative, violence, is so terrible. Like violence, nonviolence does not always succeed. Some critics even see it as 'pathology' because, they argue, it weakens motivation for violent revolution.[53] Yet nonviolence is not devoid of political success. As Wink points out:

> In 1989–90 alone, thirteen nations underwent nonviolent revolutions, all of them successful but one (China), and all of them nonviolent on the part of the revolutionaries except one (Romania, and there it was largely the secret police fighting the army, with the public maintaining nonviolent demonstrations throughout).[54]

Here I shall briefly profile three examples of nonviolence in action: the Pashtun contribution to Indian independence as an example of Muslim nonviolence, nonviolent resistance to Nazism and conciliation at the ending of the Biafra War.

Pashtun resistance to the British Raj, 1930s

Throughout most of the nineteenth century and into the twentieth, the Pashtun (or Pathan) peoples – the backbone of today's Taliban – were caught in the 'Great Game' buffer zone of the British and Russian empires. In 1893 Britain's drawing up of the Durand Line to delineate what is now modern Pakistan's north-west frontier with Afghanistan sliced through Pashtun territories. British efforts to suppress unwelcome political ideas that arose in resistance to such intrusion included collective punishments against whole communities and a gross neglect of social measures, including education.[55] Ghaffar 'Badshah' Khan (1890–1988) was a devout Muslim landowner who used his influence to open schools (*madrassas*) that would raise popular political consciousness. When his father came under pressure to rein in his son from the British Chief Commissioner, the young Khan replied that 'educating the people and serving the nation is as sacred a duty as prayer'.[56] So began a series of prison sentences for the youth, some lasting years, during which time he discoursed with prisoners of others faiths and became inspired by the teachings of Mahatma Gandhi. He concluded: 'It is my inmost conviction that Islam is *amal, yakeen, muhabat* [service, faith and love] and without these the name "Muslim" is sounding brass and tinkling cymbal.'[57]

Badshah Khan's subsequent work for Indian independence as 'the Muslim Gandhi' led him to establish the Khudai Khidmatgar – the 'Servants of God'. These were a pacifist Mujahideen who chanted slogans such as *'Allah-O-Akbar'* ('God is Great') and were derogatorily called 'red shirts' by the British on account of their uniform. The membership oath included, 'I shall always live up to the principle of nonviolence,' and, 'All my services shall be dedicated to God; they shall not be for attaining rank or for show.'[58] Khan told them:

> I am going to give you such a weapon that the police and the army will not be able to stand against it. It is the weapon of the Prophet, but you are not aware of it. That weapon is patience and righteousness. No power on earth can stand against it.[59]

In close coordination with Gandhi, the Khudai Khidmatgar invoked a spiritual jihad of civil resistance, including refusal to pay taxes, noncooperation with the Raj, boycotts and pickets, general strikes and the mass commemoration of iconic events. By 1938 Pashtun membership exceeded 100,000. Nonviolence had held fast even in the face of imprisonment, torture and the Kissa Khani Bazaar massacre in 1930. Here the British killed more than 200 civilians who were protesting the arrest of leaders, including Khan, who had just been sentenced to three years for fomenting civil disobedience. Some of the Khudai Khidmatgar sustained as many as 21 bullets in the chest as they stepped forward, peacefully to interpose their bodies between the troops and the crowd.[60] Gandhi subsequently told Khan, 'The Pathans are more brave and courageous than the Hindus. That is the reasons why the Pathans were able to remain nonviolent.'[61]

Notwithstanding the later tragedy of ethnic cleansing that resulted in the partition of India, Khan's pacifism never faltered. In 1983 he told his biographer: 'The present-day world can only survive the mass production of nuclear weapons through nonviolence. The world needs Gandhi's message of love and peace more today than it ever did before.'[62]

Norway, 1942, and Berlin, 1943

Like with violent defence, nonviolent civil defence requires awareness, commitment, training and strategy. These were largely lacking in Europe at the time of Hitler – an era that also lacked the internet, texting and other means of rapid communication through which to organize. That said, organized nonviolent resistance was developed in Denmark and Holland,[63] and here I will give examples from Norway and Berlin.

In 1942 the Norwegian 'Minister-President', Vidkun Quisling, created a fascist teachers' corporation. Membership was to be compulsory and its leader was the head of the Norwegian storm troopers. The underground called for noncompliance. Over two-thirds of the country's 12,000 teachers openly wrote letters of noncooperation. Quisling threatened them with dismissal and closed the schools, but the children's education was reorganized at home.

To set an example, the Nazis rounded up a thousand teachers and dispatched them to concentration camps. But the school children gathered to sing on railway platforms as the cattle trains passed through. Under conditions of extreme cold the Gestapo put the captured teachers through 'torture gymnastics', fed them starvation rations and issued death threats. Very few capitulated. Across Norway people signalled their revulsion by, for example, wearing a paper clip in their lapels as a way of saying, 'stick together'. Realizing that his measures were backfiring, Quisling gave in. After an eight-month ordeal the teachers were sent home. While giving a school address Quisling raged, 'You teachers have destroyed everything for me.'[64] For totalitarianism to work, it has to be total, but his veneer had cracked under the weight of superior moral authority.

In Berlin in 1943, as part of the 'Final Solution', the Jewish husbands of non-Jewish German women were rounded up and imprisoned in the Rosenstrasse. Some 6,000 wives appeared at the prison gates and, in defiance of SS guns, demanded their husbands' return. 'A decision to put one's life on the line for another can only come from the heart,' said one woman, who had expected the worst. 'One is ready, or not. One does it, or not.' Hit by embarrassment – which is always the Achilles heal of power's narcissism – the authorities negotiated. Goebbels did not want the German people's wider conformity to be jeopardized by him appearing to be in anything less than complete control. With Hitler's consent he ordered the husbands to be released. Although many were later individually rearrested, by the end of the war such intermarried Jews nevertheless comprised 98 per cent of the surviving German Jewish population that had not been driven into hiding.[65]

Both these cases show how, for oppression to succeed, it must acquire the acquiescence of the oppressed. Silence is the voice of that complicity. Nonviolent civil defence therefore seeks to break down this 'cultural invasion'[66] using truth force (*Satyagraha*) as its weapon. Hitler was very aware of this. As he said in *Mein Kampf*: 'In the long run, government systems are not held together by the pressure of force, but rather by the belief in the quality and the truthfulness with which they represent and promote the interests of the people.'[67] This is why nonviolent strategy pays close attention to the psychology of compliance, conformity and obedience.[68] It uses 'political jiu-jitsu' to throw an opponent with their own weight, in particular, unseating them from any moral high ground.[69] If Germany, and Europe generally, had been more prepared with such principles during the 1930s world history might have taken a different course. Such is the imperative for nations to teach peace.

Biafra, 1967–70

Independence from the British in 1964 left Nigeria as a fledgling state sharply divided by ethnicity and religion. Two military coups in 1966 brought brutal civil war and a violent succession bid by the Eastern Region to declare independence from federal Nigeria as the Republic of Biafra. As federal Nigerian troops and bombing destroyed Biafran military capacity, famine gripped the Ibo people. Eventually the only question was whether reconciliation might be plucked from the jaws of potential genocide. The outcome 'may have been the most extraordinary post-civil war reconciliation to have occurred in modern history'.[70] This was facilitated by Arnold Smith, secretary-general to the British Commonwealth, who called in a team of Quaker mediators headed by Adam Curle.

Curle, who in 1973 became the first Professor of Peace Studies at the University of Bradford, often distinguished between what he called 'mediable' and 'unmediable' violence.[71] For mediation to have a hope there has to be at least a possibility that both sides desire resolution. The mediator seeks to draw out and connect such desires. In the case of Biafra, it entailed:

- opening lines of communication
- reducing suspicion, misperception and fears
- advocating for negotiated settlement

Remarkably, Curle's team established the trust of both sides. The Biafran head of state, General Emeka Ojukwu, later attributed it to their 'absolute dedication to humanity' and 'an infinite capacity for neutrality'. His Nigerian counterpart, General Yakabu Gowon, said that he came to trust the mediators because: 'The basis is a belief in God and humanity ... They persisted right the way through and were accepted.'

Instead of victory celebrations, Gowon proclaimed that there were 'no victors, no vanquished'. He granted amnesties, called for three days of national prayer, and as Biafrans returned to their former federal posts he emphasized convergence on the 'three Rs': reconstruction, reintegration and reconciliation. Ignatius Kogbara, the Biafran representative in London, said that the mediators' most important contribution had been that 'they tried to resolve the hardness of the heart'. Those words lay bare the essence of spiritually informed nonviolent peacemaking.

Conclusion

Human life has an outer material expression and an inner spiritual constellation. Both are sides of the same coin. To understand war as being driven by politics, economics or science is only half the story. 'Do you know where wars come from?' asked the Indian Jesuit priest, Anthony de Mello. 'They come from projecting outside of us the conflict that is inside. Show me an individual in whom there is no inner self-conflict and I'll show you an individual in whom there is no violence.'[72]

God works on a long front. Not all its positions are visible to human view. That is why hope for peace in the world resides not just in pacifists. It also resides in principled soldiers. For example, General Sir Richard Dannatt who was until 2009 Head of the British Army and who, against the backdrop of a legally questionable war in Iraq, had the courage to tell the British people:

> Honesty is what it is about. The truth will out. We have got to speak the truth. Leaking and spinning at the end of the day are not helpful ... In the Army we place a lot of store by the values we espouse ... courage, loyalty, integrity, respect for others; these are critical things. I think it is important as an Army entrusted with using lethal force that we do maintain high values and that there is a moral dimension to that and a spiritual dimension.[73]

Neither brute violence nor naive forms of pacifism on their own can tackle the toughest issues of our times. But whatever our station on the long front, it is perilous to neglect one's spiritual life. As Dannatt concluded in delivering the Windsor lecture, we must cultivate empathy with 'something far bigger than ourselves, something bigger and deeper than we can imagine or rationalize for ourselves'. We need it, he said, because 'ships without anchors on the sea bed in turbulent times run before the prevailing wind, and the rocks can be very unforgiving'.[74]

As I have suggested, the principled soldier and the principled pacifist can find themselves occupying suprisingly similar territory. Such is the power of love that transcends the love of power.

Notes

1. When taken together with the previous chapter – 'A View From Realism' – some of the influences on either side of the 'mainstream' Just War Tradition (Chapter 4) can be seen.

2. Martin Luther King Jr, 'My Pilgrimage to Nonviolence', in Walter Wink (ed.) *Peace is the Way: Writings on Nonviolence from the Fellowship of Reconciliation* (Maryknoll: Orbis Books, 2000), pp. 64–71.

3. Based on such passages as: Thomas Merton (ed.) *Gandhi on Non-Violence: A Selection from the Writings of Mahatma Gandhi* (New York: New Directions, 1965), p. 54 (I-292).

4. Gene Sharp, *Waging Nonviolent Struggle* (Boston: Extending Horizons, 2005), inside back cover (Clausewitz/Machiavelli), pp. 19 and 111.

5. Mairead Corrigan Maguire, 'Gandhi and the Ancient Wisdom of Nonviolence', in Wink, *Peace is the Way*, pp. 159–62.

6. Merton, *Gandhi on Non-Violence*, p. 37 (I-240).

7. Cornelius Tacitus, 'Agricola', in *Dialogus, Agricola, Germania*, trans. Maurice Hutton (Cambridge: Harvard University Press, 1946), p. 221.

8. See the chapter on 'The Pleasures of War' in Joanna Bourke, *An Intimate History of Killing* (London: Granta, 2000), p. 15.

9. Alice Miller, *For Your Own Good: The Origins of Violence in Child-Rearing* (London: Virago, 1987); James Gillegan, *Violence: Reflections on a National Epidemic* (New York: Vintage, 1997).

10. Walter Wink, *Engaging the Powers: Discernment and Resistance in a World of Domination* (Philadelphia: Fortress Press, 1992).

11. Hélder Câmara, *Spiral of Violence* (London: Sheed & Ward, 1971), p. 60 – out of print, but online at http://www.alastairmcintosh.com/general/spiral-of-violence.htm

12. Except where indicated, all quotations here are from the chapter, 'The Battle of Omdurman', in Winston Churchill, *The River War: An Historical Account of the Reconquest of the Soudan* (London: Longmans, Green & Co., 1899).

13. Martin Gilbert, *The Second World War* (London: Phoenix, 2000), pp .440-441.

14. Ali Rahnema (ed.), *Pioneers of Islamic Revival* (London: Zed, 1994).

15. BBC News, 8 November 2004, 'UN Warns of Nuclear Terror Race', http://news.bbc.co.uk/1/hi/world/asia-pacific/3991305.stm, accessed 27 February 2009. The same story on Aljazeera.com at the time included his remark, 'We have to cross our fingers that nothing will happen'!

16. Rupert Smith, *The Utility of Force: the Art of War in the Modern World* (London: Penguin, 2006), p. 372.

17. Joseph Conrad, *Heart of Darkness* (Harmondsworth: Penguin, 1995), p. 62.

18. Yolanda Gampel, 'Reflections on the prevalence of the uncanny in social violence', in Antonius C. G. M. Robben and Marcela M. Suárez-Orozco (eds) *Cultures under Siege: Collective Violence and Trauma* (Cambridge: Cambridge University Press, 2000), pp. 48–69.

19. BBC News, 10 March 2009, 'Mugabe Calls Crash "Hand of God"', http://news.bbc.co.uk/1/hi/world/africa/7934931.stm, accessed 10 March 2009.
20. Alice Walker, 'S M', in *Horses Make a Landscape Look More Beautiful* (London: Virago, 1985), p. 10.
21. Dante, *Inferno*, Canto 34.
22. Sven Lindqvist, *Exterminate all the Brutes* (London: Granta, 2002).
23. Alastair McIntosh, 'Peace in the Tiger's Mouth', Fernando Enns, Scott Holland and Ann K. Riggs (eds) *Seeking Cultures of Peace: a Peace Church Conversation* (Telford and Geneva: Cascadia and World Council of Churches, 2004), pp. 215–26. Online at http://www.alastairmcintosh.com/articles/2001-basel.htm.
24. The Yearly Meeting of the Religious Society of Friends (Quakers) in Britain, *Quaker Faith and Practice* (London: The Yearly Meeting, 1994), 19:47.
25. M. K. Gandhi, *An Autobiography, or, The Story of my Experiments with Truth* (Harmondsworth: Penguin, 1982).
26. I have discussed experience of this in, 'The Power of Love: What can Nonviolence say to Violence', *Resurgence*, 219, 2003, pp. 42–4; online at http://www.alastairmcintosh.com/articles/2003—power-love.htm, accessed 15 March 2009.
27. Jolande Jacobi, *The Psychology of C. G. Jung* (London: Routledge & Kegan Paul, 1968).
28. Maurice S. Friedman, 'Hasidism and the Love of Enemies', in Wink, *Peace is the Way*, pp. 118–23.
29. John 15:1–17; 2 Peter 1:4; Romans 12:5; 1 John 4:8.
30. Adam Curle, *Tools for Transformation* (Stroud: Hawthorne Press, 1990), p. 34.
31. Wink, *Engaging the Powers*, exegesis of Romans 13:1.
32. Abdullah Yusuf Ali (trans.), *The Holy Qur'an: Text, Translation and Commentary* (Jeddah: Islamic Education Centre, 1946), p. 75.
33. Respectively Bukhari 32, Bukhari 52, Bukhari 52 and Sira 388 – see Chapter 8, 'The Prophet at War', in P. J. Stewart, *Unfolding Islam* (Reading: Garnet, 1995), pp. 75–87.
34. Stewart, *Unfolding Islam*, p. 76.
35. For example, Mohammed Abu-Nimer, *Nonviolence and Peace Building in Islam: Theory and Practice* (Gainesville: University of Florida Press, 2003); and the online resource, Glenn D. Paige, Chaiwat Satha-Anand (Qader Muheideen) and Sara Gilliatt (eds) *Islam and Nonviolence* (Honolulu: Center for Global Nonviolence, 1986), http://www.globalnonviolence.org/islam.htm, accessed 28 February 2009.
36. 'Show no pity: life for life, eye for eye, tooth for tooth, hand for hand, foot for foot' – Deuteronomy 19:21 NRSV. I quote below from translations as indicated as befits clear expression.
37. Isaiah 2:2, 4 NRSV; Micah 4:1, 3.
38. Matthew 5:38–42. See exegesis of Christ's nonviolence in Wink, *Engaging the Powers*.
39. Mark 11:15–17; John 2:13–22 NRSV.

40. See, for example, Angie Zelter and contributors, *Trident on Trial: The Case for People's Disarmament* (Edinburgh: Luath Press, 2001). For Trident Ploughshares see http://www.tridentploughshares.org.
41. Matthew 5:38–48.
42. Matthew 11:12 NRSV.
43. Mark 3:17; Luke 9:51–6.
44. Luke 22:35–8.
45. Matthew 26:52 GWT, KJV and Luke 22:51 ISV; cf. Mark 14:47 and John 18:10.
46. Luke 22:51.
47. John 18:11 WNT.
48. John 18:36 TEV.
49. Raimon Panikkar, 'Nine Sutras on Peace', in Panikkar, *Cultural Disarmament* (London:John Knox, 1995), pp.15–25, and an alternative translation from *Interculture* online at http://www.alastairmcintosh.com/general/resources/1991-Panikkar-Nine-Sutras-on-Peace.pdf, accessed 28 February 2009.
50. J. Denny Weaver, *The Nonviolent Atonement* (Grand Rapids: William B. Eerdmans, 2001).
51. Juan Mascaró (trans.), *The Bhagavad Gita* (Harmondsworth: Penguin Classics, 1962), p. 43.
52. Douglas Johnston and Cynthia Sampson (eds) *Religion, the Missing Dimension of Statecraft* (Oxford: Oxford University Press, 1994), p. vii.
53. Ward Churchill, *Pacifism as Pathology: Reflections on the Role of Armed Struggle in North America* (Winnipeg: Arbeiter Ring, 1998).
54. Wink, *Peace is the Way*, p. 1.
55. Mohammad Raqib, 'The Muslim Pashtun Movement of the North-West Frontier of India – 1930–1934', in Sharp, *Waging Nonviolent Struggle*, pp. 113–34.
56. Eknath Easwaran, *Nonviolent Soldier of Islam: Badshah Khan, a Man to Match His Mountains* (Tomales, CA: Nilgiri Press, 1999), p. 84.
57. Easwaran, *Nonviolent Soldier of Islam*, p. 63.
58. Raqib, *The Muslim Pashtun Movement*, pp. 117–18.
59. Raqib, *The Muslim Pashtun Movement*, p. 117.
60. Easwaran, *Nonviolent Soldier of Islam*, pp. 121–8; Raqib, *The Muslim Pashtun Movement*.
61. Easwaran, *Nonviolent Soldier of Islam*, p. 195.
62. Easwaran, *Nonviolent Soldier of Islam*, p. 7.
63. Chapter on 'Denmark, the Netherlands, the Rosenstrasse: Resisting the Nazis', in Peter Ackerman and Jack Duvall, *A Force More Powerful: a Century of Nonviolent Conflict* (New York: St Martin's Press, 2000), pp. 207–39.
64. Sharp, *Waging Nonviolent Struggle*, pp. 135–41.
65. Sharp, *Waging Nonviolent Struggle*, pp. 143–8.
66. Paulo Freire, *Pedagogy of the Oppressed* (Harmondsworth: Penguin, 1972), pp. 121–35.
67. Adolf Hitler, *Mein Kampf* (New York: Reynal & Hitchcock, 1940), p. 388.

68. Donald Pennington, Kate Gillen and Pam Hall, *Essential Social Psychology* (London: Arnold, 2001), pp. 207–43 (especially the Milgram and Zimbardo experiments).
69. Sharp, *Waging Nonviolent Struggle*, pp. 405–13.
70. Cynthia Sampson, '"To Make Real the Bond Between Us All": Quaker Conciliation During the Nigerian War', in Johnston and Samson, *Religion*, pp. 88–118.
71. Curle, *Tools for Transformation*, pp. 91–5.
72. Anthony de Mello, *Awareness: The Perils and Opportunities of Reality* (New York: Image Doubleday, 1992), p. 182.
73. Sarah Sands, 'Sir Richard Dannatt: A Very Honest General', *Daily Mail*, 13 October 2006, pp. 12–13.
74. General Sir Richard Dannatt, 'Leadership in Turbulent Times', *Windsor Leadership Trust Annual Lecture*, delivered 8 October 2009, http://www.windsorleadershiptrust.org.uk/en/1/cgsal09.html, accessed 30 October 2009.

Chapter 4

The Just War Tradition: A Pragmatic Compromise

DAVID WHETHAM

Introduction

The origins of the Just War Tradition in the West lie in a synthesis of classical Greco-Roman and later Christian values.[1] It can be seen developing and evolving all the way through to the codification of customary international law in the nineteenth and twentieth centuries, and beyond as it continues to grapple with the challenges of the new millennia. Influential contributors to its development have included Plato, Aristotle, Cicero, Augustine, Aquinas, Vitoria, Suárez and Grotius, all the way up to James Turner Johnson and Michael Walzer in the twentieth century. Although the Tradition is often associated with Western or even Christian traditions, it contains a broad resonance with ideas, cultures and religious principles found all over the world. For example, 'Islam, like Judaism, starts from its earliest history to develop restrictions on conduct in war'.[2] As such, the Just War Tradition represents a common language for discussing and debating the rights and wrongs of conflict.

How did it develop?

The type of thinking that leads to the Just War Tradition fits in between the two very different attitudes towards war, examples of which can be seen in Chapters 2 and 3: realism and pacifism or nonviolence. In many ways, the Just War Tradition is a compromise between these two positions. As we have seen, pacifism or nonviolence contains a deep presumption against war. It argues that while evil should be opposed in the world, resorting to war and violence is always wrong. The position sees a straightforward contradiction between the imperative to act morally, and the deliberate taking of life that fighting a war

65

requires. The act of fighting is itself unjust and one cannot do something that is unjust whatever the reason: this is an absolute deontological position (see Chapter 1). One cannot do evil even to prevent evil.
This position can be motivated by religious, political or even pragmatic
concerns.[3] For example, to illustrate each in turn: for the first three
centuries AD, Christianity was, broadly speaking, a religion that
spurned violence, taking to heart Christ's clear injunctions towards
non-violence found in the New Testament; in the late 1940s, Mahatma
Gandhi's efforts to expel the British from India provides an example of
pacifism that was politically inspired, managing to achieve results
through non-violent campaigns; finally, moving into the nuclear age,
Bertrand Russell articulated very practical grounds on which war was
simply no longer a viable or realistic policy option because even a
small-scale conflict between two or more nuclear powers would be
likely to escalate into a nuclear exchange. Whatever the cause of a possible dispute, it simply could not be as important as the mutual interest
the two sides have in not destroying each other: 'Neither side can
defeat the other except by defeating itself at the same time ... The first
and most important of their common interests is survival.'[4] The logical
consequence of such reasoning is that in the nuclear age, faced with the
very real threat of Mutually Assured Destruction, war itself must be
avoided.

Clearly advocates of pacifism or nonviolence have found different
ways of expressing and arguing their case throughout history. It is an
approach that provides a principled challenge to those who see violence
as 'the' answer. It can also provide a very profound alternative definition of the real meaning of strength. However, consistent or absolute
pacifism, eschewing force in all circumstances, also poses some profound difficulties. Taking each of the cases above, despite early state
persecution, the early Christians did not face an existential threat from
the outside because they lived within the security provided by the
Roman Empire. After the conversion to Christianity of the Emperor
Constantine (following his victory of 312 at Milvian Bridge that helped
him secure its leadership), Rome itself became a Christian empire and
Christians began to take up arms to defend it in significant numbers
(see Augustine below).[5] Mahatma Gandhi and his followers could rely,
at least to a certain extent, on the restraint of those they opposed. If
they had been living in a genuine totalitarian state, where 'opponents of
the regime disappear in the middle of the night and are never heard
from again', would their actions have been possible?[6] The pacifist who
bases the argument on pragmatic grounds can be very persuasive, but
there can be a heavy cost to applying such a principled stance, such as
when one chooses to stand by and watch tens or even hundreds of
thousands of innocent people die in a preventable genocide.

Coming from an alternative perspective, as we saw in Chapter 2, the realist has a very different view on the use of force and attempts to restrain it. The realist does not have to go as far as to say that such ethical norms and values mean absolutely nothing – after all, if understood and manipulated they can be powerful instruments for furthering the ends of the state. However, fundamentally, it is the state that makes such values possible for those who live within its boundaries and under its protection; moral or ethical thinking, or concepts such as justice are simply rendered meaningless in the world between states. Thomas Hobbes suggested that the absolute authority of the sovereign state could be likened to a Leviathan: a mighty social contract in which the individuals of a society give up a portion of their freedom in return for security provided by the state. Part of that security is the laws and values that we cherish within our society but they only exist because of the ability of the state to protect them. As there is no 'Leviathan' in the international arena, it makes no sense to talk about values, morality or ethics beyond those existing within each state.[7] The only justice that exists in international affairs is equated to power. Thucydides famously captured this idea in the Melian Dialogue with the Athenians arguing: 'the strong do what they have the power to do and the weak accept what they have to accept'.[8]

Advocates of realism often argue that it is an amoral position rather than an immoral one – a straightforward description of how the world is rather than how we feel it ought to be. However, it can itself have a strong normative core. For example, if 'war is hell', as Sherman famously stated, seeking to tame it with artificial rules is not only unworkable and misguided, it can be seen as ethically wrong: precisely because war is such a terrible thing, the best thing to do is to get it over as quickly as possible without applying misguided ethical notions or legal limitations that simply prevent a speedy victory and therefore cause more suffering in the long run.

However, as conceded in Chapter 2, realism, whether descriptive or prescriptive, can itself be decidedly unrealistic if it naively claims that strategy is most effective when it ignores the normative dimension. The distinction that Clausewitz makes between the nature of war (immutable but therefore purely theoretical) and its character (what war actually looks like in when translated into a particular time and place) acknowledges that real war is a social phenomenon and cannot simply be divorced from this context.[9] Short-term military success and longer-term political success are not necessarily the same thing and successful strategy must provide a linkage between the two. A realism that accepts constraints, even in a purely instrumental way to acknowledge the social context, is moving towards a compromise position.

A pragmatic compromise

The Just War Tradition agrees with the pacifist view that war is a terrible thing. However, rather than accepting the pacifist's presumption against war, it contains a presumption against injustice. It argues that in some (but by no means many) circumstances, war might be preferable to the injustice that will result without it. It is a view summed up well by President Carter as he received the Nobel Peace Prize in 2002: 'War may sometimes be a necessary evil. But no matter how necessary, it is always an evil, never a good.'[10] Therefore, the Just War Tradition agrees with the realist position that it is sometimes necessary to do terrible things, *up to a point*. However, very few reasons justify going to war, and even in the rare circumstances in which fighting a war might be justified, it must still be fought within strict limits. All cultures, civilizations and religions have accepted the principle that war needs to be restrained in some way, and this agreement has not happened by chance. This is because the purpose of a rational war is to gain a better peace: military victory makes no sense unless it can be transformed into political success and that can only be hampered by ignoring the normative dimension of conflict.

How has the Just War Tradition developed?

This chapter will now provide a brief history of the development of Just War thinking and the emergence of its key principles, before moving on to the contemporary understanding of the framework and the way those principles might be applied in the current international system.

Elements of the Just War principles can be found in historical practice all around the world, but ancient Greece provides a useful starting point for seeing the development of some of the Tradition's key ideas.[11] Until the late fifth century BC, broad strategies based on the destruction of an enemy's social and economic system were effectively banned by the informal system of rules and Hellenic customs. Accepted tradition required that war needed to be formally declared, truces should be respected (particularly during important events such as the Olympic Games), and fighting battles should be restricted to certain times of the year so as not to disrupt agricultural life. Noncombatants were not to be deliberately targeted and any prisoners that resulted from military action were to be ransomed rather than killed.[12]

It is obviously that there was a very pragmatic angle to these considerations. After all, clearly it was not in the interests of either side to fight during the harvest time and there is no incentive to cease fighting if one is going to be killed if one surrenders or not. However, more profound than this is the idea that a soldier is a legitimate object of attack precisely because he or she poses a threat as an instrument of the state or political entity against which you are engaged in hostilities. Once a soldier is no longer fighting you, because they are wounded or they have surrendered, they are no longer posing a threat. At that point, they cease to be a legitimate target. Because the civil population is not fighting you, they never become legitimate targets in the first place.

While allowing monuments to celebrate and record military victories, Hellenic custom also banned the use of stone as a building material for them as this would not deteriorate over time and would therefore remain as a permanent reminder of the discord. Even maintaining wooden monuments was prohibited to ensure that they deteriorated over time, with the intention that the memory of the conflict would also fade and allow an eventual successful reconciliation of the two sides.[13]

This system of rules was effectively undermined during the Peloponnesian Wars between Athens and Sparta when ideology became a factor in the quarrel and the expansionist nature of the Athenian Empire meant that national survival itself became an issue for some of the states involved.[14] This 'tragedy of epic proportions' is recorded by Thucydides.[15] Following the terrible damage inflicted on Greek society by this war, there were attempts to rethink and understand the rules that had once been in place, protecting Hellenic life. In the *Republic*, Plato records Socrates as he sets out his idea of what a perfect society would look like. Part of this is his view about warfare and the limits that should be observed in the disputes that might take place between Greeks, harking back to the customs that had governed intra-Greek conflict before the Peloponnesian War and explaining their benefits. Rather than being considered as true warfare (such as that conducted against the non-Greek 'barbarians', involving enmity and hatred), 'Greeks ... are still by nature the friends of Greeks when they act in this way, but that Greece is sick in that case and divided by faction.'[16] Because of this attitude, conflicts between the Greek peoples needed to be conducted with the intention of achieving the eventual reconciliation that must follow. Therefore, certain limits had to be respected. For example, the 'other side' should never be enslaved and their homes and farms should not be destroyed.[17] While taking food was acceptable (after all, the army had to support itself), one should not destroy the means of producing that food in the process because one needed to discriminate between those who were actually at fault for the quarrel, and those who were not. Condemning the civil population to starvation was hardly a way of promoting long-term reconciliation.

Building on the ideas of Plato, Aristotle stated some of the most important principles of the Just War Tradition, even using the exact term 'Just War' for the first time. He saw force used as a means of achieving higher goals such as peace as 'not without virtue'.[18] He also set out what he considered to be five legitimate grounds for conducting war: self-defence, vengeance against those who have caused injury, to aid an ally, to gain an advantage for one's own community and to maintain authority over those who could not rule themselves.[19] Only self-defence and the defence of an ally fit comfortably with modern ideas on the subject of just cause. The fifth criteria looks particularly out of place compared to contemporary Just War principles, but illustrates Aristotle's understanding of the natural order in which everyone had their own position: some people were simply not capable of governing themselves and therefore a war that could restore or maintain order might be a legitimate undertaking for everyone's benefit.[20]

The Roman Republic's attitude to war was firmly grounded in the idea that Rome needed to satisfy certain trial-like procedures in order to ensure that her wars could be portrayed as defensive and that the gods would therefore be on her side. Before any actual hostilities could begin, the *fetial* priests (effectively, diplomats responsible for the correct conduct of international relations) would publicly set out Rome's grievance and demand redress from the other party. If the other side were prepared to acknowledge the injury they had caused and make suitable recompense within 33 days, peaceful relations could be maintained and war avoided.[21] If (and only when) redress was not forthcoming then warfare was the legitimate next step for Rome to pursue the disagreement further. Even when Republic changed to Empire and the formal demands became increasingly outrageous to the point where war and subsequent defeat could actually be considered equal to or even better than the terms and total humiliation being offered, it should not be underestimated the degree to which the psychology of the process was important for those taking part in it. It was essential to the people of Rome that her wars were cast in the form of 'legally defensive and therefore morally justifiable acts even when they might have been launched on what would be regarded today as mere pretexts'.[22]

The Roman orator and statesman Marcus Tullius Cicero (106BC–43BC) provided philosophical justifications for many of Rome's practices and procedures, such as formal declaration of war only following a failure to make amends, making war a last resort only once the alternatives had been exhausted. The rules were important because one had to keep one's word even in times of war: 'our concern should always be for a peace that will have nothing to do with treachery'.[23] He insisted that any declaration must be made by a legitimate authority. This was partly because it was essential to maintaining the correct legal status of the procedure, and partly it was a way of minimizing the potential for civil wars. Today, states still jealously guard their monopoly on the use of legitimate violence. Cicero was also concerned that one had the correct intention for engaging in a conflict, arguing that: 'Wars, then, ought to be undertaken ... that we may live in peace, without injustice.'[24] In this way, Cicero argued that even wars fought for the glory of Rome rather than its survival (what today might be called discretionary conflicts) must still be motivated by this desire to live in peace. Like Aristotle before him, Cicero argued that expansionist wars of this type could be justified because they could 'enlarge the boundaries of peace, order and justice', thereby bringing about greater peace, prosperity and happiness than would otherwise have been the case, a theme that appears less convincing today.[25] However, more in tune with contemporary thinking, he also argued that prisoners of war should be treated fairly by their captors: 'once victory has been secured, those who are not cruel or savage in warfare should be spared'.[26] He also drew attention to the distinction between legitimate and illegitimate combatants: 'it is not lawful for one who is not a soldier to fight the enemy'.[27]

Greek and Roman thinking on war became fused with Christian ideas as they were transmitted and shaped by the writings of the early Christians. Most influential of these was St Augustine (353–430), bishop of the North African city of Hippo. Augustine was faced with the very real problem of reconciling the clear message of pacifism found in the New Testament with the fact that Rome itself, now a Christian Empire, was under very real threat. The Visigoths had sacked Rome in 410 and when Augustine died, in 430, Hippo itself was under siege. It was this terrifying reality that prompted Augustine to become what has been described as 'the reluctant just war theorist'.[28] The key to reconciling the tension between the two positions was one's intention. While Christians should, of course, still be prepared to turn the other cheek if *they* were themselves attacked, they could still employ violence to defend the innocent from evil. It was not the actual killing that made war sinful (after all, this would only shorten the lives of people who were going to die anyway). The sin occurred only if one were motivated by an unworthy intention – a love of violence, revengeful cruelty, hatred, greed and lust for power.[29] As long as one's actions were motivated by love or charity – the defence of the innocent – they could be conducted without fear of sin.

Injustice was considered a greater evil than war and Augustine argued that it was correct to carry out a lesser evil if it could prevent a greater one: clearly taking up arms to defend Christendom and the innocent men, women and children within it could actually therefore be seen as a duty rather than a sin. All rulers had an obligation to maintain the peace, and it was this obligation that gave them the right or even positive duty to wage war in order to maintain that peace: as Paul the Apostle makes clear in Romans 13, no government can exist and rule the people unless God wills it.[30] Therefore, as long as such a legitimate authority declared war, citizens were *obliged* to participate in it. This obligation meant that it was the ruler that would face any Divine retribution for waging an unjust war rather than the soldiers who participated in it, and this was true even if they had substantial doubts about the justness of their cause (a view challenged by Vitoria and one that is still contentious today).[31]

The next major contributor to the development of the Just War Tradition was St Thomas Aquinas (1225–1274). Aquinas developed and simplified the ideas of Augustine in particular, as well as other contributors such as the twelfth-century monk Gratian.[32] In doing this, he also identified right authority, just cause and right intention as the key moral criteria for assessing the moral legitimacy of the resort to the use of force. These still provide the 'basic architecture according to which discussions of just war are carried on even today'.[33] Given that he is considered so influential in the development of Just War thinking, Aquinas appears to say surprisingly little about what we think of as *in bello* considerations – that is, what can legitimately be done during a war and who it can be done to. However, this would be to do him a disservice. There is 'conceptual economy' that allows both Aquinas, and Augustine before him, to identify a relatively small number of key principles – but those key principles carried with them meanings that have, to some extent, been lost today.[34] Nowhere is this more evident than the case of right intention – motivation – the spirit in which one acts. This is not a principle that necessarily sits well with us today; it can be seen as somewhat abstract and impossibly subjective due to its essentially internal character. However, for Aquinas, right intention stood for something far more palpable; a 'certain moral character (comprising habits, attitudes, sentiments and prejudices) that disposed belligerents to limit both their recourse to war and their conduct of war'.[35] That was why this was such an important element. If it was correct, then one's conduct would automatically be affected by it. As long as one were trying to do the right thing, then the wrong people were not going to get hurt because one would make a distinction between those who were at fault and those who were not, and one would ensure that one's actions were proportionate to the injury being defended against. However, if one had the wrong intention from the outset, then no additional principles or rules would be sufficient to prevent the terrible 'descent into the moral abyss of war'.[36]

Aquinas also helped resolve a very real moral issue for Christians –

killing in personal self-defence – by clarifying what has become known as the Doctrine of Double Effect. This is the idea that individuals are not necessarily morally responsible for the foreseeable, yet unintended side effects of an otherwise legitimate action:

> Nothing hinders a single act from having two effects, only one of which is intended, while the other is beside the intention. Now moral acts get their character in accordance to what is intended, but not from what is beside the intention, since the latter is incidental ... Accordingly, the act of self-defence may have a double effect: the saving of one's own life, on the one hand, and the slaying of the attacker, on the other. Since saving one's own life is what is intended, such an act is not therefore illicit.

Of course, in doing this, one may only use as much force as is actually necessary. To use disproportionate force would be to 'exceed the limits of a blameless defence'.[37]

While Aquinas is concerned here with proportionality at the *jus in bello* level (what can legitimately be done within a war), Vitoria, writing in the mid-sixteenth century, captured the prudential aspects of the principle of proportionality at the *jus ad bellum* level (what is required to justify going to war) by explaining: 'if the recovery of one city is bound to involve the commonwealth in greater damage, for instance the devastation of several cities, heavy casualties, or rivalry between princes and the occasion of further wars, there can be no doubt that the prince should cede his right and abstain from war'.[38] This proportionality consideration was not purely limited to one's own damages, but to the total costs to everyone involved.

The important moral division between the *ad bellum* and the *in bello* level of war was also clarified and explained by Vitoria, who demonstrated that the distinction between the two levels could have profound implications for the long-term defence of the state:

> if subjects can not serve in war except they are first satisfied of its justice, the State would fall into grave peril and the door would be opened to wrongdoing ... if subjects in a case of doubt do not follow their prince to the war, they expose themselves to the risk of betraying their State to the enemy, and this is a much more serious thing than fighting against the enemy despite a doubt.[39]

In the absence of clear evidence to the contrary, soldiers on both sides of a dispute are required to give their own leaders the benefit of the doubt. However, even while coming down on the side of the state here as far as selective conscientious objection goes, Vitoria goes further than Augustine by arguing that there are still limits as to how far soldiers can wash their hands of their moral responsibilities for the decision to go to war and their ensuing participation in it. In particular, there may be: 'arguments and proofs of the injustice of war so powerful, that even citizens and subjects ... may not use ignorance as an excuse for serving as soldiers'.[40] This important qualification 'represents a succinct account of the legal arguments regarding the limitations on obedience and the duty of dissent raised during the Nuremberg war crimes trials following World War II'.[41] The burden on the most senior of military commanders is somewhat heavier than that for the rank-and-file soldier, for they have a professional responsibility to examine the justice of a war and through their advice to the ruler, avert a conflict that is unjust. There is a clear responsibility here for those of senior rank to 'speak truth to power' (even if those arguments are ultimately over-ruled).[42] While Vitoria was writing nearly 500 years ago, this is a theme that still has clear contemporary relevance.

Even where giving the benefit of the doubt is not an issue, error may still induce a belligerent to believe that they are in the right, even when they are actually 'squarely at fault. This gives rise to a situation in which the guilty party (sincerely) believes itself to be innocent.'[43] After all, very few people would deliberately fight for a cause they knew was wrong and this results in a moral equality of combatants.[44] Because of the lack of metaphysical certainty, it cannot matter who 'started it'; Vitoria demonstrates clearly why both sides are obliged to afford their opponents some respect and conduct their conflict within limits.[45] He also expressly ruled out the long-standing notion that 'the Church or empire had a universal right to wage war (a pivotal idea in the crusades), the claim that wars of conversion were just, and the argument that non-believers had fewer rights than believers'. As Bellamy points out, up until this point in history, this was a rare case in which a 'public intellectual criticised official policy'.[46]

Fransisco Suárez (1548–1617) considered himself to be a disciple of both Aquinas and Vitoria. He echoes many of the arguments already made and drew a distinction between defensive wars that were a reaction to armed attack, and offensive wars that were seeking redress for an injury or injustice.[47] Particularly when undertaking the latter kind of conflict, a ruler must ensure that he has a reasonable expectation of victory: 'if the expectation of victory is less apt to be realized than the chance of defeat, and if the war is offensive in character, then in almost every case that war should be avoided'. However, he argued that a defensive war, regardless of the chance of success, was 'a matter of necessity, whereas the offensive war is a matter of choice'.[48]

There are many other great names associated with the further development of the Just War Tradition. Hugo Grotius, and his magnificent *De jure belli pacis* (On the Law of War and Peace) is probably the most significant of these, marking the real foundation of international law. However, in many ways, this was really the summation of the ideas already encapsulated within the Just War Tradition, the key principles of which were clearly well established by the early modern period (although, of course, there was then and still is now much healthy debate as to how exactly such principles should be applied). As it became increasingly difficult to challenge the justice of the cause of Europe's absolute monarchs, Grotius attempted to 'constrain the sovereign's discretion by emphasizing that a war could only be just if it was launched in a procedurally correct fashion and ... justly conducted'.[49] Many other thinkers sought to develop and refine these principles over the next centuries. At the same time, the Enlightenment project sought to replace religion with reason and provided an increasingly secular basis for the laws of war. Particularly following Vitoria's emphasis on the moral equality of combatants, there was a growing interest in bolstering *in bello* concerns, looking at who and what were legitimate objects of attack and how this could be done. This paved the way for the eventual codification of customary practice into such bodies of rules as the 'Lieber Code' of 1863, the Hague and Geneva Protocols. Chapter 5 explores the international legal framework that emerged from this period, and developed on into the twentieth and twenty-first centuries. Given such developments, the contemporary Just War Tradition is now 'substantially secular', expressed in 'philosophical and legal rather than religious idiom'.[50] This highlights the importance of historical context when seeking to understand the development of the Tradition. James Turner Johnson, who can be credited with laying the foundations for the wider appreciation of Just War thinking in the twentieth century, is particularly aware of this factor, charting the way that its key principles have emerged and been interpreted in light of the issues that were pertinent in each age. As a result, today's Just War Tradition represents a 'fund of practical moral wisdom, based not in abstract speculation or theorization, but in reflection on actual problems encountered in war as these have presented themselves in different historical circumstances'.[51]

Contemporary Just War categories: *jus ad bellum, jus in bello* **and** *jus post bellum*

As can be seen above, the moral distinction between the two levels of conflict allows us to draw a line between the decision to go to war and the actual conduct of that war. Soldiers are not necessarily responsible for the decision to send them to war (except as part of the electorate of a democratic state), but they are responsible for its conduct. Very senior military officers may straddle this line, but, as Walzer points out, this gives us a pretty good idea of where that line should be drawn.[52] As Vitoria established in the mid-sixteenth century, every soldier has a moral and legal duty to disobey any illegal or manifestly immoral order, including an order to take part in a war that is clearly unjust. How far this affects the actual participation of an individual in a war that is demonstrably (or at least arguably) legal but still widely considered immoral is certainly a profound question of personal integrity with no easy answers.

While an unjust cause cannot be made better by fighting a war well, one can certainly undermine a just cause by conducting a war badly. Therefore, the two levels of war are clearly related in a number of ways, but they also remain distinct. It is this separation that allows a nation's public to support their military forces even in an otherwise unpopular conflict. As will be seen below, there is also an emerging third category, *jus post bellum*, concerned specifically with the aftermath of conflict.

We will now look at the principles contained within the contemporary Just War categories in turn.

Jus ad bellum: **just cause**

The clearest example of a just cause is when a state acts in self-defence due to a direct attack against its people or its territory. This is enshrined in Article 51 of the UN Charter, which affirms the inherent right of self-defence possessed by every state: 'Nothing in the present Charter shall impair the inherent right of individual or collective self-defence if an armed attack occurs.' Thus, an action taken in direct self-defence requires no additional authority to sanction it. Defending a neighbour or an ally, or responding to requests to protect the weak against the strong can also be examples of acting in defence, thus the Persian Gulf conflict of 1990–1, where the international community, almost without exception, supported the defence of Kuwait following its invasion by Iraq, provides a good example of a just cause.

As we have seen, Augustine established 1,600 years ago that defence of the innocent was then considered to be the primary justification for the use of force, and some contemporary legal arguments are, in many ways, returning to such a position. The Kosovo intervention in 1999 was justified in terms of preventing gross human rights violations and the emerging 'responsibility to protect' is based upon the idea that if and when a state proves itself to be unwilling or unable to carry out its core responsibility to look after its own population, that responsibility should be transferred to the international community so that it can act instead to prevent massive human rights abuses. Of course, the international community must still employ peaceful means where possible, with military force only being employed as a last resort.[53] Whether this idea will have a practical rather than merely a theoretical impact on state behaviour is unclear, although indications are that so far there is little appetite for intervention 'despite there being no lack of humanitarian crises to provide opportunities for the international community to give meaning to a responsibility to protect'.[54]

Although brought to fore by the Bush doctrine in recent years, the question of the legitimacy of pre-emptive or even preventive action is not actually a new issue. As far back as the thirteenth century, Raymond of Peñafort argued that killing an ambusher before they strike, 'if there is no other way to counter the threat', was considered lawful.[55] Sometimes a threat must be anticipated if it is to be successfully defended against, but how far can this be taken? The importance of demonstrating true necessity before striking first was clearly considered important even then. Morally, the problem with pre-empting a threat is that it can all too easily turn the defender into the attacker (the legal background and implications of this are explored in Chapter 5). The portrait painted by National Security Advisor, Condoleeza Rice, in 2002 suggested that traditional notions of self-defence were no longer adequate for the new security environment of uninhibited actors with potential access to weapons of mass destruction (WMDs). Clearly one cannot wait until the appearance of the mushroom cloud to act in such circumstances.[56] However, if you just decide that somebody or something, at a future time and date as yet unspecified, theoretically *might* pose a threat to you even though currently they do not, attacking them clearly cannot be considered an act of self-defence, either legally or morally. The key to legitimacy in such a case is getting this balance right.[57]

Jus ad bellum: **right intention**

Most people would accept that the motivation of an act has a bearing on whether or not it can be judged morally good or bad. While satisfying the just cause criteria can help establish that one is doing the right thing, having the right intention is necessary to ensure that one is doing it for the right reasons. Creating, restoring or keeping a just peace, righting wrongs and protecting the innocent would all clearly qualify as right intentions, while seeking to expand one's territory, enslave or convert others to one's religion, hatred or revenge would not.[58] Motivation shapes action and if a conflict is motivated by these last two emotions in particular, it becomes all too easy for the enemy to be regarded as less than an equal. Such a climate makes it far more likely that war crimes and atrocities will be committed. Of course, good intentions are often mixed up with those that are less so and, anyway, how does one really know what motivates a state to act? It would be naive to assume that one can always find a single, pure motivation for an action, particularly in the messy world of international politics. However, this does not diminish the importance of this principle as it recognizes that wars fought *primarily* for the wrong motives will invariably lead to an unjust peace. This, in turn, is likely to sow the seeds for further conflict in the long run.

Jus ad bellum: **legitimate authority**

Acting in direct self-defence requires no further authority. However, anything that does not fall into this category requires a declaration by a legitimate authority. Going all the way back to the Roman legal procedural requirements, when an injury has been suffered, the offending party has to be told what it is they have done wrong, and also what it is they can do to restore the situation and therefore prevent war. Some have questioned the continued usefulness of such a declaration, particularly given that one does not actually tend to declare war itself any more.[59] However, the clear declaration of what has been done wrong before embarking on a conflict, whether it is legally called war or not, is more than purely procedural and remains pertinent if the other side is to have any chance of rectifying the situation or making redress. Demonstrating the way that the Just War principles can become intertwined, it would be very hard to demonstrate that one had reached the point of last resort if this declaration had not yet happened.

What qualifies as a legitimate authority in contemporary terms? In Roman times it was the Emperor; in medieval Europe, it was the nobility; while the absolute monarchs of the early modern period were very clearly sovereign authorities. States had a free reign to authorize war when they saw fit from the eighteenth century through to the early twentieth. However, following the formation of the United Nations, the answer is not quite as straightforward. Article 2(4) of the UN Charter declares: 'All Members shall refrain in their international relations from the threat or use of force against the territorial integrity or political independence of any state, or in any other manner inconsistent with the Purposes of the United Nations.' Everything other than actions taken in immediate self-defence requires the prior authority of the Security Council, effectively so that an exception to the rules can be agreed upon.[60] However, in practice, the actual power to do this still appears to rest with states or their regional alliances rather than exclusively with the UN.[61] Even here though, states still have legal procedures for committing their armed forces into a conflict and these need to be adhered to before any such deployment can be considered to have legitimate authority. For example, unless the Unites States is already under attack, the President can only send its military into action abroad with the authorization of Congress. Even if the US has been attacked, military forces cannot remain indefinitely committed without the approval of Congress.[62] All states have some kind of constitutional mechanism for ensuring that their armed forces can only be legitimately employed under certain conditions. It is arguable that legitimate authority does not have to be state-centric, but it must be possible to at least demonstrate a 'significant measure of popular support ... It is worth noting that this aspect of the condition means that even some states cannot justify going to war.'[63]

Jus ad bellum: **a goal that is proportional to the offence**

While war might be legitimate to right certain wrongs, not all wrongs can legitimize war – one must therefore ask whether the overall harm likely to be caused by the war is less than that 'caused by the wrong that is being righted'.[64] Is war a proportionate response to the injury received, or not? Of course, this can be a very subjective criteria to fulfil – after all, what value does one put on something as intangible as national honour? This question demonstrates why this principle cannot be purely a prudential matter, for we must take into account the total cost of the war, not just the cost to ourselves. Clearly this is an enormously difficult assessment to make in advance of military action, particularly due to the inevitable unintended consequences and knock-on effects that may be impossible to accurately predict. However, the Just War Tradition asks that at least a credible attempt is made to answer this question before committing to the use of force.

Jus ad bellum: **reasonable prospect of success**

Some people may regard it as noble to die fighting for a hopeless cause. However, even while recognizing that war involves at least some degree of evil, most accept that it is unethical to sacrifice life and cause unnecessary pain and suffering if there is not at least a reasonable chance that it will change anything. Clearly this is a prudential calculation.[65] Of course, asking this question also means that one must have a clear idea of what success actually is in this situation. Few people would deny that it is sound strategic planning to ask what a war is trying to achieve before embarking upon it. The link between Just War thinking and sound strategy can be demonstrated in this area by Clausewitz, who stated that:

> no-one starts a war – or rather, no-one in his senses ought to do so – without being clear in his mind what he intends to achieve by that war and how he intends to conduct it. The former is a political purpose; the latter its operational objective.[66]

By establishing a clear and realistic objective at the outset, mission creep can be avoided and war can be kept firmly as an instrument of policy rather than its master. Outright victory through the categorical defeat of the other side may not be the definition of success that is required. For example, in 1939 Finland defended itself against the Soviet Union even though it could clearly not win. On the face of it, this would not satisfy the reasonable prospect of success criteria. However, the Finns almost certainly obtained significantly better terms, when they did finally capitulate five months later, than they would have achieved if they had not fought at all.[67]

Jus ad bellum: **last resort**

Has every rational nonviolent alternative been tried before armed force is employed? Such alternatives could include diplomacy, international political pressure and economic sanctions, among many other options. Of course, it is always possible to do *something* that is nonviolent even if it will have absolutely no chance of succeeding, so how does one genuinely know that the point of last resort has been reached? The last resort simply requires that all other *practical* options that might achieve success have been exhausted before military action is initiated. Given that the context may be very time sensitive, it may of course be that some options are simply unavailable.

Jus in bello: **discrimination**

As we have seen above, determining who or what are legitimate objects of attack has been a concern of the Just War Tradition throughout its history. Given the huge number of casualties in the wars of the twentieth century, and the rising percentage of civilian deaths as a proportion of overall casualties, this is an issue that is possibly even more pressing now.[68] Reflecting over two millennia of the Just War Tradition, contemporary law provides a clear separation between the two groups of people:

> Only combatants are permitted to take a direct part in hostilities. It follows that they may be attacked. Civilians may not take a direct part in hostilities and, for so long as they refrain from doing so, are protected from attack.[69]

Civilians have the right never to be intentionally attacked by military forces. On their part, they have a duty not to take up arms, except in direct self-defence.[70] This should be seen as an absolute principle – even extreme military necessity cannot override the prohibition on deliberately targeting noncombatants as it can only require us to do that which is legal in the first place (see Chapters 6 and 7).

Jus in bello: **proportionality**

Closely related to the idea of discrimination is proportionality: just as the war itself must be a proportional response to the injury suffered, the means employed to pursue the war must also be proportionate. The principle requires that the damage, losses or injury resulting from any military action, not just to one's own side but considered overall, should not be excessive in relation to the expected military advantage.

Under the principle of proportionality, would it be 'fair' to employ standoff PGMs against an opponent who cannot defend against them? This is not the appropriate question – the principle of proportionality is *not* about being fair, it is about not using more force than is necessary to achieve the required ends. To use the same 'dumb' weapons as a less sophisticated opponent, thereby inflicting more collateral damage and causing more unnecessary destruction and loss of life than necessary out of some sense of fair play, would, obviously, be obscene.

It was in line with the idea of proportionality that the St Petersburg Declaration of 1868 prohibited the use of incendiary or explosive projectiles below a certain size and weight – the parties accepted that they could achieve the same effect with a traditional solid round, so why cause additional and unnecessary suffering?[71] Going back much further in history, the Laws of Manu prohibited Hindus from employing poison arrows, and both the Greeks and Romans, likewise, prohibited the use of poison and poisoned weapons in their warfare.[72] The use of biological and chemical weapons is banned today for the same reason and one is still prohibited from poisoning water and food supplies.[73] Compared to the military advantage achieved by their use, some types of weapon or methods of war are simply considered too inhumane due to the degree of suffering they can inflict.

The doctrine of double effect

Underpinning both discrimination and proportionality is the doctrine of double effect. As Aquinas explains (above), individuals are not necessarily morally responsible for a foreseeable, yet unintended side effect of an otherwise legitimate action. However, the foreseeable side effects of any military action, even while not directly intended, must still be proportionate to the expected military utility of the target. Part of this calculation requires that any noncombatant casualties are to be avoided as far as is possible. Because of this, the doctrine of double effect can only justify military activities up to a point. For example, it cannot be used to defend the use of weapons of mass destruction against an area containing a civilian population, as these weapons are so indiscriminate that the resulting casualties cannot be regarded as merely a secondary result.[74] This is reinforced by the principle of proportionality and also the idea of *mala in se*, which recognizes that some methods of war are simply evil in themselves, and cannot be justified under any circumstances. This might include the use of rape, genocide or torture as an instrument of war: 'we do not have to do a cost–benefit analysis to determine whether such are impermissible in warfare: we already judge such acts to be heinous crimes because of their very nature'.[75]

This illustrates the way that the Just War Tradition is not only a compromise between realist and pacifist positions, but also accepts both consequentialist and deontological reasoning. Clearly, a certain level of regrettable side effect may sometimes be permissible, but the ends cannot be used to justify *any* means because there are absolute lines that cannot be crossed. As Chapter 1 points out, the Just War Tradition provides a useful framework for balancing this type of ethical consideration because it represents a set of principles that have emerged over time, in part, through a dialogue between deontological and consequentialist reasoning. However, does it really matter if one is killed intentionally or unintentionally? In many practical ways, of course, it does not. Nonetheless, there is a difference between getting into a car and being involved in an accident that hurts somebody, and the conscious act of getting behind the wheel of a car with the intention of running someone down. Even though the effect may be the same, one is a tragic accident, the other is murder. If it turns out that the driver was talking on a mobile phone and having an animated discussion at the same time as the accident, the driver's negligence will have contributed to the accident. In the same way, bad staff work that results in otherwise avoidable deaths is effectively negligence. If such negligence becomes routine, the injuries and deaths caused can no longer be considered accidents and it is difficult to see how anyone is supposed to tell the difference, 'least of all those who have lost loved ones'.[76] The cumulative affect of perhaps individually justifiable actions must also be taken into account:

> Press reports suggest that over the last three years drone strikes have killed about 14 terrorist leaders [in Pakistan]. But, according to Pakistani sources, they have also killed some 700 civilians. This is 50 civilians for every militant killed, a hit rate of 2 percent ... every one of these dead noncombatants represents an alienated family, a new desire for revenge, and more recruits for a militant movement that has grown exponentially even as drone strikes have increased.[77]

The way that one applies the doctrine of double effect is also likely to be affected by the character of the conflict one is involved with. For example, under the banner of the UN, many militaries find themselves involved in humanitarian operations. These are motivated and justified by the desire to aid and protect civil populations in need of support. It would seem inappropriate and even perverse to move the burden of risk on to that same population, accepting a high degree of collateral damage, for example, in an effort to minimize one's own casualties when the very purpose of being there is to protect those people who are now being put at additional risk.[78] The (often erroneous) idea that military operations motivated by humanitarian concerns are somehow less dangerous than conventional military operations should therefore not be allowed to produce a 'radical force protection' mentality when the purpose of being there is to protect other people's lives.

Jus post bellum

Depending on how one views this, it is either a subset of the existing categories, making explicit that which is already implicit in the *ad bellum* and the *in bello*, or it is a separate category in its own right that is overlooked to the detriment of sound strategic thinking. While 'making a better peace' has been the goal of Just War thinking throughout the history of the Tradition, following the widely held view that the US-led coalition that invaded Iraq in 2003 failed to think through the latter stages of the conflict, there has been a resurgence in attention paid to the idea of *jus post bellum*, or justice after war. For example, in *The Morality of War*, Brian Orend sets out a 'general blue-print' for post-war justice.[79] This approach considers factors such as the legitimate ends of a Just War: the settlement must be publicly declared and proportionate to the initial justification for the conflict; it must recognize and vindicate the rights of everyone involved, not just the victor; it must discriminate between those who are morally cul-pable and those who are not; provide appropriate punishment for those (on both sides) who may have violated both *ad bellum* and *in bello* principles; consider compensation that does not sow the seeds of future conflict; and, finally, allow rehabilitation or reform of state institutions that are requiring of it.

Some people argue that these elements can be found in extant *ad bellum* and *in bello* principles already – for example, the reasonable chance of success test requires a definition of success and this must take into account the long-term, post-war situation or it is meaning-less.[80] One can certainly argue, as Kant himself did, that 'both *jus ad bellum* and *jus in bello* are fundamentally constrained by a vision of what shall happen after the war'.[81] However, there is also no doubt that focusing attention on more than just the 'shooting war' at the start of a conflict has got to be a positive development. Once the longer-term implications of military action are really appreciated *before* the decision to use force is made, the chances of military force being employed lightly have got to be reduced, while the chances of it being used appropriately in pursuit of a better peace might be improved.

Conclusion

Statesmen and women who wish to be successful must keep in mind that war must only be considered as a means towards achieving a better peace, rather than as an end in itself. This was something that was recognized by Plato in the fourth century BC and has remained a constant idea throughout the development of the Just War Tradition.[82] This consideration needs to shape and guide both policy and conduct at all the different levels of war.

The Just War Tradition (and the legal and moral norms that it repre-sents) is not necessarily about providing a set of answers. It is not a formula which somehow generates a successful outcome as long as the right things are put into it. However, it can help to structure decision-making as the factors it asks us to consider should be taken into account before and during any use of armed force. It provides a frame-work for distinguishing between justifiable military action within an ethical framework, and murder on a massive scale.[83] It also provides a common language within which the rights and wrongs of conflict can be discussed and debated.

Violating the basic principles contained in the Just War Tradition will, more often than not, prove counterproductive in the long run by making the return to peace harder, thus delivering only hollow military success rather than genuine political victory. Ignoring those principles can only be to one's moral and strategic detriment.

Notes

1. See James Turner Johnson, *The Just War Tradition and the Restraint of War* (Princeton, NJ: Princeton University Press, 1981).
2. Richard Sorabji and David Rodin (eds) *The Ethics of War: Shared Problems in Different Traditions* (Aldershot: Ashgate, 2007), p. 5. See also N. Allen, 'Just War in the *Mahābhārata*' in the same work. For further examples, see Paul Robinson (ed.) *Just War in Comparative Perspective* (Aldershot: Ashgate, 2003).
3. The following three example are taken from David Whetham, 'Ethics and the Enduring Relevance of Just War Theory in the 21st Century', in John Buckley and George Kassimeris (eds) *The Ashgate Research Companion to Modern Warfare* (Aldershot: Ashgate, 2010), pp. 242f.
4. Bertrand Russell, *Common Sense and Nuclear Warfare* (London: George Allen & Unwin, 1959), pp. 18–21.
5. There were some Christians in the Roman military before this date, but there is no evidence of Christian participation in the Roman military before 173. See Alex J. Bellamy, *Just Wars: From Cicero to Iraq* (Cambridge: Polity Press, 2006), p. 21.
6. George Orwell, cited in Michael Walzer, *Just and Unjust Wars: A Moral Argument with Historical Illustrations*, 2nd edn (New York: Basic, 1992), p. 332.
7. The position results in a form of ethical relativism as a result. See Chapter 1.
8. Thucydides, *The History of the Peloponnesian War*, trans. Rex Warner (Harmondsworth: Penguin, 1972), p. 402.
9. See Paul Cornish, 'Clausewitz and the Ethics of Armed Force: Five Propositions', *Journal of Military Ethics* (Volume 2, Edn 3, 2003), pp. 213–26.
10. http://nobelprize.org/nobel_prizes/peace/laureates/2002/carter-lecture.html
11. For example, see P. Christopher, *The Ethics of War and Peace: An Introduction to Legal and Moral Issues* (Englewood Cliffs, NJ: Prentice-Hall, 1994), pp. 9f.
12. Josiah Ober, 'Classical Greek Times', in Michael Howard, George J. Andreopoulos and Mark R. Shulman (eds) *The Laws of War: Constraints on Warfare in the Western World* (New Haven: Yale University Press, 1994), p. 13.

13. Coleman Phillipson, *The International Law and Custom of Ancient Greece and Rome* (London: Macmillan, 1911), p. 296.
14. See Ober, 'Classical Greek Times', pp. 17ff.
15. Bellamy, *Just Wars*, p. 17.
16. Plato, 'Republic V', line 470, Edith Hamilton and Huntington Cairns (eds) *The Collected Dialogues of Plato* (Princeton, NJ: Princeton University Press, 1989), p. 709.
17. Deuteronomy (20:10, 19) instructs the Jews to refrain from destroying fruit-bearing trees, echoing the Greek ideas of restraint.
18. Aristotle, 'Politics', from J. L. Ackrill (ed.) *A New Aristotle Reader* (Oxford: Clarendon, 1990), I, 6.1255a5–25 and VII, 14.1333a30–5.
19. Bellamy, *Just Wars*, p. 18.
20. M. Hamburger, *Morals and Law: The Growth of Aristotle's Legal Theory* (New Haven: Yale University Press, 1951), pp. 172–5.
21. Alan Watson, *International Law in Archaic Rome: War and Religion* (Baltimore: Johns Hopkins University Press, 1993), pp. 4ff. and 43.
22. David Whetham, *Just Wars and Moral Victories: Surprise, Deception and the Normative Framework of European War in the Later Middle Ages* (Leiden: Brill, 2009), p. 36.
23. Cicero, *On Duties*, bk 1, sections 34–41. From Gregory M. Reichberg, Henrik Syse and Endre Begby (eds) *The Ethics of War: Classic and Contemporary Readings* (Oxford: Blackwell, 2006), p. 52.
24. Cicero, *On Duties*, bk 1, sections 34–41.
25. Such arguments have a certain resonance with the neo-conservative agenda today.
26. Cicero, *On Duties*, bk 1, sections 34–41. Cicero remarks that he would have preferred if Rome 'had not destroyed Carthage and Numantia'.
27. Cicero, *On Duties*, bk 1, sections 34–41.
28. Reichberg, *Ethics of War*, p. 71.
29. Augustine, *Against Faustus the Manichean*, bk XXII, ch.74. In Ernest L. Fortin and Douglas Kries (eds) *Augustine: Political Writings*, trans. Michael W. Tkacz and Douglas Kries (Indianapolis: Hackett, 1994), p. 222.
30. Augustine, *Against Faustus*, bk XXII, ch.75.
31. Bellamy, *Just Wars*, p. 28.
32. See Whetham, *Just Wars and Moral Victories*, ch. 2.
33. Reichberg, *Ethics of War*, p. 169.
34. Anthony Coates, 'Culture, the Enemy and the Moral Restraint of War', in Sorabji and Rodin, *The Ethics of War*, p. 215.
35. Coates, 'Culture, the Enemy and the Moral Restraint of War'.
36. Coates, 'Culture, the Enemy and the Moral Restraint of War'.
37. Aquinas, *Summa Theologica*, Qu.64, Art.7, pp. 1465f.
38. Vitoria, *On the Law of War*, Qu.2, Art. 3. In Reichberg, *Ethics of War*, p. 323.
39. Vitoria, *De Indis De Jure Belli*, Part III.31. From Walzer, *Just and Unjust Wars*, p. 39.
40. Vitoria, *On the Law of War*, Qu. 2, Art.1.1. In Reichberg, *Ethics of War*, p. 318.

41. George R. Lucas, 'Advice and Dissent: the Uniformed Perspective', in D. Whetham and D. Carrick (eds) *Journal of Military Ethics*: 'Saying No: Selective Conscientious Objection', Special Edition, 8, Edn 2, June 2009). This article provides an excellent summary of the arguments for and against various forms of selective conscientious objection.

42. McMaster argues that there was an abdication of responsibility by American military leadership during the Vietnam War. H. R. McMaster, *Dereliction of Duty: Johnson, McNamara, the Joint Chiefs of Staff, and the Lies that Led to Vietnam* (New York: Harper Perennial, 1998).

43. Vitoria and Aquinas. From Reichberg, *Ethics of War*, p. 317.

44. This moral equality is beautifully articulated on the Kemal Atatürk Memorial in Gallipoli: 'There is no difference between the Johnnies and the Mehmets to us where they lie side by side now here in this country of ours.' It has been challenged recently by McMahan among others, who question why those on the 'wrong' side should be afforded the same rights as those who are innocent of waging aggressive war, etc. See J. McMahan, *Killing in War* (Oxford: Oxford University Press, 2009).

45. As Chapter 1 argues, accepting the moral equality of some combatants is not easy in many contemporary operations, but neither is it a requirement for the adherence to the other Just War principles.

46. Bellamy, *Just Wars*, pp. 51–2.

47. Fransisco Suárez, *Disputation XIII*, Section I. From Reichberg, *Ethics of War*, pp. 340–2.

48. Suárez, *Disputation XIII*, Section IV. From Reichberg, *Ethics of War*, pp. 352–3.

49. Bellamy, *Just Wars*, p. 76.

50. Sorabji and Rodin, *The Ethics of War*, p. 2.

51. James Turner Johnson, *Can Modern War be Just?* (New Haven, CT: Yale University Press, 1984), p. 15.

52. Walzer, *Just and Unjust Wars*, p. 39.

53. See the International Commission on Intervention and State Sovereignty. http://www.iciss.ca/

54. Steven Haines, 'Humanitarian Intervention: Genocide, Crimes against Humanity and the Use of Force', in John Buckley and George Kassimeris (eds), *The Ashgate Research Companion to Modern Warfare* (Aldershot: Ashgate, 2010), pp. 322-325. See also Chapter 5 in this volume.

55. Raymond of Peñafort, *Summa de casibus poenitentiae*, II, 17–19. From Reichberg, *Ethics of War*, p. 140.

56. http://transcripts.cnn.com/TRANSCRIPTS/0209/08/le.00.html

57. Whetham, 'Ethics and the Enduring Relevance of Just War Theory', p. 248.

58. While the Greco-Roman and Christian tradition took until Vitoria's time to reject the idea of forced conversion being a legitimate motivation for military action, the Islamic faith embraced this idea long before this. 2: 256 of the Qur'an states unequivocally that 'there is no compulsion in religion'.

59. See Ian Holliday, 'When is a Cause Just?', *Review of International Studies*, 28 (2002), p. 565.

60. Peacekeeping and the majority of Peace Support Operations will generally be conducted under the legal authority of a UN Security Council Resolution.

61. For example, the authority for the Kosovo intervention came from the agreement by the countries of NATO and was then retrospectively legitimized by the UN.

62. http://avalon.law.yale.edu/20th_century/warpower.asp.

63. Holliday, 'When is a Cause Just?', p. 567.

64. Bellamy, *Just Wars*, p. 123.

65. Johnson argues that three prudential criteria have been 'attached' to the classical *ad bellum* criteria: proportionality, reasonable prospect of success and last resort. These are obviously important, but are also 'at best supportive concerns having more to do with the wise practice of government' and effectively cede priority to the other, deontological criteria. See James Turner Johnson, *The War to Oust Saddam Hussein: Just War and the New Face of Conflict* (Lanham, MD: Rowman & Littlefield, 2005), p. 36.

66. Carl von Clausewitz, *On War* (Oxford: Oxford University Press, 2007), VIII: 2, p. 223.

67. Walzer, *Just and Unjust Wars*, p. 70.

68. For a particularly compelling argument for upholding civilian rights, see Hugo Slim, *Killing Civilians: Method, Madness and Morality in War* (London: Hurst & Co, 2007). See also http://www.correlatesofwar.org/

69. *The Manual of the Law of Armed Conflict* (Oxford: Oxford University Press, 2004), p. 24. Referring to Additional Protocol I, Art. 43 & 51.

70. Both the idea of protected persons, and that this protected status can be forfeited by taking up arms and fighting alongside soldiers can be found in many of the religious Just War traditions. For example, see John Kelsay, 'Arguments Concerning Resistance in Contemporary Islam', in Sorabji and Rodin, *The Ethics of War*, p. 65.

71. Kelsay, 'Arguments Concerning Resistance in Contemporary Islam', p. 109.

72. Adam Roberts and Richard Guelff (eds) *Documents of the Laws of War*, 2nd edn (Oxford: Oxford University Press, 1995), p. 29.

73. For example, the Geneva Gas Protocol of 1925, which prohibited 'the use in war of asphyxiating, poisonous or other gases, and of all analogous liquids, materials or devices'. See *The Manual of the Law of Armed Conflict*, p. 11.

74. Whetham, 'Ethics and the Enduring Relevance of Just War Theory', p. 253.

75. Brian Orend, *The Morality of War* (Toronto, Ontario: Broadview Press, 2006), p. 123.

76. David Whetham, 'Killing Within the Rules', *Small Wars and Insurgencies*, 18(4), December 2007, p. 727.

77. David Kilcullen and Andrew McDonald Exum, 'Death From Above, Outrage Down Below', *New York Times*, 17 May 2009.

78. Whetham, 'Ethics and the Enduring Relevance of Just War Theory', p. 253.

79. Orend, The *Morality of War*, pp. 160–219.
80. For a survey of the different positions, see Doug McCready, 'Jus Post Bellum and the Just War Tradition', *Journal of Military Ethics*, 8(1) (2009), pp. 66–78.
81. Reichberg, *Ethics of War*, p. 539.
82. Plato, *Laws I*, p. 1230.
83. Bellamy, *Just Wars*, p. 1.

Chapter 5

War Law and Its Intersections

CHRISTOPHER P. M. WATERS

Introduction: law's bedfellows

One of the aims of this book is to bring together in one place three subjects normally treated apart: ethics, law and military strategy. Since this is a chapter on law, the bulk of it will be on what the law on the use of force actually is. It will also address how the law is evolving and where controversies arise in the law (though to be clear from the start, controversies arise much more in the application of law to facts than with respect to the law itself). Before getting to that, however, it will be useful to briefly return to the nature of the relationships between law and ethics and between law and strategy that have been raised in the Introduction. The former relationship has seen much ink spilt, but the latter is only coming to the fore now. Both provide a necessary back-drop to this chapter.

With respect to law and ethics, questions that have long been asked include: is there an internal morality to law? Should wicked laws be obeyed? Should wicked laws – say in the Nazi or apartheid eras – even be considered 'law'? To caricature a complex philosophical debate, one extreme of the debate might be considered the 'law is law' approach, which sees ethics or morality as irrelevant to the law's content.[1] The other would give law little autonomy and consider it 'law' only when it maps on – substantively (genocide is wrong) or procedurally (laws should be clear and not retroactive) – to ethics or morality. This peren-nial debate has sometimes been labelled a 'positivist' conception of law versus a 'natural law' view of law. What is clear to most observers is that while ethics and law are distinct concepts, there are intersections between the two. For example, few in the 'law is law' camp would deny that questions of 'What ought the law to be?' are relevant to law-making, even if strictly different from questions of 'What is the law now?' Whether or not the use of cluster munitions is universally illegal is a different question from whether or not they ought to be made illegal. Before moving on then, let us simply note that what is 'legal' and what is 'ethical' hopefully, but not necessarily, resemble each other. Simply obeying the law as a guide to operational conduct does not, perhaps unfortunately, necessarily relieve commanders and others of undertaking a separate ethical calculus in all circumstances.

Unlike the relationship between law and ethics, the relationship between law and strategy is only now being explicitly and fully teased out. One dimension of this relationship is the idea that adherence to international law (and in particular the branch of that law called international humanitarian law [IHL]) helps win wars. This is particularly true in counter-insurgency wars, where breaches of IHL – unjustified destruction of civilian property or prisoner abuse – can lead to public disenchantment with counter-insurgency forces and corresponding support for insurgents. This idea is now largely established at the doctrinal (if not practical) level for many militaries, including the UK and the US,[2] although there are some who argue that breaches can help to win conflicts, including the 'war on terror', by, for example, providing useful intelligence through coercive techniques.[3] Personally I find the detractors to be wrong on this point – evidence obtained under torture is unreliable and torture only serves to make bitter enemies – but undertaking this kind of calculus does demonstrate the dangers in providing instrumental or strategic reasons for obeying the law. Strategy should not trump law when the two depart. Another dimension to the law–strategy relationship is that law may be self-consciously used as a strategic tool. In the US, the term 'lawfare' has gained in popularity since the 2002 Iraq invasion.[4] It is generally seen as the use of courts and legislation by asymmetric enemies to restrain superior Western militaries. Attempts to use the courts to close the detention facility at Guantanamo Bay have been described as an example of lawfare in action. Conversely, Western militaries will use law strategically, by justifying actions through an enabling legal framework. For instance, civilian casualties may be cast as an unfortunate but inevitable and acceptable outcome because they are not disproportionate to the military objective of an operation. As David Kennedy puts it in his masterful examination of the subject: 'Law now offers an institutional and doctrinal space for transforming the boundaries of war into strategic assets as well as a vernacular for legitimating and denouncing what happens in war.'[5]

Law's pervasiveness and relevance

One of the things that recent analyses of the law–strategy axis reveal is that legal questions have pervaded questions surrounding the military use of force. Rather than being a bit-player, law is now seen as central to the conduct of military operations. One only need think of the rise of the legal adviser in Western militaries – vetting bombing targets in Kosovo and subsequent operations – to see that this is true. Some have resisted this legal creep, in the UK calling it 'legal encirclement' and suggesting that operational effectiveness is reduced as a result.[6] Others, including the author of this chapter, have suggested that law-making with respect to the military is an interactive process; the military is a player in determining, for example, the nature of the Armed Services Act or the relationship with humanitarian organizations in the delivery of emergency food aid.[7] Whether one welcomes or loathes the pervasiveness of law in the conduct of military operations – and indeed with respect to various aspects of military life including recruitment and discipline during peacetime – it appears the law as one framework for military decision-making is here to stay. Not surprisingly given the pervasiveness of legal questions in Western military practice, some British military officers have expressed the view that they have insufficient legal training and would like to receive greater familiarity with international law and its domestic implementation.[8] The growing perception of law's relevance has various causes. One has been alluded to already and that is the strategic need to justify or defend operational decisions. Another is to the need to explain to oneself and subordinates, as well as, outside of the military, civilians (both at home and in theatres of operation) and the press, the legal basis for a mission. The threat of personal liability for one's actions (or those of subordinates under the rubric of command responsibility) is also a natural motivator in wanting to know the law. To these reasons might be added the fact that law and order and rule of law reform are often central aspects of a mission. Officers may need to be familiar with the legal framework to successfully pursue mission goals. Unfortunately, at present, it would appear that insufficient training – and I would argue failures in leadership in terms of setting out in unambiguous terms the need for strict compliance with law – have contributed to some misperceptions about what the law actually is. One need only think of the unwarranted fear and confusion in some quarters over the International Criminal Court's mandate to illustrate this point.[9] This is a good place then to turn to the law itself.

As a preliminary matter, it should be noted that the focus of the discussion will be on international law rather than on domestic civilian or military law. There are numerous introductory writings on international law and no more than a superficial sketch can be given here of the nature of this body of law.[10] Suffice it to say for present purposes that although linked with domestic legal systems, international law represents a separate and distinct legal system. It has different sources, actors, substantive rules, methods of interpretation and enforcement.

International law is primarily created between states which are sovereign and legally equal and may be made in one of two ways. The first is by way of treaties, or as they are often called, conventions, which are binding agreements akin to contracts. Nowadays multilateral treaties cover many spheres of international life, from postal exchange to war crimes and from air travel to trade. The second, customary international law, is more difficult to grasp but remains an enduring and evolving source of law even in an era where it is sometimes overshadowed by the rise of treaty law. It is created through state practice, which is largely constant and uniform, combined with an acceptance that the practice is law. Both the objective actions, which constitute state practice, and the subjective element (often referred to as *opinio juris*) – the understanding that the state practice is governed by law and is not merely habit – are needed for a practice to be considered law. For instance, while naval vessels may salute each other at sea with some consistency, there is no sense that there is a 'legal' obligation to do so and therefore no customary international law is created on the subject. By contrast, allowing free passage of ships on the high seas is both state practice and perceived as a binding legal norm by states themselves. The free passage rule thus represents customary international law binding even states which have not signed the UN Convention on the Law of the Sea. In addition to treaties and customs, it should be noted that general principles of law from around the world and the resolutions and practices of international organizations also contribute to our understanding of what international rules exist.

It will be evident that unlike the prototypical domestic legal system, which can be categorized as vertical (a legislature centrally passing binding laws on all citizen-subjects), international law is created 'horizontally' between states. There is, for example, no international legislative body capable of passing laws which will be automatically enforced by an international police force. Although the UN apparatus may play some of the roles played by domestic governance institutions, international law is primarily set, interpreted and enforced by states themselves. This is, in one sense, the law's weakness. But as will be discussed below, international law is enforced through different means and preconceptions about what a legal system looks like by analogizing from the domestic sphere have to be put aside.

International law on the use of force

There are two generally recognized branches of 'war law', commonly referred to by their Latin terms. These terms derive from the Just War Tradition (see previous chapter) though it should be noted that their legal meaning – especially in the post-UN Charter era – is distinct from the ethical tradition. The first branch is the *jus ad bellum* (law on the use of force). It is concerned with whether resort to force is legal or illegal. The second is the *jus in bello* (the law in war). This branch – IHL essentially – deals with how hostilities can be conducted when force is being used, regardless of whether or not the overall use of force is legal or illegal. Practically speaking the two categories are far from watertight. Participants in, and observers of, armed conflict do in fact – if not self-consciously – let perceptions of the justness of the fight impact on their perceptions of the justness of a particular kind of action or use of a particular kind of weapon (think of justifications for the use of the atomic bomb against Japanese civilian targets in 1945 as an extreme example). Conceptually, however, the *jus ad bellum* is broadly distinct from the *jus in bello* and will be treated as separate sections of this chapter. Before moving on, it has also become fashionable to speak of a *jus post bellum*, a branch of law which deals with peace agreements and transitional justice.[11] That branch will only be addressed tangentially here when discussing enforcement.

Jus ad bellum

There are pre-UN Charter antecedents which attempted to limit states' discretion in opting for war as an instrument of their foreign policy. Notably, the Covenant of the League of Nations allowed for a 'cooling off' period before resort to arms was permissible and the 1928 Kellog–Briand Pact explicitly sought to outlaw war. The latter treaty stated: 'The High Contracting Parties solemnly declare in the names of their respective peoples that they condemn recourse to war for the solution of international controversies, and renounce it, as an instrument of national policy in their relations with one another.'[12] The ultimate failure of the League and the Pact are obvious, however, and, while of historical interest, only provide the backdrop to the post-1945 regime of the UN Charter. The Charter was adopted explicitly to prevent further world war and it not only seeks to prohibit the threat and use of 'force' (a broader notion than the prior, formal term of 'war' used in the Kellogg–Briand Pact) but provides for a centralized response to breaches of the prohibition. The basic rule on the use of force is Article 2(4): 'All states shall refrain from the threat or use of force against the territorial integrity or political independence of any State, or in any other manner inconsistent with the Purposes of the United Nations.' Although there is some ambiguity in the language of Article 2(4) which some states have sought to explore (Argentina in the Falklands dispute claiming it was not in breach of 2(4) because it was

reclaiming its own 'territorial integrity'), the Article has been interpreted as a broad prohibition on the use of force. Thus, violations of
international borders or the use of armed groups to make incursions
into another state would both contravene the article.[13] Conversely it
should be noted that not only is force prohibited under the Charter,
but also states are under a positive duty to seek to resolve their disputes peacefully. It should always be borne in mind that the peaceful
settlement of disputes – through negotiation, mediation, arbitration,
courts and 'good offices' – is the usual course of action in the vast
majority of international disputes which arise.

The Charter, which is ultimately a treaty albeit one which has a
quasi-constitutional character, has been ratified by all members of the
UN. It is also widely recognized that Article 2(4) represents customary
international law on the subject of the use of force. Thus, if a state
which is not a member of the United Nations were to embark on an
aggressive war, such action would still be illegal. There are, however,
two recognized exceptions to the prohibition on the use of force. The
first is the use of force pursuant to the collective security measures
established under the Charter. When the 15-member Security Council
decides that there has been a threat or beach of the peace it may order
states to act or desist from acting in a certain manner. If its orders are
disobeyed, measures 'not involving the use of armed force' may be
imposed.[14] These measures can include economic or political sanctions
(increasingly becoming 'smarter' and more targeted against individuals
or elites within a state) or other measures such as a weapons inspection
regimes or the establishment of an international criminal tribunal (such
tribunals were established for both the former Yugoslavia and
Rwanda). If, however, those measures are deemed inadequate, the
Security Council may 'take such action by air, sea, or land forces as
may be necessary to maintain or restore international peace and security'.[15] As the UN as currently constituted has no standing forces, the
Security Council may delegate enforcement action to a regional security organization (such as the North Atlantic Treaty Organization
[NATO] or the Economic Community of West African States
[ECOWAS]) or to a coalition of states. During the Cold War, with
deadlock among the five veto-wielding permanent members on the
Security Council, the enforcement provisions were largely 'dead letter'.
However, with the end of the Cold War, more robust Security Council
action was possible. Beginning with the grant of authority to the US-
led 'coalition of the willing' following the 1990 invasion of Kuwait by
Iraq, an era of Charter-sanctioned forceful intervention (sometimes
called peace enforcement) was ushered in as demonstrated through,
among other examples, intervention in the Balkans.

The most controversial use of Security Resolutions to justify the use of force came with the 2003 invasion of Iraq by US/UK-led forces. Both the US and UK purported to find legality for their operations under Security Council grants of authority to the first Gulf War coalition which, they argued, were automatically reinstated by Iraq's failure to comply with the weapons inspection regime put in place after the 1990 invasion of Iraq.[16] This reasoning has been condemned by most international lawyers in the UK, who have argued that, among other things, such reasoning is contrary to the purposes and principles of the United Nations and represents a turning of backs on the historic 'transatlantic commitment to international law'.[17] What is clear is that while the dispute over the interpretation of the Security Council Resolutions was a body blow to the collective security system, it was not a lethal blow as was feared in many camps in 2003. While the taste for open-ended 'peace enforcement' missions has subsided, the Security Council maintains a busy agenda and has taken action on such matters as terrorist financing and peacekeeping (which, unlike peace enforcement, involves the consent of the host states).

The second exception to the use of force prohibition is the right to self-defence, which the Charter itself refers to as an 'inherent' right. Article 51 provides that:

> Nothing in the present Charter shall impair the inherent right of individual or collective self-defence if an armed attack occurs against a Member of the United Nations, until the Security Council has taken measures necessary to maintain international peace and security. Measures taken by Members in the exercise of this right of self-defence shall be immediately reported to the Security Council and shall not in any way affect the authority and responsibility of the Security Council under the present Charter to take at any time such action as it deems necessary in order to maintain or restore international peace and security.

This passage provides that there must be an armed attack, meaning, among other things, that political or economic pressure would not invoke a right to forceful self-defence. Questions might also be asked whether a terrorist attack by nonstate actors – or at least when there is support by a government – might be considered an 'armed attack' for the purposes of self-defence. This was claimed by the US following '9/11' as a justification for the invasion of Afghanistan and appears to have been accepted by many states, if tacitly in some instances, as legal on that occasion. Article 51, also states that self-defence might be individual or collective, permitting for example, NATO-style agreements which provide that an attack on one state is an attack on all. Further, there are what might be called procedural aspects to Article 51 in relation to Security Council action. There is little in the Charter itself, however, to indicate the scope of the inherent right. For this we must look to customary international law. The usual starting point is the case of the *Caroline* which arose out of an 1837 rebellion in Canada. This is not really a case in the sense of a court decision, but rather an exchange of letters between the British foreign minister and his American counterpart following the firing and sending over Niagara Falls by British troops of an American vessel, the *Caroline*, which had been docked in an American port. The *Caroline* had been used to supply American nationals attacking British assets in Canada. The formulation which came out of that exchange of letters was that there had to be 'necessity of self defence, instant, overwhelming, leaving no choice of means, and no moment for deliberation' and that the response could not be 'unreasonable or excessive'.[18] In other words, to use the usual shorthand, self-defence must be necessary and proportionate. It is here that debates most often arise as to whether a response – say the US-led invasion of Afghanistan – was necessary and proportionate and often there can be no mechanical or formulaic response to that question. However, to give a clear example of disproportionate response, one can point to the 2008 Russian attacks in Georgia far beyond the initial conflict zone of South Ossetia (where its peacekeepers were initially threatened by a Georgian attack on the South Ossetian capital). As a European Union fact-finding commission put it in its 2009 report on the war:

> [The] massive and extended military action ranging from the bombing of the upper Kodori Valley to the deployment of armoured units to reach extensive parts of Georgia, to the setting up of military positions in and nearby major Georgian towns as well as to control major highways, and to the deployment of navy units on the Black Sea ... cannot be regarded as even remotely commensurate with the threat to Russian peacekeepers in South Ossetia.[19]

One further area of controversy in the area of self-defence is with respect to anticipatory self-defence. Must the attack have actually occurred, as suggested by a strict reading of Article 51, or can an attack be 'imminent'? What about peremptory attacks against nonimminent, but nonetheless deadly threats, as suggested by the Bush – and now possibly the Obama - doctrine?[20] There is a volume of scholarly literature on this topic, but state practice suggests that anticipatory self-defence in the face of imminent attack would be legal while a broader peremptory right – a 'preventative' attack on an installation in State A which might be making weapons to be used at some point in the future against State B – would be illegal.[21] Thus, the Israeli strike on the Iraqi nuclear reactor at Osiraq in 1981 was broadly condemned as illegal. Quite clearly claims of a broad peremptory right of self-defence are open to unilateral abuse and threaten the basic Charter regime on the use of force.

While there are some other proposed exceptions to the prohibition on the use of force (one of the more outlandish ones is the use of force to promote democracy), the main contender might be said to be 'humanitarian intervention'. That the Security Council can intervene on humanitarian grounds is now uncontroversial – it has found the 'trigger' for its action in the threat to international peace and security caused by refugee flow and regional instability inherent in humanitarian crisis – and it has done so on several occasions. What, however, if the Security Council is deadlocked or disinterested in a humanitarian crisis, one caused either by civil strife or by natural disaster, where the state is unwilling or unable to help its people, or indeed where the state itself is the perpetrator of atrocities? This situation arose in Kosovo in the late 1990s. Faced with some atrocities being committed against Albanian Kosovars by 'Yugoslav' security and paramilitary forces, the perceived intransigence of the Milosovic regime and in light of a probable Russian and Chinese veto in the Security Council, NATO countries undertook a bombing campaign against 'Yugoslav' targets. This bombing campaign ultimately resulted in a ceasefire and the deployment of a peacekeeping mission. Significantly, most NATO states did not claim to be acting under self-defence, Security Council authorization (though several pointed to the fact that the Security Council had previously passed resolutions noting the existence of a humanitarian emergency) or a new, stand-alone 'humanitarian intervention' exception to the basic prohibition on the use of force. Most claimed that the Kosovo crisis was an extra-ordinary situation where an exception was warranted to protect a civilian population. This position was essentially an argument that the NATO action was 'technically' illegal but legitimate nonetheless. Kosovo, therefore, cannot be used as a strong precedent for the creation of a customary legal norm of humanitarian intervention. What Kosovo did do, however, was set off a soul-searching moment about how the international community should react to humanitarian catastrophes. What has gradually emerged to take the place of humanitarian intervention (where the emphasis seemed to be on 'intervention') is the idea of a 'responsibility to protect' ('R2P'). The emphasis of the R2P doctrine is on the responsibility of each individual state to protect its population. Where, however, that state fails to protect its population, the doctrine provides for the international community to 'step up to the plate'. As pledged by UN General Assembly members at the 2005 World Summit:

[w]e are prepared to take collective action, in a timely and decisive manner, through the Security Council, in accordance with the Charter, including Chapter VII, on a case-by-case basis and in cooperation with relevant regional organizations as appropriate, should peaceful means be inadequate and national authorities are manifestly failing to protect their populations from genocide, war crimes, ethnic cleansing and crimes against humanity.[22]

Unlike resolutions of the Security Council, which are binding on all states, General Assembly resolutions are not binding and are sometimes referred to as 'soft law'. Nonetheless, the R2P framework suggests the possibility of a coherent, principled approach to states where governments fail to protect their own people. Non-governmental organizations (NGOs) have taken the principle up with gusto in their efforts to pressure governments into responding robustly to international crises in places such as Darfur, and R2P will likely be an important feature of the legal and political discourse on the use of force for some time to come.

It is difficult to know how to conclude on the topic of legal constraints on the use of force. Some, from the classical realist perspective (as evident in Chapter 2) view this area of the law with cynicism. Yet it is worth noting that no state has claimed to be exempt from the Charter regime on the use of force. When states do use force, justification is always sought under an exception to the basic prohibition. The Charter regime has shaped the way states perceive their options. For example, aggressive war for territorial conquest is now a nonstarter. The influence on state behaviour may or may not be affected by perceptions of legitimacy, or the law's 'compliance pull'. It may be mostly about states' perceptions that their long-term self-interest is tied up with a predictable and stable world order. One other potential reason for the general success of the Charter regime, as suggested by the reference to NGOs in the preceding paragraph, is the increasing engagement of nonstate actors with international legal issues. Governments are now pressured to abide by the rule of law in international affairs by 'norm entrepreneurs', NGOs which lobby and challenge government actions in court. These nonstate actors include the population at large. One need only think of the street demonstrations in London opposing the Iraq war, partly on the basis of its legality, or the fact that the war's legality became an election issue in Britain, to realize the truth of this.

One other organization which had been previously unconcerned with the *jus ad bellum* is the military. While traditionally an active participant in the creation and development of the *jus in bello*, Western militaries under the doctrine of civilian supremacy have essentially not concerned themselves with the question of whether or not a conflict is legal. This appears to be slowly changing, as evidenced by reports that the British Chief of Defence Staff at the time of the Iraq invasion insisted on a government legal opinion that the war was legal before committing his armed forces to action.[23] And, at a 'grass roots' level, legality matters as well, as suggested by calls for the Military Covenant to be rethought. For example, in an open letter to the Prime Minister published in the *Independent on Sunday*, the signatories – who included family members of active and deceased service people – demanded 'the right [of British service people] to expect any war to be lawful'.[24]

Jus in bello

The ideas underlying IHL (also called the law of war or the Law of Armed Conflict) find resonance in ancient notions such as chivalry and a warrior's honour.[25] Most cultural and religious traditions can be plumbed for examples of 'proto-IHL' protection afforded to civilians (such as women and children) or special classes of fighters (such as those carrying a white flag). The origins of its modern and multilateral incarnation, however, can be traced to the latter half of the nineteenth century in Europe.[26] In fact, the story of IHL is most often – somewhat simplistically so – grounded in a particular European battlefield. In 1859, during the Battle for Solferino in the war for Italian unification, a travelling Swiss businessman, Henri Dunant, witnessed battlefield carnage on a massive scale. Together with local citizens he collected and cared for the wounded who had been left on the battlefield. Touched by what he had witnessed and convinced of the need for action, Dunant wrote a tract entitled *A Memory of Solferino,* in which he suggested the need to create a relief group to address the inadequacy of army medical services. He also asked the militaries of various countries whether they could formulate 'some international principle, sanctioned by a convention and inviolate in character, which, once agreed upon and ratified, might constitute the basis for societies for the relief of the wounded in the different European countries?'[27] The group formed in 1863 to continue the agenda suggested by Dunant was the International Committee of the Red Cross (ICRC), an organization which remains the guardian of much of IHL. At the urging of the ICRC, the Swiss government agreed to convene a diplomatic conference which resulted in the 1864 Geneva Convention for the Amelioration of the Condition of the Wounded in Armies in the Field and set the stage for a multilateral treaty regime which continues to evolve to this day.

The most important IHL treaties today are the four Geneva Conventions (GCs) of 1949 and the additional protocols to the GCs.[28]

The GCs are the most universally ratified (ratification is a process whereby a state formally agrees to be bound) treaties with 194 states party. They cover victims of land (GC I) and sea (GC II) warfare, prisoners of war (GC III) and civilians (GC IV). In 1977 two additional protocols (APs) were created to fill gaps left by the 1949 set of treaties and to recognize the evolving character of warfare. AP I more clearly addresses the conduct of hostilities (for example, prohibiting weapons which cause superfluous injury) and additional types of combat (notably aerial warfare). AP II addresses civil wars (or to use the arcane language of IHL, 'non-international armed conflicts') and supplements the minimal protections provided for victims of civil wars in the GCs themselves.[29] Given that a clear majority of the victims of warfare are victims of civil wars, AP II is perhaps particularly important, though it has been ratified by fewer states than AP I and certainly the GCs themselves. AP III of 2005 was made to deal with the discrete issue of the emblem of protection. The red crystal – a symbol without possible religious connotations – was adopted to stand beside the red cross and red crescent as internationally recognized symbols to be used on, among other things, medical transport vehicles. There are roughly 20 additional important IHL treaties – on issues ranging from child soldiers and cultural property through to laser-blinding weapons – including, most recently, the Convention on Cluster Munitions of 2008. They have been subscribed to with varying degrees of support. In addition to the IHL treaty regime, customary international law also provides IHL content. The ICRC in 2005 completed an exhaustive review of customary international law – by, among other things, surveying military manuals – to determine the customary rules.[30] Finally, it is worth briefly noting that international human rights – rights guaranteed to individuals vis à vis governments through a separate though overlapping treaty regime from IHL – may also apply. Rights do not automatically cease to exist in times of armed conflict. For example, while in national emergencies states may derogate from some rights, others, such as freedom from torture, are nonderogable.[31]

Mastering the body of IHL and related areas is a challenge for any military or civilian lawyer and, clearly, nonlegally trained officers and enlisted personnel are not expected to have a detailed knowledge of the law. The following chapter on the operational aspects of the law will detail how IHL is distilled into, among other things, rules of engagement and points to the availability of legal resources (advisers and manuals) to answer legal questions. Suffice it to say here that while the legal detail is voluminous and complex – and that special training is required to master any particular area, such as say, setting up a prisoner of war camp – the rudiments of IHL are distillable into some basic, key concepts of which all military personnel should be aware.

While quite properly insisting that a summary cannot act as a substitute for the text of the treaties themselves, the ICRC has put forward seven basic rules which go a long distance towards sketching the crux of IHL. They are as follows:[32]

1. Persons *hors de combat* [in other words those taken prisoner or wounded/injured] and those who do not take a direct part in hostilities are entitled to respect for their lives and their moral and physical integrity. They shall in all circumstances be protected and treated humanely without any adverse distinction.
2. It is forbidden to kill or injure an enemy who surrenders or who is *hors de combat.*
3. The wounded and sick shall be collected and cared for by the party to the conflict which has them in its power. Protection also covers medical personnel, establishments, transports and equipment. The emblem of the red cross or the red crescent [and now the red crystal] is the sign of such protection and must be respected.
4. Captured combatants and civilians under the authority of an adverse party are entitled to respect for their lives, dignity, personal rights and convictions. They shall be protected against all acts of violence and reprisals. They shall have the right to correspond with their families and to receive relief.
5. Everyone shall be entitled to benefit from fundamental judicial guarantees. No one shall be held responsible for an act he has not committed. No one shall be subjected to physical or mental torture, corporal punishment or cruel or degrading treatment.
6. Parties to a conflict and members of their armed forces do not have an unlimited choice of methods and means of warfare. It is prohibited to employ weapons or methods of warfare of a nature to cause unnecessary losses or excessive suffering.
7. Parties to a conflict shall at all times distinguish between the civilian population and combatants in order to spare civilian population and property. Neither the civilian population as such nor civilian persons shall be the object of attack. Attacks shall be directed solely against military objectives.

Many points of clarification can be made here, but let us content ourselves with just one, albeit a clarification that goes to the heart of the internal tensions within IHL. Underlying several of the seven rules is the principle of distinction; the notion that fighters should distinguish between civilians and combatants and between civilian and military objects. To the extent that this means that civilians should never be specifically targeted, the rule is unambiguous. What if, however, in pursuing a legitimate military target, civilians will be harmed? The rule is that the harm to civilians cannot be unnecessary or excessive – in other words *disproportionate* – to the military importance of the military objective. To put it another way, military logic or military necessity is to be balanced against the principle of distinction, with 'proportionality' acting as the fulcrum. When one civilian night-watchman will be killed in an attack on a major munitions dump, the harm to civilians will obviously not generally be considered disproportionate. Similarly, an air strike which is expected to kill dozens of civilians as the price for killing a mid-ranking enemy officer will be disproportionate. The problem comes in the grey areas, with different militaries taking different approaches. What is clear for most observers is the essential permissiveness of IHL itself as it now stands (see Chapter 7 for an exploration of military necessity). As the British MOD *Manual on the Law of Armed Conflict* puts it, 'The Law of Armed Conflict is consistent with the economic and efficient use of force. It is intended to minimize the suffering caused by armed conflict rather than impede military efficiency.'[33] While some specific acts or weapons are prohibited, in general the 'balancing' is often tipped towards military necessity, at least in current practice. The generally permissive nature of IHL is one major reason why ethics remains an important part of the military decision-making calculus; what is legally permitted may be unethical or simply, especially in the context of counter-insurgency, imprudent.

Having sketched out the nature of IHL, a fair question remains: is there any way of enforcing it? Yes, though these methods are far from comprehensive. The first way is through individual criminal accountability. Thus war crimes trials of individuals can be conducted by national authorities or by an international *ad hoc* tribunal based on the Nuremburg precedent (such as the International Criminal Tribunal for the former Yugoslavia and the International Criminal Tribunal for Rwanda). Those committing war crimes, genocide or crimes against humanity may also be tried by the International Criminal Court (ICC), a permanent, independent court created by a 1998 treaty with 110 states participating. If there is a national or territorial connection of a suspected individual criminal to a state party, and that state has proven unwilling or unable to investigate or prosecute, the ICC may take that case over. As important as the ICC is in the fight against impunity, however, it only has a mandate to go after 'big fish'; in the words of the treaty establishing the court, '[t]he jurisdiction of the Court shall be limited to the most serious crimes of concern to the international community as a whole'. Other ways of enforcing IHL include ICRC monitoring and reporting (confidential except in the most extreme of circumstances), fact-finding by international organizations (such as the

UN's 'Goldstone Report' on the Gaza conflict[34]), 'naming and shaming'-type activities by non-governmental organizations, judicial complaints against states by individuals or other states and, as a practical matter, long-term self-interest.

To focus on the enforcement of IHL, however, risks overlooking the larger enterprise of IHL implementation. In this regard training and dissemination of IHL is key. So too is clear and consistent leadership from the top. The mixed signals sent from the civilian and military leadership with respect to the treatment of detainees, has, for example, set the state for prisoner abuse scandals which have not only legal – and obviously ethical – ramifications, but strategic ones as well. As the noted international law specialist Adam Roberts has been quoted as saying, 'Some of the biggest coalition failures in Afghanistan and Iraq have come from failures of the coalition to observe basic norms on certain matters, especially with regard to the treatment of prisoners.'[35]

Conclusion

When the rule of law in international relations fails, the results can be deadly. However, the core rules themselves are generally clear – both with respect to the (il)legality of a conflict and the conduct of hostilities – and breaches of the rules are actually much rarer than commonly perceived. International law in this area seeks to minimize the use of force in international affairs. If that fails, it seeks instead to provide the tools which can be used to protect the victims of that armed conflict. It is clear that there is a significant overlap between legal and strategic concerns and, returning to themes set out in the Introduction, it is also clear that in many areas, a purely legal calculus does not and cannot displace the necessity of an ethical one as well.

Notes

1. An excellent entry point into the debate is the now classic Hart–Fuller debate carried out in the pages of the Harvard Law Review and elsewhere. See H. L. A. Hart, 'Positivism and the Separation of Law and Morals' (1958) 71, *Harvard Law Review*, p. 593, and Lon L. Fuller, 'Positivism and Fidelity to Law: A Reply to Professor Hart' (1958) 71, *Harvard Law Review*, p. 630. For a more recent and controversial work on the subject (one which argues that law is divorced from morality despite 'window dressing' to the contrary), see Costas Douzinas, *Justice Miscarried: Law and Aesthetics in Law* (New Jersey: Prentice Hall, 1995).
2. See, for example, US Army and Marine Corps, *The US Army*

Counterinsurgency Manual (Chicago: University of Chicago Press, 2007), pp. 7–37.

3. For a discussion of this, see Robert Barnidge, 'Should National Security Trump Human Rights in the Fight Against Terrorism?' (2007) 37, *Israel Yearbook on Human Rights*, p. 85.

4. Charles J. Dunlop, 'It Ain't No TV Show: JAGS and Modern Military Operations' (2003) 4, *Chicago Journal of International Law*, p. 479.

5. David Kennedy, *Of War and Law* (Princeton: Princeton University Press, 2006), p. 116.

6. Jenny Booth, 'Military Top Brass Attack Soldier Prosecutions' (14 July 2005), available at http://www.timesonline.co.uk/tol/news/uk/article544087.ece.

7. Christopher P. M. Waters, 'Is the Military Legally Encircled?' (2008) 8, *Defence Studies*, p. 26.

8. W. G. L. Mackinlay, 'Perceptions and Misperceptions: How are International and UK Law Perceived to Affect Military Commanders and Their Subordinates on Operations' (2007) 7, *Defence Studies*, p. 111.

9. On widespread misunderstanding of the Court – and its British implementing legislation – see Mackinlay, 'Perceptions and Misperceptions'.

10. For a clear introductory account of international law, see Vaughan Lowe, *International Law* (Oxford: Oxford University Press, 2007). For further readings on the use of force and IHL, see the chapters by Christine Gray and Christopher Greenwood, respectively, in Malcolm D. Evans (ed.) *International Law* (Oxford: Oxford University Press, 2006).

11. See Carsten Stahn, '"Jus ad bellum", "jus in bello" ... "jus post bellum"? Rethinking the Conception of the Law of Armed Force' (2006) 17, *European Journal of International Law*, p. 243.

12. General Treaty for the Renunciation of War (1928), Article 1. The Treaty, named after the American Secretary of State and his French counterpart at the time, was also known as the 'Pact of Paris'.

13. See generally, the principles adopted by the UN General Assembly, by consensus, in the *Declaration on Principles of International Law Friendly Relations and Co-operation Among States in Accordance with the Charter of the United Nations*, UNGA Res. 2625 (XXV), 1970.

14. UN Charter, Article 41.

15. UN Charter, Article 42.

16. In brief, the argument goes that Resolution 678 (1990) authorised force against Iraq in part to 'restore peace and security to the area'. Resolution 687(1991) set out ceasefire conditions which included Iraq's compliance with a weapons inspection regime. When those ceasefire conditions were breached by Iraq's noncompliance, as recognized by Security Council Resolution 1441 (2002), Resolution 678 was revived. For more detail see 'The Advice of the United Kingdom Attorney-General, Lord Goldsmith, on "The Legal Basis For the Use of Force Against Iraq"' (17 March 2003), available at http://www.number-10.gov.uk/output/Page3287.asp.

17. Philippe Sands, *Lawless World* (London: Allen Lane, 2005), p. 225.

18. 'The Caroline Case', 29 BFSP, pp. 1137–8, 20 BFSP, pp. 195–6.

19. Independent International Fact-Finding Mission on the Conflict in Georgia, Volume I (September 2009) at 24, available at http://www. ceiig.ch/pdf/IIFFMCG_Volume_I.pdf.

20. For the 'Bush Doctrine' see 'National Security Strategy of the United States' (September 2002), available at http://georgewbush-whitehouse. archives.gov/nsc/nss/2002/index.html. The 'Obama Doctrine' is more opaque at this point. In receiving his 2009 Nobel Peace Prize, President Obama said: 'To begin with, I believe that all nations – strong and weak alike – must adhere to standards that govern the use of force. I – like any head of state – reserve the right to act unilaterally if necessary to defend my nation. Nevertheless, I am convinced that adhering to standards, international standards, strengthens those who do, and isolates and weakens those who don't.' See Nobel Lecture of 10 December 2009, available at: http://nobelprize.org/nobel_prizes/ peace/laureates/2009/ obama-lecture_en.html.

21. See, for example, James A. Green, *The International Court of Justice and Self-Defence in International Law* (Oxford: Hart, 2009), pp. 96–101, and Niaz Shah, 'Self-Defence, Anticipatory Self-Defence and Pre-Emption: International Law's Response to Terrorism' (2007) 12, *Journal of Conflict and Security Law*, p. 95.

22. Para 139. In addition, the international community embraced a responsibility to *prevent* by assisting states under stress.

23. Antony Barnett and Martin Bright, 'British Military Chief Reveals New Legal Fears over Iraq War', *Observer*, 1 May 2005, available at www.guardian.co.uk/politics/2005/may/01/uk.iraq.

24. T. Judd, *et al.*, 'The Betrayal of British Fighting Men and Women', *The Independent*, available at http://news.independent.co.uk/uk/politics/ article2347537.ece.

25. On some of the antecedents in medieval Europe, see David Whetham, *Just Wars and Moral Victories: Surprise, Deception and the Normative Framework of European War in the Later Middle Ages* (Leiden: Brill, 2009).

26. Though the first modern codification of humanitarian law is often credited to the Lieber Code issued to Union troops during the US civil war.

27. ICRC, 'From the Battle of Solferino to the Eve of the First World War' (2004), available at http://icrc.org/web/eng/siteeng0.nsf/html/57JNVP.

28. Convention (I) for the Amelioration of the Condition of the Wounded and Sick in Armed Forces in the Field. Geneva, 12 August 1949; Convention (II) for the Amelioration of the Condition of Wounded, Sick and Shipwrecked Members of Armed Forces at Sea. Geneva, 12 August 1949; Convention (III) relative to the Treatment of Prisoners of War. Geneva, 12 August 1949; Convention (IV) relative to the Protection of Civilian Persons in Time of War. Geneva, 12 August 1949; Protocol Additional to the Geneva Conventions of 12 August 1949, and relating to the Protection of Victims of International Armed Conflicts (Protocol I); 8 June 1977, Protocol Additional to the Geneva Conventions of 12 August 1949, and relating to the Protection of Victims of Non-

International Armed Conflicts (Protocol II), 8 June 1977; Protocol additional to the Geneva Conventions of 12 August 1949, and relating to the Adoption of an Additional Distinctive Emblem (Protocol III), 8 December 2005. The text of these and all other IHL treaties can be found on the ICRC's website: www.icrc.org.

29.   Common Article 3 of the GCs has been described by the International Court of Justice as setting out 'elementary considerations of humanity' (The Corfu Channel case [*United Kingdom* v. *Albania*], *ICJ Reports 1949*, p. 22). It requires that those not taking part in hostilities – including noncombatants and detainees – should be treated humanely. Acts such as torture, murder and the passing of sentences without guarantees of judicial fairness are specifically prohibited. The United States Supreme Court has held that Common Article 3 is the appropriate IHL framework to be applied by American forces in the 'War Against Terrorism': *Hamdan* v. *Rumsfeld* 126 S.Ct. 2749 (2006).

30.   Jean-Marie Henckaerts, 'Study on Customary International Humanitarian Law: A Contribution to the Understanding and Respect for the Rule of Law in Armed Conflict' (2005) 87, *International Review of the Red Cross*, p. 175.

31.   A good primer on the overlap of the two regimes is ICRC, 'International Humanitarian Law and International Human Rights Law: Similarities and Differences' (2003), available at http://www.icrc.org/web/eng/siteeng0.nsf/html/57JR8L. Another debate has been over the exact territorial reach of human rights? Does, for example, the European Convention of Human Rights apply to the actions of British soldiers in Iraq? The House of Lords recently held that European Convention Rights did apply in a case where an Iraqi civilian was killed while in British custody, though not where civilians were killed by British soldiers on patrol: *R (on the application of Al-Skeini and others)* v. *Secretary of State for Defence*, 13 June 2007 (HL).

32.   ICRC, Basic rules of the Geneva Conventions and their Additional Protocols (1988), available at http://icrc.org/Web/Eng/siteeng0.nsf/htmlall/p0365/$File/ICRC_002_0365_BASIC_RULES_GENEVA_CONVENTIONS.PDF!Open.

33.   UK Ministry of Defence, *The Manual of the Law of Armed Conflict* (Oxford: Oxford University Press, 2005), p. 21.

34.   Report of the United Nations Fact-Finding Mission on the Gaza Conflict ('The Goldstone Report'), UN Doc. A/HRC/12/48, 25 September 2009.

35.   As cited in R. Norton-Taylor and C. Dyer, 'International Laws Hinder UK Troops – Reid', *Guardian*, 4 April 2006, p. 1.

Chapter 6

Law at the Operational Level

PHILIP McEVOY

Introduction

The complexities of the current operational environment, including advances in global communications and weapons technology, makes it very difficult to define discrete levels of warfare. Doctrine[1] describes three levels of warfare – strategic, operational and tactical. General Sir Rupert Smith in *The Utility of Force*, describes a fourth – political.[2] Decisions taken, and the general conduct of troops at the tactical level, can have considerable consequences at the operational, strategic and political levels. This chapter will examine the impact of law at all four levels by addressing the issues from the point of view of the commander and the units, sub-units and soldiers they command. While, due to the author's background and experience, this chapter is informed by a British Army perspective, the actual issues raised are of wider relevance beyond the UK experience.

The British military have been involved in one domestic and six major expeditionary campaigns in the past quarter-century – Northern Ireland, the Falklands, the First Gulf War, the Balkans, Kosovo, Iraq and Afghanistan.[3] Those campaigns, each with their own characteristics and complexities, have seen an increase in, and development of, the impact of international, national and human rights law at all four levels of warfare but perhaps most significantly at the tactical and operational levels. There is therefore a greater need for training and education not only of the military in operational law,[4] but also a wider understanding among those at the strategic and political levels on the role of the armed forces and the demands placed upon them in the contemporary operating environment.

Background

As acknowledged in previous chapters, all wars or conflict, whether they are classified in legal terms as domestic, international or noninternational armed conflicts, or military operations other than war, and no matter whether of long or short duration, are capable of being perceived at least by one of the participating parties as being cruel. They inevitably result in death and destruction. The idea that there can be any laws governing something as cruel as war may be regarded as surprising but as Chapter 1 notes, rules or regulations relating to behaviour in conflict can be traced back to nearly all the great civilizations – the Romans, the Chinese, the Greeks. Over the centuries those rules have been encapsulated into the laws of war.

The majority of the world's armed forces have an obligation to train in and comply with the Law of Armed Conflict[5] – the Hague Law and the Geneva Law. However, the legal dimension has expanded beyond this area significantly. For example, the European Court of Human Rights became a forum to challenge UK military procedures in Northern Ireland. In 1998 the United Kingdom introduced the Human Rights Act and in 2001 the Rome Statute of the International Criminal Court was implemented into United Kingdom legislation by the International Criminal Court Act. The European Convention on Human Rights and the Human Rights Act continues to be used as a mechanism for challenging military activity in non-European conflicts such as the Falklands conflict and in Iraq.[6] The United Kingdom is a signatory to treaties regulating the use of weapons and weapon systems such as anti-personnel mines and cluster munitions. It seems on the face of it that warfare is becoming over-regulated and this may have prompted Lord Boyce, a former Chief of Defence Staff, to comment in the House of Lords:

> The Armed Forces are under legal siege and are being pushed in a direction that will see such an order being deemed as improper or legally unsound. They are being pushed by people not schooled in operations but only in political correctness. They are being pushed at a time when they will fail in an operation because the commanding officer's authority and his command chain has been compromised with tortuous rules not relevant to fighting and where his instinct to be daring and innovative is being buried under the threat of liabilities and hounded out by those who have no concept of what is required to fight and win.[7]

This was the view of a former member of the armed forces. What was perhaps more surprising was the contents of a speech made at the Royal United Services Institute on 3 April 2006 by Dr John Reid, when Secretary of State for Defence, in which he said:

I believe we need now to consider whether we, the international community, in its widest sense need to re-examine these conventions. If we do not we risk continuing to fight 21st Century Conflict with 20th Century Rules.[8]

At first glance both the quotes from Lord Boyce and Dr Reid might be seen as challenging the relevance of the laws of war. Lord Boyce was addressing the possible impact of the Human Rights Act and the International Criminal Court Act upon the exercise of command and Dr Reid was looking more at the question of international terrorism and armed intervention from a state perspective rather than conduct on the battlefield. But here we have both a military and political view expressing some doubt about the impact and effectiveness of some aspects of both domestic and international law. The quotes must, of course, be read in context and neither suggests that there is no place for law on the battlefield.

The contemporary operating environment

What, then, might have caused Lord Boyce and Dr Reid to challenge or at least debate the efficacy of the law as it is applied today? The answer may lie not in the defence adopted by Cicero in his representation of Titus Milo – 'silent enim leges inter arma' (in war, the law is silent) – but, given the complexities of the modern battlefield, to quote from the Chief Justice of the United States, William Rehnquist, in 1998 'The laws will thus not be silent in time of war but they will speak with a somewhat different voice.'[9] The character, some argue even the nature, of warfare has changed. Rarely now do we have the concept of two neatly formed and uniformed armies facing each other in a clearly defined battlespace. The start of a conflict may be comparatively easy to define, but defining the end of the same conflict is far more difficult. Even if the start point is distinct, whether conflict is regarded as purely internal or of international or noninternational nature is complex for politicians or their legal advisers. Multinational coalition operations involving different countries with different legal backgrounds add to the complexity. The contemporary operating environment has many definitions but is characterized by being complex, demanding, chaotic, challenging, irregular, unpredictable and uncertain. An army must be trained to meet the challenges and demands of any type of operation from war-fighting to peace support, in a range of physical environments and against all kinds of threats – simultaneously.

The application of law in this contemporary operating environment is equally challenging and demanding. Like the contemporary operating environment it is also uncertain and unpredictable. The soldier requires and deserves a degree of certainty and clarity which, regrettably, it is often not possible to provide.

Command and control: the relationship between politics and the military

From a commander's perspective the start point is the mission itself. British Defence Doctrine lays down three key principles for the use of military force. These are:

- A decision to use force must be based on necessity and legitimacy, and have clear objectives.
- The use of force must be under democratic control at all times. This control is provided by the elected government answerable to Parliament.
- An organized structure is required to produce and apply force.[10]

A commander should therefore have a clear and unambiguous directive stating what his mission is and implicit in this statement should be that the mission is lawful. From this flows the powers he needs to accomplish the mission and any restrictions imposed upon those powers. Those restrictions may be legal restrictions or restrictions imposed by policy. Legal restrictions are those imposed by domestic or international law, whereas policy restrictions are those imposed by national caveats or, in some circumstances, imposed as a result of international UN or NATO restrictions. These powers include, at one end of the spectrum, the authority to use force either offensively or, at the other end of the scale, the authority to interfere in the daily life of members of the civilian population such as the power to stop, detain, question and arrest. Put simply, although the issue is complex, in a war-fighting situation the Law of Armed Conflict applies. In peace support operations the commander will follow a mixture of international and national law, including human rights and local law. There may also be a status of forces agreement or memorandum of understanding (MOU) with the host nation which further sets out permissions and restrictions.

The decision to commit forces is taken at the political level. Implicit in the political decision to commit military forces is the need for that military force to act either only in self-defence or more offensively. The parameters within which force may be applied are also decided at the political level and the task of the commander is to determine when and how to apply that force within those parameters. Ultimately the military may be required to use lethal force. Rupert Smith sums this political/legal relationship in the following way,

> A regular force is employed to serve a political purpose decided upon by a lawful government, which instructs the military, as a legally sanctioned and formed body answerable to that government, to apply the force. A regular force is therefore legal, deadly and destructive.[11]

The military trains for war-fighting, yet the type of operation it is likely to face in the contemporary operating environment is a mixture of high-intensity fighting, counter-insurgency, peace enforcement and peacekeeping, famously described by General Krulak as 'the Three Block War'.[12] This presents a dilemma for both politicians and commanders. In announcing the decision to commit British troops to the International Security Assistance Force (ISAF) mission in Afghanistan in January 2006, the Secretary of State may have hoped that the aim of the mission was to establish 'Provincial Reconstruction Teams ... building up Afghan's [sic] capacity'.[13] The reality was far different, and probably understood by military commanders in the request for combat troops supported by artillery and attack helicopters. The politician will probably be more hesitant in their approach considering public opinion at home and abroad and not wishing to paint a picture that might portray them as warmongering. The military commander will be more concerned at the planning stage that he has political support – thereafter his focus will be on what his mission is and whether he will have sufficient resources provided to him to enable him to effectively carry out that mission. Beyond the planning stage and as the operation develops there is still a constant need to review the political support and to ensure all aspects of the operation are conducted in accordance with the political mandate. This can only be achieved through trust – trust and confidence in the military commander and the trust that the commander has that he will be supported by the political leadership.

This sometimes uneasy political relationship is not new. In other conflicts politicians have been unable or unwilling to give clear direction. The Falklands conflict started in April 1982 with the invasion of the Falkland Islands and South Georgia by Argentina and ended with the Argentine surrender in 14 June 1982. The conflict was a result of a diplomatic confrontation over the sovereignty of the Islands. The United Kingdom launched a task force consisting of 110 ships and 28,000 men. The conflict lasted 74 days and claimed the lives of 255 British and 649 Argentine servicemen. The UK lost 34 aircraft and six ships, HM ships *Sheffield, Ardent, Antelope, Coventry*, the RFA *Sir*

Galahad and the MV *Atlantic Conveyor*. To the commanders, soldiers, sailors and airmen on the ground, at sea and in the air the conflict was a war, yet neither the United Kingdom nor Argentina considered itself at war. In this conflict, whether the United Kingdom or Argentina were at war made little practical difference – captured prisoners were treated as prisoners of war and the provisions of the Geneva Conventions were applied.

The same could be said about the first Iraq war. Again there was no formal declaration of war but the US coalition acted under UN Resolution 678, passed on 29 November 1990, to 'use all necessary means to uphold and implement Resolution 660 (1990) [demanding that Iraq withdraw from Kuwait] and all subsequent resolutions and to restore international peace and security in the area'.[14] From a UK perspective this was clear and unambiguous. The difficulties encountered by General Sir Peter de la Billiere and identified in his account of this conflict in *Storm Command* appear to have centred on the justification to Parliament of force levels, the practical difficulties of 'jointery' and the application of ROE in multinational operations.[15]

The Kosovo campaign was, in legal terms, more difficult. At the end of 1998 Serbian forces continued their build-up in Kosovo and there were numerous reports of atrocities being committed against native Albanians. Peace talks had failed and the Former Republic of Yugoslavia had refused to sign the US-drafted Rambouillet Agreement.[16] On 24 March 1999 NATO commenced its ten-week bombing campaign without the authority of the UN, relying upon the concept of 'humanitarian intervention'.

The legitimacy of the second Iraq conflict was surrounded by controversy. In this case there was no clear UN mandate but a compelling argument was advanced relying upon UN Resolutions 660, 678 and 687 and resolution 1441. In the days preceding the conflict there was talk about the need to obtain a further resolution and debate as to the military's task, in particular whether they would be an occupying power. Occupation, as can be seen from the aftermath of the Second World War, brings with it responsibilities which the coalition seemed unwilling or at least ill prepared to accept.

There was unprecedented media challenge, which must have been of concern to not only commanders but also the soldiers who saw headlines such as 'Weapons of Mass Deception'[17] and 'Blood on his Hands'. In his account of the conflict in *Soldier*, General Sir Mike Jackson, who at this time was Chief of the General Staff, stated:

> A lot of people had asked me in casual conversation Are you sure you're OK on this one? The Chiefs of Staff discussed the matter and collectively agreed that we needed to be sure of our ground. So

Mike Boyce, on behalf of us all sought the Attorney General's assurances on the legality of the planned military action.[18]

There was then controversy over the decision to release the Attorney General's advice. How much simpler would it have been from the commander's and soldier's perspective if clear unequivocal advice could have been given from the outset? However, does the soldier really need to know? Army Doctrine Publication (ADP) Land Operations addresses this issue in Chapter 7, 'The Moral Component':

> 0711 The morality of actions within a military operation is affected by the morality of the operation itself as well as the more immediate circumstances. Few soldiers may have sufficient access to the relevant context to judge whether a military operation is lawful. The decision to deploy UK forces to engage in armed conflict will always be taken at the highest political and military level and the decision will be subject to scrutiny by Law Officers and others. In the United Kingdom we can have the highest degree of confidence that our armed forces will not be launched into operations that are unlawful.[19]

Recent events, especially the debate over the legality of the second Iraq conflict, might cause some commanders to question the accuracy of the quote. The United Nations Charter prohibits the threat or use of force unless acting under individual or collective self-defence or under the authority of the UN itself. The decision to deploy forces will be taken not by the military but by politicians. The commander needs to understand what he can do and what restrictions will be imposed upon him. Even where the commander has a clear mandate, the mandate itself may not go into the specific details of the means to be employed, instead using such phrases as 'all necessary measures'. This phrase is capable of either a broad or narrow interpretation – but in the contemporary operating environment it will be the politician and not the commander who makes the interpretation.

Thus, military doctrine recognizes the need to have clear political direction to support the military mission. This has never been more relevant than in the present contemporary operating environment. A force sent on a mission for which it is not trained or equipped has a high risk of failure or, to quote from Rupert Smith, it has no 'utility of force'.

Command and control: the strategic corporal

General Krulak termed the 'strategic corporal' as the low-level unit leader able to take independent action and make decisions with major implications. In this complex legal environment and in the full glare of the media, any isolated and inappropriate actions of comparatively junior ranks can cause untold damage to the mission. This is illustrated by the conduct of Charlie Company of the US 11th Light Infantry Brigade in the village of My Lai, Vietnam, on 16 March 1968. Here 150 soldiers, under the command of a lieutenant, massacred several hundred civilians, mostly women, children and old men. It was a massacre that would haunt the conscience of the US military and the American people for years to come. Those with the moral courage to intervene and prevent further bloodshed, helicopter pilot Warrant Officer Hugh Thompson and his door gunner, Specialist Lawrence Colburn, were awarded the Soldier's Medal in 1998 – 30 years after the incident.

In March 1993, Shidane Arone, a teenage Somali prisoner, was tortured and beaten to death by members of the Canadian Airborne Regiment serving on a peacekeeping mission in Somalia. One soldier was found guilty of the torture and sentenced to five years' imprisonment. This incident led to cover-ups, media saturation, courts martial, the disbandment of the Airborne, suicide attempts, resignations and a full-scale public inquiry.

The British military are sadly not immune from these incidents. Baha Musa, a 26-year-old hotel worker, was taken into British military custody in Iraq in September 2003. After 36 hours in custody he died from asphyxiation. He had been subjected to torture and on his body were 93 separate injuries. Although nine soldiers faced trial by court-martial only one was found guilty – a corporal who pleaded guilty to inflicting inhumane treatment on Baha Musa and his fellow detainees.[20] In May 2008 the government announced a public inquiry into the incident. This inquiry has yet (at the time of writing) to hear evidence, but the circumstances surrounding the deaths of both Shidane Arone and Baha Musa bear some striking similarities. One analysis of the Somalia incident concluded that Arone's death could be traced to the fact that 'Canada's people, government, military, and specifically its army have failed to keep real soldiers, combat effectiveness, and traditional military leadership at the centre of the Canadian army.'[21] The British Army commissioned its own internal investigation into the death of Baha Musa and other incidents and concluded:

> It is not possible for any organisation to prevent criminal activity or disgraceful behaviour absolutely. It is, however, possible to create the conditions which make the commission of criminal or disgraceful acts less likely. That requires leadership, education and training (including the clear articulation of the requisite doctrine to

support it), and the effective operation of all aspects of a criminal justice system. The small number of instances of deliberate abuse highlighted in this report have exposed some failings in all of those areas.[22]

Rules of engagement

Rules of engagement are defined as directions for operational commanders that set out the circumstances and limitations under which armed force may be applied by armed forces to achieve military objectives for the furtherance of government policy. ROE are thus issued as a set of parameters to inform commanders of the limits of constraint imposed or of freedom permitted when carrying out their assigned tasks. They are designed to ensure that any application of force is appropriately controlled.[23] It should be noted that there is, quite properly, a political element to ROE. Regular forces are, to quote from Rupert Smith, 'employed to serve a political purpose decided upon by a lawful government, which instructs the military, as a legally sanctioned and formed body answerable to that government, to apply the force'.[24] It follows that the lawful government can decide when and how force can be applied and in doing so it should take into account what the military commander requires at the operational level. At first sight this might seem to be politicians interfering in a purely military aspect of the mission and to some extent it is, although, when analyzed, it is understandable. At the tactical level commanders may decide to issue not a rule of engagement, but a fire control order – 'Do not shoot until you see the white of their eyes' is an example attributed to the General William Prescott at the battle of Bunker Hill in the American War of Independence in 1775. That order was given for sound military reasons, probably to preserve dwindling ammunition stocks. Governments may not be so concerned about the use of ammunition, but will be concerned to see that commanders apply force appropriately. This is especially important in the build-up to conflict where a commander will wish to know what his ROE are, that his soldiers are appropriately aware of them and that they have been trained in them.

General Sir Peter de la Billiere makes frequent reference to ROE and their importance to soldiers in the context of the first Gulf War. He notes that 'to soldiers, sailors and airmen they are literally a matter of life and death, for they govern one's response to hostile action and lay down the circumstances in which one may or may not open fire on the enemy'.[25] A military commander seeks the most liberal ROE and the delegated authority to apply them when the situation permits. The politician may wish to exercise restraint, especially in the build-up to conflict, and may choose not to delegate except in self-defence. General de la Billiere describes the emphasis he placed on clear and concise ROE, the challenge faced by the divergence of ROE within the coalition and the 'constant debate and argument at both national and international level'. He formed the view that may well be shared by commanders today that 'back in the United Kingdom people were taking a cool, detached attitude and worrying more about the possible legal consequences of making our ROE too generous than about the chances of our losing a ship or aircraft in a pre-emptive air strike'.[26] Matters came to a head on 10 January 1991 before the commencement of hostilities. HMS *Gloucester* was a type 42 destroyer with 300 men on board. History had shown the damaging effect that a single strike with an Exocet missile could inflict on such a warship (in the Falklands conflict HMS *Sheffield*, a sister ship of the same class, was hit on 4 May 1982 by an Exocet missile fired from an Argentinean Super Etendard at a range of 20–30 miles, with the loss of 20 crew), and in 1991, the Iraqi airforce had just such a capability. At 10.00hrs HMS *Gloucester* was patrolling when her captain reported to General de la Billiere that eight enemy aircraft were approaching HMS *Gloucester* in attack formation. The Captain requested authority to fire weapons and authority was granted. The decision was taken instantly. In the event, the aircraft turned away and the incident, which could have led to an immediate commencement of hostilities, was avoided. According to General de la Billiere's account of the incident, the Secretary of State's office were in 'a panic' and demanding a 'flash explanation'. It is not clear whether, if time had permitted, the response from the Secretary of State would have been any different to that of General de la Billiere, but the incident illustrates the degree of political oversight.

The sinking of the *Belgrano* in the Falklands conflict was a similar incident. On 29 April 1982 Task Group 79.3, consisting of the *General Belgrano* accompanied by two destroyers, were patrolling the Burdwood Bank, south of the Falkland Islands. On the following day she was detected by the British nuclear-powered hunter-killer submarine HMS *Conqueror*. The submarine approached over the following day. Although outside the British-declared Total Exclusion Zone of 370km (200 nautical miles) radius from the islands, the British decided that the group was a threat. After consultation at Cabinet level, the Prime Minister, Margaret Thatcher, agreed that Commander Chris Wreford-Brown should attack the group.

On 2 May, at shortly after 20.00hrs, *Conqueror* fired three conventional Mk 8 mod 4 torpedoes, each with an 800lb (363kg) Torpex warhead, two of which hit the *General Belgrano*. The ship began to list to port and to sink towards the bow. Twenty minutes after the attack, at 20.24hrs, Captain Bonzo ordered the crew to abandon ship.

Inflatable life rafts were deployed and the evacuation began without panic. Argentine and Chilean ships rescued 770 men in all from 3 May to 5 May. In total 323 were killed in the attack, 321 members of the crew and two civilians who were on board at the time.

The sinking of the *Belgrano* was a significant event in the Falklands War. Significant because it was the first engagement in the mission to regain control of the Islands. Significant in that it was controversial, the *Belgrano* being outside the British-imposed exclusion zone and significant in that the British Prime Minister Margaret Thatcher and her cabinet played a crucial role in the decision to engage the *Belgrano*.

The challenge in producing effective ROE is to balance the competing interests. They must give the commander sufficient authority to use force to enable him to achieve his mission within the political constraints. The commander must understand the political context; the politician must understand the military context. The military commander, however, must ensure that the rules are not so rigid as to be utterly proscriptive, and to permit soldiers to utilize their initiative and judgement, with the rules of engagement providing them with suitable guidelines. The 'Yellow Card' used in Northern Ireland was often criticized for being too restrictive, yet it is almost identical to that used in Iraq and Afghanistan without difficulty. What has changed is the approach to ROE judgemental training, emphasizing what the soldier can do rather than what they cannot do and encouraging them to make the judgement call in the knowledge that they will be supported in their decision. In Northern Ireland the difficult decision was deciding when to open fire. In Iraq and Afghanistan the difficult decision is often deciding when not to open fire – normally because of the risk of civilian casualties. In his personal account of his operational tour in Iraq, *Sniper One*, Sgt Dan Mills describes the visit of a senior officer to Al Amarah in Iraq to discuss ROE. That officer explained a less restrictive interpretation of the ROE:

> The date of his visit was 6th June, the sixtieth anniversary of the D Day landings. It was fitting because what he said was fitting for us too. It was exactly what was needed. Of course the relaxation didn't mean that we went straight out to drop a load of twelve year olds for chucking stones at us but it did give us the ability to blunt a few of the enemy's subsequent attacks; attacks after all that had only one intention, to kill us.[27]

Whatever ROEs are in operation in a theatre, the inherent right of self-defence remains. It is therefore necessary to think in terms of ROE and self-defence rather than self-defence ROE. It provides a slightly different legal justification for the use of force in a conflict situation as it is based upon domestic law rather than IHL. This also makes it the legal basis for many military actions in areas that are not classified as international armed conflicts. What can be covered by self-defence can and does vary from state to state (for example, defence of property is not considered self-defence as far as the UK is concerned, but can be for the US: that does not mean that UK forces cannot defend property, but the legal justification would need to be based upon the vulnerability of people inside it or the essential function it carried out, etc., rather than just the bricks and mortar). In UK law, an act of self-defence requires that the person carrying out the act (or the person calling in the fire) has 'an honest belief' that the action is required. It is not absolutely necessary that the defendant be attacked first. As Lord Griffith said: 'A man about to be attacked does not have to wait for his assailant to strike the first blow or fire the first shot; circumstances may justify a pre-emptive strike.'[28] A person who is being attacked is also not necessarily expected to 'weigh to a nicety the exact measure of his necessary defensive action'. In practice, this means that taking valuable time to make detailed casualty assessment calculations may not be required for actions taken in direct self-defence, as long as all reasonable precautions not to injure anyone other than the intended target are taken.

Command and control: accountability

The circumstances surrounding the shooting of the unarmed Brazilian Jean Charles de Menezes in London on 22 July 2005 illustrate the difficulties of responding to a threat in a highly charged and fast-moving operational environment. This single operation faced by the Metropolitan Police, although not without its complexities, is a situation similar to incidents faced daily by soldiers on operations. The incident illustrates the need for all those involved in an operation, at each level in the chain of command, to know their individual authority and responsibilities. Each individual must have professional competence but needs to have trust – trust in subordinates and trust in superiors. This can only be achieved through effective training.

The details of Operation Kratos remain confidential, but it seems to relate to a 'shoot to kill' policy apparently designed to deal specifically with the threat of a terrorist incident or a suicide bomber, obviously recognizing the very difficult judgement a law enforcement officer may be required to make in a very short space of time. Often they will be unable to make that judgement themselves and may have to rely upon information relayed to them by a third party. To put this in the military context this may range from the order to drop a missile from an unmanned aerial vehicle controlled from thousands of miles away, to a request for artillery or air support through the medium of a forward air controller or a sniper operating under the direction of his observer. In these scenarios the firer/pilot/gunner is simply the person delivering the munition and the person exercising the judgement is the person who would be held to account. The law on this point is not clear in that when acting in self-defence, including in defence of others, the judgement in deciding whether to use force should be that of the person actually applying the force. Can a soldier rely upon the superior orders in this context? If the soldier acts in good faith on the basis of orders given to them by others who were privy to other intelligence or information, and the order was not manifestly illegal, then it would be a nonsensical to make the firer responsible for those actions.

Some authority for this position can be obtained from the European Court in the case of *McCann* v. *UK*. On 6 March 1988, Mairaid Farrell, Danny McCann and Sean Savage were shot dead by an SAS undercover team in Gibraltar. As appears to have been the case for de Menezes, the three IRA suspects were not given the chance to surrender but were killed outright, the soldiers believing that the terrorists had the ability to blow up a car bomb remotely. The court accepted:

> that the soldiers honestly believed, in the light of the information they had been given ... that it was necessary to shoot the suspects in order to prevent them from detonating a bomb and causing serious loss of life. The actions which they took, in obedience to superior orders were thus perceived by them to be absolutely necessary in order to safeguard innocent lives.[29]

In the McCann case the actions of the individual soldiers who fired were vindicated but there was criticism over the planning and control of the operation. Both the McCann and de Menezes case highlight the difficulties of dealing with the suicide bomb threat and calls for an almost impossible judgement from the soldier's perspective. The soldier will make that judgement based upon a combination of the threat level, their intelligence brief, the general operational conditions and the specific picture in the minutes or seconds leading to their decision – together with their training and experience. If the soldier gets the decision right they will have averted a threat with the potential to cause loss of life; if they get it wrong they have taken an innocent life and will have to account for their actions.

Targeting

Targeting is broadly defined as 'the process of selecting targets and matching the appropriate response to them taking account of operational requirements and capabilities'.[30] There are many other definitions, some more complex and recognizing the comprehensive approach to targeting, but from the military lawyer's perspective targeting is usually, but not exclusively, taken to mean kinetic, as opposed to nonkinetic targeting.[31] In war-fighting under the Law of Armed Conflict the principles of distinction, proportionality, necessity and humanity govern the conduct of any targeting. In the contemporary operating environment these principles have equal weight and provide sound guidance for the conduct of any operation. Like ROE, targeting will comprise mixtures of law and policy set in the operational context. ROE authorize the use of force and targeting is concerned with applying that force.

The commander must first ensure that the object is 'a legitimate target'. This includes enemy combatants, unlawful combatants[32] and military objectives. Second, they must protect civilians and civilian property. The fact that an attack on a legitimate target might inflict civilian casualties does not automatically render the attack unlawful. However, the collateral damage caused by that attack must not be disproportionate to the concrete and direct military advantage anticipated by the attack. Where the commander has a choice of method of attack and means of attack they must use the one that is best designed to avoid or reduce collateral damage.

Targeting is not an exact science and targeteers are only required to take all reasonable steps to ensure their targets are legitimate, decisions being made in the circumstances existing at the time the decision was made. To do this they will rely upon the intelligence picture, but there needs to be a process by which that picture is validated to ensure reliability.

The NATO bombing campaign in the Balkans lasted from 24 March 1999 to 9 June 1999. A number of specific incidents took place which were described as 'problematic'. These included the attack on a railway bridge in the Grdelica Gorge on 12 April 1999, the bombing of the RTS Radio Station in Belgrade on 23 April 1999 and the attack on the Chinese Embassy on 7 May 1999.

A bridge, in simple military parlance, is a potential target of attack if it is being used as a supply route. Where the bridge is being used for both military and civilian traffic it is a dual use target.[33] At 11.40am on 12 April 1999 NATO dropped a bomb on the Leskovac Bridge at the exact time that a passenger train was crossing. There were civilian casualties – ten killed and 15 injured. While the damaged train was still on the bridge a second bomb was dropped. There was no evidence to suggest that the train was deliberately targeted. The intended target was the bridge on the supply route. In military legal terms the targeting of the bridge, and the method and manner in which the attack was carried out, was justified. The attack on the RTS Station in Belgrade was a planned attack aimed at disrupting the Command Control and Communications network. RTS is state owned and while used for commercial purposes, was also an integral part of the military communications system and as such was dual use. The attack took place at night and 16 civilians were killed. This incident should be compared with that of the attack on the Radio Milles Collines during the Rwandan Conflict, but in the latter case the justification was based more upon the use of the radio station to incite violence rather than as a command and control network. In any event, like the attack on the Leskovac Bridge, this attack was justified.

The final incident was the attack on the Chinese Embassy. Three Chinese citizens were killed, approximately 15 injured and extensive damage was caused to the building. The Chinese Embassy was not the intended target for attack. The Embassy had been wrongly identified as the Yugoslav Federal Directorate for Supply and Procurement, which played a key role in military procurement and would therefore have been a legitimate target. However, the building actually hit was clearly a civilian object and not a legitimate military objective. The fault in this case was not that of the aircrew involved who hit their intended target but the accuracy and reliability of the intelligence used to identify the target.

The conduct of the NATO bombing campaign in the Former Yugoslavia was referred to the Prosecutor of International Criminal Tribunal for the Former Yugoslavia to determine whether the conduct of the campaign established violations of international humanitarian law. The committee established to investigate the incidents described above concluded that while objects of attack might be subjected to legal debate and that while mistakes occurred and that there were errors of judgement, that no further investigation should be commenced. The report by the committee is important in that it recognizes the complexities of the current operating environment, especially with regard to dual use targets. The legal principles of distinction, proportionality and necessity were sufficient to enable targets to be prosecuted. What caused the controversy was not the legal basis for directing the attacks but the inevitable media and political fallout when, in the event, and despite the best of intentions, such attacks go wrong or have unintended consequences. Managing those consequences is a considerable challenge to commanders as can be evidenced by, in the case of the Leskovac Bridge incident, the Supreme Allied Commander Europe (SACEUR) personally justifying the actions of the pilot.

In the contemporary operating environment, defining what is a military objective or assessing 'the concrete and direct military advantage' to be gained by the attack, even with state-of-the-art ISTAR assets, is problematic.[34]

As can be seen from the examples referred to from the Balkans, successful targeting is a war-winning exercise. Do it well at the operational and tactical level and it bodes well for the strategic level. Do it badly, relying upon poor intelligence and causing unnecessary collateral damage, it can cause a disproportionate effect at the tactical level with the consequence of restricting a commander's ability to respond by imposing greater levels of control over the use of force. The German commander's decision to destroy two oil tankers that had been hijacked by the Taliban in Afghanistan's Kanduz province in 2009 provides an illustration of the type of political fallout that that can follow from an action that resulted in a large number of unintended civilian casualties.[35] This led to several high-level resignations and repercussions throughout German society as the nature of the country's involvement in the operation was debated.[36] To withdraw from a particular mission or to divert a missile away from the intended point of impact because of the risk of causing higher than anticipated civilian casualties may be seen by some as a tactical failure – but in the longer term it may lead to strategic success.

Combatants and noncombatants

Rupert Smith employed the phrase 'war amongst the people' and, in doing so, identified 'the people' themselves as part of the terrain on the modern battlefield.[37] This causes untold problems for both commanders and soldiers alike. Under the Geneva Conventions, lawful combatants include members of the regular armed forces of a state party to the conflict; militia, volunteer corps and organized resistance movements belonging to a state party to the conflict, which are under responsible command, wear a fixed distinctive sign recognizable at a distance, carry their arms openly and abide by the laws of war; and members of regular armed forces who profess allegiance to a government or an authority not recognized by the detaining power. They are entitled to prisoner of war status upon capture, and are entitled to 'combatant immunity' for their lawful pre-capture warlike acts. Combatant immunity is a doctrine rooted in the customary international law of war, and forbids prosecution of soldiers for their lawful acts committed during the course of armed conflicts against legitimate military targets. For example, Article 57 of the Lieber Code of 1863, which governed the conduct of war for the Union Army during the American Civil War and which served as the basis for the modern law of war treaties, provided that 'so soon as a man is armed by a sovereign government and takes the soldier's oath of fidelity, he is a belligerent; his killing, wounding, or other warlike acts are not individual crimes or offenses'.[38] Thus far so good. If we conveniently forget, for the sake of argument, the status of members of private security companies, members of the British Army are entitled to combatant immunity for lawful acts committed during and in the course of their duty. In traditional war-fighting soldiers are entitled to target and kill enemy combatants without warning even if those combatants pose no immediate threat. The combatant could be asleep, eating or 'off duty'. The only combatants he is not allowed to kill are those who have surrendered or are *hors de combat*. In the war among the people, however, the enemy may be part of an organized group, they may have a command structure, they may carry arms openly but they will rarely wear uniform and may rely upon insurgency tactics. Under their ROE the soldier, unable to identify the enemy, will respond to the threat. While this is good for the enemy, it places the soldier at a distinct disadvantage and is yet another challenge he or she has to face in the contemporary operating environment. The exception to this is intelligence-led operations where, if the enemy can be identified by reliable information, then by this means the status can be determined.

Captured persons

The treatment of captured persons has been one of the most difficult and challenging issues to have arisen in the contemporary operating environment. In traditional conflict the enemy, once captured, were treated as prisoners of war and steps taken to deal with them in accordance with the Geneva Conventions. The treatment of prisoners of war in the Falklands worked well from both British and Argentinean perspectives. Status was not in doubt, the International Committee of the Red Cross were involved from the outset, no distinction was made in medical treatment between those of different nationality and prisoner transfers were conducted in a proper and dignified manner. The deaths of two Argentine prisoners of war were investigated by Boards of Inquiry conducted in accordance with the principles set out in the Geneva Conventions. The only practical difficulty arose over the status of Captain Alfredo Astiz who was captured on 23 April when South Georgia was retaken by Royal Marines. Astiz, perhaps somewhat foolishly in the circumstances, signed the document of surrender. Astiz was wanted by both the French and Swedish authorities for the torture, disappearance and murder of two French nuns in 1976 and a Swedish tourist in 1976. He was not, therefore, repatriated but brought back to the UK while the British authorities considered the requests for extradition. In the event he was not extradited but was interviewed by the French and Swedish authorities. He was returned to Argentina on 10 June 1982.[39]

Similarly there were few difficulties in the first Gulf War. There were practical difficulties with the registration of prisoners and allegations made that personal property, including large sums of money, had been stolen. One of the more interesting events (which reveals how seriously the UK applies the Geneva Conventions) occurred on the first day of hostilities. At this time there were 35 members of the Iraqi military who were lawfully in the UK. The majority were attending educational courses. These prisoners were arrested and taken to Rollestone Camp on Salisbury Plain, which became a prisoner of war camp. The prisoners received appropriate treatment in accordance with the Geneva Conventions. Some disputed their status and Boards of Inquiry were held in accordance with the Determination Regulations and three were released.

Dealing with prisoners of war is a military requirement and the cost in terms of resources – including manpower and equipment – must be included in the commander's planning for operations. In the first Gulf War the arrangements for POW handling were not finalized until January 1991, when the Prisoner of War Guard Force was established, comprising 1st Battalion (Bn) Coldstream Guards, 1st Bn Kings Own Scottish Borders and 1st Bn Royal Highland Fusiliers under the command of a full colonel.

Although expensive in manpower terms, dealing with prisoners of war in traditional war-fighting is comparatively straightforward. In the contemporary operating environment forces are more likely to be dealing with civilians who will fall into the category of either security internees or criminal detainees. The military have the power to detain prisoners of war for the duration of the war-fighting and have power to intern when an occupying force. When acting under a Security Council mandate or with the consent of a host nation, the power to detain and intern will depend upon the authority granted by the UN or the receiving state. The treatment of internees is governed by the Geneva Conventions but there are no specific rules for the treatment of criminal detainees and precise arrangements will vary from theatre to theatre. At the very least, however, captured personnel of whatever status should receive basic humanitarian treatment equivalent to those afforded to prisoners of war or internees. In particular, they should be protected from attack and not subjected to torture or any inhuman or degrading treatment.

Private security companies

With the current overstretch of military forces engaged in operations in both Iraq and Afghanistan, the use of private security companies and private military companies to undertake activities short of war-fighting is attractive. Since the end of the Cold War there has been a huge growth in the number of commercial organizations who can offer a service to the military on operations. The difficulty in law, however, is the status of the individuals and the companies that employ them with an unfortunate and unhealthy comparison often being made with mercenaries. While an argument can be advanced that a private military company could constitute a state's 'armed forces' within the meaning of Article 43(1) of Additional Protocol 1, this is unlikely to be the case for the UK in the absence of specific regulation which would bring the company under the 'responsible command' of the military once deployed. The absence of this command responsibility would leave members of private military companies in the highly undesirable position of being virtually immune from any jurisdiction. This can lead to two tiers of soldiers – those who are accountable at all times and those who fall into a grey area. The ongoing arguments as to which jurisdiction applies to the Blackwater private security contractors alleged to be involved in the deaths of Iraqi civilians in 2007 demonstrates the international nature of this accountability issue.[40]

Journalism and the media

If modern conflicts are fought among the people and in the presence of the media, what challenges does this present? We have already noted the strategic corporal; is there also a 'strategic cameraman'?

A key element of influence operations (intended to affect the perceptions and behaviours of leaders, groups, or even entire populations) is media operations and a key tenet of the comprehensive approach is the maintenance of political and public support for the military activity which supports Her Majesty Government's strategic objectives.[41] Included in the factors for media operations is the need for accountability. In this respect media accountability is inextricably linked to accountability under the law – some may say more so. Virtually all conflicts are conducted under the spotlight of the media. There is a thirst for knowledge and conflict is news. It is UK policy to provide the media with a range of facilities to enable first-hand reporting, in addition to an accurate, objective and timely information service. Those journalists who choose to accompany the force become 'embedded'. The pool of embedded journalists covering the Iraq conflict was 158 compared to 29 covering the Falklands conflict. In the first Gulf War General de la Billiere addressed all embedded journalists, stating that the military would offer every opportunity to give them safe passage and send their stories back, but in return he wanted full cooperation: security of the mission and the lives of soldiers was paramount. An example of accountability to the media was the images of prisoner abuse at Abu Ghraib. This led to an immediate inquiry and investigation. The British were not immune and, soon after the breaking of the Abu Ghraib incidents, images of British troops abusing detainees at Camp Breadbasket appeared in the newspapers. One of the challenges in dealing with the media when they report atrocities such as Camp Breadbasket is that the damage is done as soon as the story breaks. Media accountability is therefore instant. When the *Daily Mirror* published photographs in May 2004 allegedly of British soldiers beating an Iraqi prisoner, General Sir Mike Jackson felt obliged to say 'If proven, the perpetrators are not fit to wear the Queen's uniform. They have besmirched the good name of the Army and its honour.'[42] On that occasion the photographs were fakes. Legal accountability, on the other hand, takes time. The legal process is lengthy and time is required to investigate and, where appropriate, to bring perpetrators to justice.[43] Often legal accountability only serves to reopen old wounds many months or years after the initial incident. Like the actions of the strategic corporal, adverse media reporting can cause immeasurable damage to the campaign. The temptation to cover up bad news must be resisted at all costs. The potential damage caused by the My Lai incident in 1969 and the death of Shidane Arone in Somalia in 1993 was immediately apparent to lower-level commanders and attempts were made to conceal the truth. With the intense media and legal interest in operations today the story will emerge at some time. If the unfortunate happens, the proper and only course of action is to confront it straight away by investigating the circumstances.

Public opinion is crucial to the success of a military operation. It is difficult to classify conflict as 'popular' or 'unpopular', but an unpopular war is clearly difficult to justify. Politicians risk losing their mandate, but for a military commander the challenges run much deeper. Journalists, like the weather, are unpredictable – one day they will offer a story that provides absolute support, the next day the headlines could present a damning account of impropriety or incompetence. However, before we condemn journalists it is best to remember that the incidents giving rise to these headlines often come from the cameras or mobile phones of the soldiers themselves. Morale, a key element of the moral component, is influenced, shaped and developed through training and leadership and can be irreparably damaged by poor public opinion influenced through adverse media. The US General George Marshall noted that:

Morale is a state of mind. It is steadfastness and courage and hope. It is confidence and zeal and loyalty. It is élan, esprit de corps and determination. It is staying power, the spirit which endures to the end – the will to win. With it all things are possible, without it everything else, planning, preparation, production, count for naught.[44]

Multinational operations

Multinational operations are not new. Wellington commanded a coalition of six countries at Waterloo and the Normandy landings in the Second World War were a multinational operation including Americans, British, Canadians and Free French ground forces, supported by Polish, Dutch, Australian and New Zealand air forces and navies. It has probably more to do today with modern communication systems, but Wellington was free to commit his forces without political approval or scrutiny. The difficulty today goes back to the problem of state as opposed to individual responsibility. As noted above with respect to domestic self-defence legislation, those forces taking part in multinational operations will come from different legal backgrounds. While almost all will have adopted the Geneva Conventions, not all will be signatories to the Additional Protocols, the Rome statute of the International Court or the Ottawa Convention on anti-personal mines or the Dublin Convention on cluster munitions. While there is very little disagreement about the application of IHL in international armed conflicts, the application of different laws to other types of conflict are not necessarily consistent between countries. For example, the UK chooses to apply Geneva Conventions I–IV to internal conflicts as well as international ones. This is a matter of policy, not law, and provides a level of consistency across the board for the UK but can lead to different coalition partners applying different legal standards while operating alongside each other.

In practice, while this sounds a recipe for disaster, this has not proven to be an insurmountable problem for the conduct of effective multinational operations. What causes more difficulty at the operational and tactical levels is the interpretation of UN Security Council Resolutions, for example with the interpretation of phrases such as 'all necessary measures'. ROE will not be identical, with some countries being more robust and others adopting a conservative approach. For example, US forces may apply lethal force in the protection of property while no such provision applies to UK forces (see above). Again, in practice, these issues tend not to result in practical difficulties. However, what does cause some difficulty is the targeting process, where, in coalition operations, approval may have to be sought from two or three different chains of command: the national chain of command of the troops occupying the area in which the strike is taking place; the national chain of command of the aircraft delivering the munition; and the chain of command of the organization having overall responsibility for the mission itself. What also is not clear is whether, for example, the commander of a NATO operation is ultimately responsible to NATO or to his home nation.

Interrogation and tactical questioning

Following the issues highlighted in the Baha Musa case, there is to be a public inquiry in the UK. Whether his death had anything to do with interrogation is still a matter for debate; however, the treatment of prisoners in Guantanamo and the deliberate decision to place prisoners outside of the jurisdiction of the United States adds fuel to the controversies in the broader area of interrogating prisoners.

The questioning of prisoners is, itself, perfectly legitimate. However, the Geneva Conventions prohibit the use of physical or mental torture, or any other form of coercion on prisoners of war to secure from them information of any kind. Prisoners who refuse to answer questions may not be threatened, insulted or exposed to any unpleasant or disadvantageous treatment.[45] Similar provisions apply to members of the civilian population who may be interned by members of the security forces.[46] The prohibition on torture and inhuman or degrading treatment is reinforced under the UN Convention against Torture and other Cruel Treatment and under the Human Rights Act. Yet the interrogation of prisoners is a legitimate and vital part of any military operation: as with any civil police investigation, lawful techniques or practices may be applied to get the prisoner or suspect to talk. Clearly, effective interrogation requires respect and professionalism.

Conclusions

There can be little doubt that the contemporary operating environment is tough and demanding for all those involved, from the private soldier to the commander. The levels of scrutiny, responsibility and accountability have never been higher. This is not just from a legal perspective but also from political and public levels, the latter generally being applied through the media. The degree of professionalism required to operate in this environment has never been greater. At the tactical level the soldier has to be trained to enable him or her to respond to a myriad of scenarios from periods of high-intensity combat to routine patrolling. At the operational level the commander will have access to vast quantities of information or intelligence from a variety of sources – patrol reports, human intelligence, signal intelligence and surveillance assets. They will have at their disposal a vast array of weaponry deliverable by land, sea or air and possibly by different contributing nations each subject to national restrictions and with differing rules of engagement. In that battlespace, and operating alongside them, will be representatives from other government organizations and agencies. All of this takes place in a fast-moving and dynamic environment. A high level of competence is required, together with mutual trust. The doctrine of mission command, that promotes decentralized decision-making, emphasizing speed of action and the taking of initiative while acting within the commander's intent, is a key element of British Army Doctrine and is vital for success in this environment. In this environment, does the law constrain the prosecution of effective military operations or, to put it in the words of Dr Reid, are twentieth-century rules still appropriate for twenty-first-century conflict? The law recognizes that conflict will occur, that armed force may be used, that the armed force will result in death, injury and destruction of property and, while the conflict should be directed against the opposing armed forces, civilians will inevitably be drawn into the conflict. These permissions were applicable in twentieth-century conflict and they remain applicable in twenty-first-century conflict. Therefore, from a legal perspective, with one exception, nothing has changed. If a soldier or commander feels constrained it is usually governmental policy that restricts the freedoms of action. It may be the case that force can be applied legally, but the question for the specific circumstance is: is it appropriate to apply that force? If we look at the changes in the structure of a commander's headquarters in these last 25 years of conflict then we notice the creation of two branches – those of the legal and political advisors. It is also of note that these advisors usually work closely together and are themselves sometimes in conflict, balancing the freedoms that the law gives with the expediency of applying those freedoms in the particular circumstances. It is not suggested that this political control is in any way detrimental to the effective conduct of operations. It is simply a reality of the modern battlefield.

What then is the exception? Perhaps it is no more than a sign of the times and an example of the litigious society in which we live, but the present conflict in Iraq has seen a number of civil cases brought against the UK's Ministry of Defence. These cases cover a multitude of issues and at the heart of these cases is the application of the European Convention on Human Rights to military operations overseas. This is not new and there were similar challenges relating to the conflict in

Northern Ireland. Turkey and the Russia find themselves in a similar position. In the United Kingdom these challenges have met with only limited success but underlying these challenges are allegations of murder, manslaughter, assault and other improper behaviour. As identified in the Aitken Report, there is, regrettably, some substance to some of these allegations. Some coroners are using the forum of their courts to openly criticize not only the Ministry of Defence but also individual commanders, applying a commercial health and safety standard to the complexities of the modern battlefield. Mistakes are bound to occur in this environment but the reality is that the civilian victim in conflict may be awarded more in compensation terms than the injured soldier. All these cases have the potential to undermine morale and have one aspect in common: they are often judged years after the event in the comfortable surroundings of the courtroom and with the benefit of hindsight. Soldiers who commit wrongful acts should be investigated and, where appropriate, prosecuted. To set the contemporary operating environment into a legal context it would be well to remember the doctrine of combat immunity set out in the 1940 case of *Shaw Savill and Albion Company Ltd* v. *The Commonwealth*.[47] The case, decided by the High Court of Australia, was a claim in damages brought by the plaintiff companies alleging that HMAS *Adelaide* was sailing too fast, that it failed to keep a proper look out and that it was not navigated in a proper and seaman like manner. The defence was that, at the relevant time, the *Adelaide* was part of the naval forces of Australia and was engaged in active naval operations against the enemy.

The High Court of Australia accepted that, in principle, the defence of combat immunity was open to the state. Dixon J, in the course of his judgment, said this:

> It could hardly be maintained that during an actual engagement with the enemy or a pursuit of any of his ships the navigating officer of a King's ship war was under a common-law duty of care to avoid harm to such non-combatant ships as might appear in the theatre of operations. It cannot be enough to say that the conflict or pursuit is a circumstance affecting the reasonableness of the officer's conduct as a discharge of the duty of care, though the duty itself persists. To adopt such a view would mean that whether the combat be by sea, land or air our men go into action accompanied by the law of civil negligence, warning them to be mindful of the person and property of civilians. It would mean that the Courts could be called upon to decide whether the soldier on the field of battle or the sailor fighting on his ship might reasonably have been more careful to avoid causing civil loss or damage. No one can imagine a court

undertaking the trial of such an issue, either during or after a war. To concede that any civil liability can rest upon a member of the armed forces for supposedly negligent acts or omissions in the course of an actual engagement with the enemy is opposed alike to reason and to policy. But the principle cannot be limited to the presence of the enemy or to occasions when contact with the enemy has been established. Warfare perhaps never did admit of such a distinction, but now it would be quite absurd. The development of the speed of ships and the range of guns were enough to show it to be an impracticable refinement, but it has been put out of question by the bomber, the submarine and the floating mine. The principle must extend to all active operations against the enemy. It must cover attack and resistance, advance and retreat, pursuit and avoidance, reconnaissance and engagement. But the real distinction does exist between active operations against the enemy and other activities of the combatant services in time of war.

This Ruling was given not in the context of the contemporary operating environment, but in 1941. Although it refers to civil liability it is a timely reminder that the fog of war existed then and still today.

Notes

1. Army Doctrine Publication Land Operations. Army Code 71819.
2. General Sir Rupert Smith, *The Utility of Force* (London: Allen Lane, 2005), p. 10. British Defence Doctrine (JWP 0-10), also refers to four levels of warfare, namely grand strategy, military strategy, operational level and tactical level.
3. Sierra Leone, East Timor and the UN commitment to Cyprus are not referred to. That is not to say they were trivial or insignificant. All military deployments are unique and have, from a legal perspective, their own complexities.
4. Operational law is not defined but includes the myriad of Treaty and Customary law known collectively as the Law of Armed Conflict, the applicable domestic law, and the policies and regulations which support those laws. This includes the procedures for the treatment of detainees, interrogation and tactical questioning, rules of engagement, targeting, the investigation of operational incidents and the planning for operations
5. Also known as International Humanitarian Law.
6. *Luisa Diamantina Romero de Ibanez and Roberto Guillermo Rojas* v. *UK* (ECHR 2000), Al Skeini and Al Jeddah.
7. *House of Lords Official Report,* 14 July 2005; Vol. 673, c. 1236.
8. http://news.bbc.co.uk/1/hi/uk/4873856.stm.
9. William Rehnquist, *All the Laws But One: Civil Liberties in Wartime* (New York: Alfred A Knopf, 1998), pp. 224f.

10. JWP 0-10 British Defence Doctrine.
11. Smith, *The Utility of Force*, p. 7.
12. Charles C. Krulak (1999). 'The Strategic Corporal: Leadership in the Three Block War', available at http://www.au.af.mil/au/awc/awcgate/usmc/strategic_corporal.htm.
13. John Reid: 'British Task Force has a Vital Job to do in Southern Afghanistan', MOD website 26 January 2006. Cited in http://www.kosb-edinburgh-branch.co.uk/articles.php?id=402&table=articles&o=350.
14. UN Security Council Resolution 678, adopted by the Security Council 29 November 1990, available at http://daccessdds.un.org/doc/RESOLU-TION/GEN/NR0/575/28/IMG/NR057528.pdf?OpenElement.
15. Peter de la Billiere, *Storm Command: Personal Account of the Gulf War* (London: HarperCollins, 1995).
16. The Rambouillet Agreement called for the NATO administration of Kosovo with a force of 30,000 NATO troops to maintain order.
17. *Guardian*, 28 March 2006.
18. Mike Jackson, *Soldier: An Autobiography* (London: Corgi, 2008).
19. Directorate of Land Warfare, *ADP Land Operations* (London: Ministry of Defence, 2005).
20. 'British Soldier Admits War Crime', available at http://news.bbc.co.uk/go/pr/fr/-/1/hi/uk/5360432.stm.
21. David Bercuson, *Significant Incident: Canada's Army, the Airborne, and the Murder in Somalia* (Toronto: McClelland & Stewart, 1996), pp. 238–9.
22. *The Aitken Report: An Investigation into Cases of deliberate Abuse and Unlawful Killing in Iraq in 2003 and 2004* (MOD: 25 January 2008), p. 10.
23. JSP 398.
24. Smith, *The Utility of Force*, p. 7.
25. De la Billiere, *Storm Command*, p. 140.
26. De la Billiere, *Storm Command*, p. 193.
27. Dan Mills, *Sniper One: The Blistering True Story of a British Battle Group Under Siege* (Harmondsworth: Penguin, 2008), p. 188.
28. *R v. Beckford* [1988] AC 130. See www.bailii.org/uk/cases/UKPC/1987/1.html.
29. Mark W. Janis *et al.*, *European Human Rights Law* (Oxford: Oxford University Press, 2008), p. 141.
30. For example, US Marine Corps, *Tactics, Techniques, and Procedures for Field Artillery Target Acquisition* FM 3-09.12, 21 June 2002.
31. The effects-based approach to operations (EBAO) includes all targeting, kinetic and nonkinetic. For the purposes of this chapter the author distinguishes between 'soft' and 'hard' effects.
32. The term 'unlawful combatants' is not defined precisely under customary or convention law but it is widely used and subject to many different interpretations. This is a further example of the complexities of the contemporary operating environment.
33. Military objectives are limited to those objects which by their nature, location, purpose or use make an effective contribution to military

action and whose total or partial destruction or neutralization ... offers a definite military advantage. Additional Protocol 1 Article 52.

34. Intelligence, surveillance, target acquisition and reconnaissance.

35. 'Germany's Top Soldier Quits over Afghanistan Raid', available at http://news.bbc.co.uk/1/hi/world/south_asia/8380226.stm.

36. 'German Ministers Face Kunduz Air Strike Inquiry', available at http://news.bbc.co.uk/1/hi/8415305.stm.

37. Smith, *The Utility of Force.*

38. Instructions for the Government of Armies of the United States in the Field (Lieber Code). 24 April 1863, Section III, Art. 57, available at http://www.icrc.org/ihl.nsf/5a780f680129b33841256739003e6367/fc13 7ca8da6a5a65c12563cd00514d35!OpenDocument.

39. Astiz was convicted, in absentia, by the French Courts for his role in the disappearance and torture of the two nuns Alice Domon and Leonie Duquet and sentenced to life imprisonment.

40. Alissa J. Rubin and Paul von Zielbauer, 'Blackwater Case Highlights Legal Uncertainties', *New York Times*, 11 October 2007, available at http://www.nytimes.com/2007/10/11/world/middleeast/11legal.html?_r= 2&oref=slogin.

41. The Comprehensive Approach refers to the 'commonly understood principles and collaborative processes that enhance the likelihood of favourable and enduring outcomes within a particular situation' (Joint Discussion Note 4/05). It is based on the expectation that military forces usually work alongside other government departments, international organizations (IOs) and NGOs and that such action is most effective when each organization contributes its strengths, with those of others, in an holistic approach. See http://www.mod.uk/NR/rdonlyres/BEE7F0A4-C1DA-45F8-9FDC-7FBD25750EE3/0/dcdc21_jdn4_05.pdf.

42. Paul Byrne, 'Shame of Abuse by Brit Troops', *The Mirror*, 1 May 2004.

43. David Whetham, 'Killing Within the Rules', *Small Wars and Insurgencies*, 18(4), December 2007, p. 729.

44. Address at Trinity College Hartford, Connecticut, June 1941. John M. Collins, 'The Principles of War', in Arthur F. Lykke Jr (ed.) *Military Strategy: Theory and Application* (Carlisle Barracks, PA: US Army War College, 1 May 1984), pp. 3–9.

45. GC III Article 17.

46. GC IV Article 31. 'No physical or moral coercion shall be exercised against protected persons'.

47. *Shaw Savill and Albion Company Ltd* v. *The Commonwealth* [1940] HCA40; (1940) 66 CLR 344.

Chapter 7

Civilian Protection and Force Protection

HENRY SHUE[1]

At the heart of any effort to maintain some minimal level of civilization is a commitment to attempt, insofar as possible, to settle disputes without the use of force. Our best attempts, however, may fail, sometimes requiring, if the stakes are high enough, military force to be used. But even during the resort to force, not all restraint is abandoned by civilized societies: 'In any armed conflict, the right of the Parties to the conflict to choose methods or means of warfare is not unlimited.'[2] The three basic limits are necessity, discrimination and proportionality, as is widely understood.[3] The devil, as always, is in the details. What do these three concepts mean? And how do they relate to each other?

Military necessity

The early landmark catalogue of the customary law of war, the Lieber Code (1863), recorded the following: 'Military necessity, as understood by modern civilized nations, consists in the necessity of those measures which are indispensable for securing the ends of the war, and which are lawful according to the modern law and usages of war.'[4] The 2004 British manual says: 'Military necessity permits a state engaged in an armed conflict to use only that degree and kind of force, not otherwise prohibited by the law of armed conflict, that is required in order to achieve the legitimate purpose of the conflict, namely the complete or partial submission of the enemy at the earliest possible moment with the minimum expenditure of life and resources' (2.2). The manual adds in further explanation: 'Conversely, the use of force which is not necessary is unlawful, since it involves wanton killing or destruction' (2.2.1). These definitions make three elements of necessity quite clear but leave a fourth obscure.

First, an action's being a physically necessary means to the end pursued in the war is a logically necessary condition, but not a logically

135

sufficient condition, of its being justified. A means is permitted *only if* it is physically necessary – only if, in Lieber's term, it is 'indispensable'. This is an entirely different matter from its being the case that *if* it is physically necessary, then it is justified. 'If it is physically necessary, then it is justified' is a rationale for ruthlessness; this formula says that if one would need to do it in order to succeed in attaining one's goal, one is justified in doing it. That amounts to saying that one's end justifies absolutely any means that it takes to reach it; in other words, there are no limits on acceptable means except whether they, in fact, serve the end. The claim that there are no limits on acceptable means is obviously completely incompatible with the requirement that what is necessary also be 'not otherwise prohibited by the law of armed conflict', including the laws requiring discrimination and proportionality.

Military necessity is a matter of *only if*, not of *if*. A means is justified only if it is of a kind that is physically necessary, indispensable, or unavoidable in pursuit of the goal of the war. Military necessity does not mean that one may do whatever it takes to win. On the contrary, both the Lieber and the British formulations make it perfectly clear that necessity is itself constrained by the Law of Armed Conflict. If an act is prohibited by the Law of Armed Conflict, it may not be done. If it were the case that one could attain one's end in the fighting only by doing what was prohibited by the Law of Armed Conflict, one would have to accept failure. Fortunately, by conscious design over the centuries, the Law of Armed Conflict does not, in fact, prohibit all actions that are sufficient to attaining a 'legitimate purpose'. This means that 'otherwise, we will not win' is rarely an applicable – and never a good – reason for violating the law of war.

That one may be confronting an emergency has already been taken into account in the writing of the rules. This was put eloquently in a decision in a US Military Tribunal following the Second World War:

> It is an essence of war that one or the other side must lose and the experienced generals and statesmen knew this when they drafted the rules and customs of land warfare. In short, these rules and customs of warfare are designed specifically for all phases of war. They comprise the law for such emergency. To claim that they can be wantonly – and at the sole discretion of any one belligerent – disregarded when he considers his own situation to be critical, means nothing more or less than to abrogate the laws and customs of war entirely.[5]

In sum, military necessity is not a positive authorization saying that 'one may do whatever it takes (legal or illegal)'. Military necessity is fundamentally a prohibition – a prohibition on wasteful and pointless destruction – and may thus best be captured in the British manual's negative formulation: 'The use of force which is not necessary is unlawful, since it involves wanton killing or destruction.'

The second point, already evident, is that the critical restraint on necessity is law. Necessity is subordinate to legality: first, an action must be in accord with the laws of war, and then it must in addition be indispensable or necessary. That an illegal act looks as if it might be helpful, or even essential, to one's cause counts for nothing. Necessary military actions must be chosen from among legal military actions.

Third, to be necessary an action must, as also already mentioned, serve a 'legitimate purpose'. That actions might be indispensable to the attainment of a purpose one has no business pursuing counts for nothing as well. Military necessity consists of actions that are legal and indispensable to the attainment of a kind of purpose that can be justified. It most certainly does not consist of 'whatever it takes to get whatever one wants'!

But, fourth, the British manual rightly raises an issue that cannot be fully handled by reference to necessity alone, with its final phrase: 'with the minimum expenditure of life and resources'. The basic question is: whose lives and resources, ours, theirs, or both? The difficulty is that possible trade-offs often arise between force protection and civilian protection. Reducing the risks of combat to civilians will sometimes require members of the forces to take greater risks upon themselves. This raises the question of the meaning of the principles of discrimination and proportionality, to which we must turn in order to tackle the issue of the relative weight to be given to force protection and civilian protection.

Discrimination and proportionality

These principles are two complementary components of the protection of civilians. The current authoritative handbook of the customary law of war presents as Rule 1: 'The parties to the conflict must at all times distinguish between civilians and combatants. Attacks may only be directed against combatants. Attacks must not be directed against civilians.'[6] Turning from persons to objects, the handbook reports as Rule 7 of customary law: 'The parties to the conflict must at all times distinguish between civilian objects and military objectives. Attacks may only be directed against military objectives. Attacks must not be directed against civilian objects.'[7] 1977 Geneva Protocol I combines the two distinctions concerning persons and objects respectively into a single 'Basic Rule'.[8] For strong emphasis it adds: 'In case of doubt whether a person is a civilian, that person shall be considered to be a civilian.'[9] Thus, the fundamental point is perfectly clear in both customary and treaty law: only military personnel and military objectives may ever be attacked. That is the principle of distinction. There are, for this purpose, only two kinds of people; one may be attacked and the other must never be.[10] Similarly, there are only two kinds of objects; one may be attacked and the other must never be.

The principle of distinction is, however, not only negative. It positively requires constant care. The chapter of 1977 Geneva Protocol I on 'Precautionary Measures' begins: 'In the conduct of military operations, constant care shall be taken to spare the civilian population, civilians and civilian objects.'[11] It is, of course, understood that even if one does take constant care for the protection of all civilians, some will nevertheless be maimed or killed. The principle of proportionality governs such unintended civilian losses when they are foreseeable. Strictly speaking, then, a lack of proportionality is one kind of lack of discrimination; rather than being a completely independent standard, the principle of proportionality prohibits one especially egregious form of indiscriminate fighting, which might be called disproportionate indiscriminateness. Customary law, Rule 14, says: 'Launching an attack which may be expected to cause incidental loss of civilian life, injury to civilians, damage to civilian objects, or a combination thereof, which would be excessive in relation to the concrete and direct military advantage anticipated, is prohibited.'[12] That is the principle of proportionality.

Eminent theorist Michael Walzer has put the point well: 'Simply not to intend the death of civilians is too easy ... What we look for in such cases is some sign of a positive commitment to save civilian lives.'[13] What does 'positive commitment' mean concretely? Walzer continues: 'If saving civilian lives means risking soldiers' lives, the risk must be accepted. But there is a limit to the risks that we require ... War necessarily places civilians in danger ... We can only ask soldiers to minimize the dangers they impose.'[14] This is the answer to the question raised above by the British manual's formulation of necessity as requiring 'the minimum expenditure of life and resources'. Force protection is one duty, but the fundamental duty is civilian protection. Force protection cannot be pursued to a degree that compromises civilian protection. A well-trained, well-armed soldier is expected to take some risks with his/her own life in order to save the lives of some civilians. Fortunately, a well-trained, well-armed soldier is considerably more likely, in fact, to survive any given level of danger than the unarmed, untrained civilian. A soldier is sometimes expected to put himself in harm's way in order to keep civilians out of harm's way, but, other things equal, he/she is more likely actually to come out of it safely than they would have been without such protection. Such a moderate position about the risks that combatants are expected to run is widely accepted as an accurate account of the professional responsibility of honourable fighters.

The position just advanced treats all civilians alike. Its meaning can be sharpened by contrast with a recently advanced extreme position that rejects the Protocol I requirement of constant care for all civilians. Asa Kasher and Amos Yadlin, who controversially claim that the rules of war ought to be different when one's adversaries include terrorists, advocate the creation, through the use of a multiplicity of dichotomies, of a sharply differentiated ranking of packages of duties, which begins as follows:

(d.1) Minimum injury to the lives of citizens of the state who are not combatants during combat;

(d.2) Minimum injury to the lives of other persons (outside the state) who are not involved in terror, when they are under the effective control of the state;

(d.3) Minimum injury to the lives of the combatants of the state in the course of their combat operations;

(d.4) Minimum injury to the lives of other persons (outside the state) who are not involved in terror, when they are not under the effective control of the state.[15]

The list continues with the duties towards (d.5) those indirectly involved in terror and (d.6) those directly involved in terror.

This remarkable priority ranking contains a number of extraordinary features. For the purpose at hand, the radical proposal is that (d.3) force protection should receive priority over the protection of everyone except (d.1) fellow citizens and (d.2) non-citizen civilians who happen to be in areas under one's own control. In other words, force protection takes priority over protection for all civilians in contested territory, which will presumably include virtually all the citizens of one's adversary! This is the opposite of the 'constant care' required by 1977 Geneva Protocol I. As Michael Walzer and Avishai Margalit have noted in rebuttal, 'For Kasher and Yadlin, there no longer is a categorical distinction between combatants and non-combatants. But the distinction should be categorical, since its whole point is to limit wars to those – only those – who have the capacity to injure (or who provide the means to injure).'[16]

The international consensus on the protection of civilian objects is also under challenge. The generally accepted position is given in the handbook of customary law: 'Rule 8. In so far as objects are concerned, military objectives are limited to those objects which by their nature, location, purpose or use make an effective contribution to military action and whose partial or total destruction, capture or neutralization, in the circumstances ruling at the time, offers a definite military advantage.'[17] The broad international consensus on this two-pronged test is joined by the 2004 British manual.[18] The United States, however, has for some time been advocating its own, much more permissive definition of 'military objective'. In the first prong of the test, the US would like the language, 'make an effective contribution to military action', to be replaced by the following much broader and looser terminology: 'effectively contribute to the enemy's war-fighting or war-sustaining capability'.[19] 'War-fighting ... capability' may be largely equivalent to the generally accepted language of 'contribution to military action', but 'war-sustaining capability' would add large but indeterminate swathes of the civilian economy to the list of acceptable targets. The loosening advocated by the US would be a highly retrogressive step, and the British manual is right to resist it.

The language advocated by the US opens up indeterminate amounts of the nation beyond its military forces to attack. This appears designed to open the door to a return to aerial 'morale-bombing'. That this is indeed the goal of some of the leadership of the US military is made clear by a passage in the 2007 edition of Joint Publication 3-60, *Joint Targeting*: 'Civilian populations and civilian/protected objects, *as a rule*, may not be intentionally targeted, although *there are exceptions to this rule*. Civilian objects consist of all civilian property and activities other than those used to support or sustain the adversary's war-fighting capability. Acts of violence *solely* intended to spread terror among the civilian population are prohibited.'[20] This chilling passage not only twice emphasizes (groundlessly) that there are allegedly unspecified exceptions to the prohibition on intentionally targeting civilian objects, but it also goes out of its way to volunteer that targeting 'solely' with the intention of producing terror is prohibited, implicitly but unmistakably inviting the thought that if one can find some other pretext, one may engage in targeting that has the supposedly beneficial side effect of terrorizing the civilian population. Whether terror is one's sole intention is a complete red herring; what is illegal is having terror as one's 'primary purpose'.[21] If one targets a dual use facility with an intention to deny its benefits to the military but also with the main intention of terrorizing civilians, the denial is only a pretext and the terror is the primary purpose. In that case the attack is illegal.[22]

As is the case with the Kasher and Yadlin position on civilian persons, this US position on civilian objects challenges the fundamental purpose of the principle of distinction, which is to restrict the death and destruction of war as nearly as possible to the military forces involved. In the famous words of the *1868 St Petersburg Declaration*: 'The only legitimate object which States should endeavour to accomplish during war is to weaken the military forces of the enemy.'[23] Morale-bombing attempts to avoid the need to weaken the military forces by weakening the society's support for the forces. Not only has this indirect strategy always, in fact, failed,[24] but it is a formula for expanding the misery of war that the principles of discrimination and proportionality are intended to minimize.

Even if one does not smudge the lines between civilian and military as the two positions just criticized do, other problems arise about the prohibition on disproportionate indiscriminateness or, as it is usually known, the principle of proportionality. The heart of the proportionality standard is the requirement that civilian losses expected from an attack not 'be excessive in relation to the concrete and direct military advantage anticipated'. What does this mean? Is its meaning sufficiently determinate to provide genuine guidance? Plainly the test is relational: the expected civilian losses are to be compared to the anticipated concrete and direct military advantage, and a judgement is to be made about whether the civilian losses would be excessive in relation to the military advantage.[25] The question is whether there is any real standard for 'excessive', or whether 'excessive' and 'not excessive' are simply in the eye of the beholder.

A leading commentary on 1977 Geneva Protocol I makes the crucial point in response, but muddles it up with the confusing and unnecessary introduction of the terminology of 'subjectivity':

> As both sides of the equation are variables, and as they involve a balancing of different values which are difficult to compare the judgment must be subjective. In the final analysis, however, most decisions on the major political, economic, and social affairs of societies as well as major military decisions rest on the subjective judgment of decision makers based on the weighing of factors which cannot be quantified. The best that can be expected of the decision maker is that he act honestly and competently.[26]

Once it is conceded that 'most decisions on the major political, economic, and social affairs of societies as well as major military decisions' are judgements of the same kind as judgements about proportionality, the pointlessness of gratuitously calling them 'subjective', as if they were groundless preferences between different flavours of tea, is evident. All that is being granted is that, like virtually all important decisions, these can be well grounded and carefully reasoned, but they cannot be quantified. Since very few fundamental choices can be quantified, however, it is wildly misleading to label almost all fundamental human judgement 'subjective' as if it were inferior to some readily available 'scientific' alternative.[27] A clear-eyed qualitative judgement by a competent and experienced judge is better than a muddled quantitative calculation that has artificially collapsed fundamental distinctions in pursuit of commensurability. Yes, proportionality requires judgement, but there is no reason at all to think that good judgement based on reliable facts and sound reasons is in any meaningful sense 'subjective'.

What kinds of reasons, then, can a judgement about whether civilian losses are excessive compared to a concrete and direct military advantage appeal to? Naturally, some considerations apply to the side of the balance holding military advantage and others apply to the side with civilian losses. We may begin with military advantage. First, the principle of military necessity already does important work before one reaches the principle of proportionality. A means involving the use of force can be justified *only if* it is, in fact, necessary to the military end it is supposed to serve. Instrumental necessity is, of course, only a necessary condition of justification, not a sufficient one, as has already been noted in the initial discussion of military necessity. But it is significant that any attack that is supposed to produce a concrete and direct military advantage must, for a start, be an attack that is necessary.

Second, the simple fact that a military attack must be expected to produce a military advantage makes a difference. This may simply be a further, specific implication of the requirement of military necessity. It means that the attack must have some rational military purpose, and an attack can hardly be necessary if it has no rational purpose. This is an extremely low standard, but in view of the amount of completely senseless destruction that occurs in war, it is far from negligible. Violence that is at least not senseless is one small step forward. An attack must make some sense. It cannot consist simply of killing people and blowing things up for the sake of doing so. The killing and destruction must play some intelligible role in a coherent plan for military success. Mere high body counts as such, for example, do not constitute a military advantage. Attrition as such is not necessarily militarily advantageous, as the First World War and the American intervention in Vietnam demonstrated beyond all dispute. Whatever is done must have some reasonable prospect of actually advancing some sensible conception of military success.

Third, the firm requirement that the advantage be 'concrete and direct' is also significant. One of many objectionable features of the US proposal to treat 'war-sustaining' facilities that have no 'present connection' with military action as military objectives is the highly speculative and indirect nature of any advantage that could be expected from their destruction.[28] That some military advantage could imaginably accrue sooner or later is not good enough. One leading commentary observes:

'Concrete' means specific, not general; perceptible to the senses. Its meaning is therefore roughly equivalent to the adjective 'definite' ... 'Direct', on the other hand, means 'without intervening condition of agency'. Taken together the two words of limitation raise the standard ... A remote advantage to be gained at some unknown time in the future would not be a proper consideration to weigh against civilian losses.[29]

When one turns to the other side of the balance holding the civilian losses, the considerations are somewhat more controversial. First, I believe that the civilian losses should be calculated over the longer term and especially that they should be considered in the light of their cumulative effects.[30] This is obviously just the opposite of the requirement for military advantage, that it be concrete and direct. Why should the respective treatment of the two sides of the balance be so asymmetrical? A minor general reason is that when one is contemplating an action that one expects to be favourable to oneself, one is psychologically inclined to underestimate the costs to others. Tilting somewhat in the opposite direction by counting only the concrete and direct benefits on the side of the military advantage while counting the longer term and cumulative costs on the side of civilians may provide a useful corrective to the psychological tendency to underestimate the costs for others. Far more important is the fact that the protection of civilians is a fundamental purpose, making it essential not to misrepresent the losses that they will, in fact, suffer. In the battle zone military equipment and soldiers themselves are, for as long as replacements are available, replaced when they are lost, but civilian facilities that are destroyed in battle are normally not replaced until the war has ended, if then. Because of the lack of replacement, civilian losses tend to endure to a greater extent than military losses. It is only accurate for the more enduring character of the wartime civilian losses to be reflected in the manner of calculation.

Second, it is a generally accepted rule of customary law – designated as Rule 54 in the recent handbook – that 'attacking, destroying, removing or rendering useless objects indispensable to the survival of the civilian population are [sic] prohibited'.[31] Indispensable objects include 'foodstuffs, agricultural areas for the production of foodstuffs, crops, livestock, drinking water installations and supplies[,] and irrigation works'.[32] If this is to be meaningful, it must include sufficient electricity supplies to operate the water purification installations and irrigation works. The UK government has taken the position that the prohibition on destruction of indispensable objects applies only to destruction intended to deprive civilians.[33] This seems to me such a minimal requirement that it ought to hold for all foreseeable destruction of vital facilities; without the protection of adequate food or water, other protections of civilians are rather empty.

In the end, a judgement must be made whether any given expected civilian losses are excessive compared to the anticipated military advantage. Rather than requiring a comparison of 'apples [civilian losses] with oranges [military advantage]', as is often said, the judgement actually requires only the avoidance of exchanging a large apple for a small orange. Various military advantages can be ranked in their own terms, just as civilian losses can be ranked in their own terms. What a reasonable military commander would consider a small military advantage must then not be gained at the cost of what any ordinary person would consider a large civilian loss. This is not an exact rule but, applied in good faith, it can significantly protect civilians.

Notes

1. I have benefitted from comments on an earlier draft by Seth Lazar and anonymous reviewers, as well as conversations with Janina Dill and David Rodin. None should be assumed to agree with it all, and some definitely do not.
2. 1977 Geneva Protocol I, Art. 35(1). Also see 1907 Hague Convention IV, Annex: Regulations, Art. 22; Michael Bothe, Karl Josef Partsch and Waldemar A. Solf, *New Rules for Victims of Armed Conflicts: Commentary on the Two 1977 Protocols Additional to the Geneva Conventions of 1949* (The Hague, London and Boston: Martinus Nijhoff, 1982), p. 194 [2.2]; 1994 San Remo Manual on International Law Applicable to Armed Conflicts at Sea, Art. 38; and United Kingdom, Ministry of Defence, *The Manual of the Law of Armed Conflict* (Oxford: Oxford University Press, 2004), 2.1 [sometimes hereinafter the British manual].
3. See, for example, David Luban, 'War Crimes: The Law of Hell', in Larry May (ed.) *War: Essays in Political Philosophy* (Cambridge: Cambridge University Press, 2008), p. 275.
4. Lieber Code, Art. 14. Professor Lieber, writing at the request of President Abraham Lincoln in the midst of the American Civil War, went on to say: 'Military necessity does not admit of cruelty ... nor of torture' (Art. 16). See Richard Shelly Hartigan, *Lieber's Code and the Law of War* (South Holland, IL: Precedent, 1983), p. 45; L. Friedman (ed.) *The Law of War: A Documentary History*, Vol. I (New York: Random House, 1972), p. 158; or http://www.icrc.org/ihl.nsf/FULL/110?OpenDocument. Hartigan's edition contains commentary and other useful documents, including correspondence with Lieber. For a valuable overview, see Burrus M Carnahan, 'Lincoln, Lieber and the Laws of War: The Origins and Limits of the Principle of Military Necessity', *American Journal of International Law*, 92 (1998), p. 213.

5. US Military Tribunal at Nuremberg, *US* v. *Alfred Krupp et al.*, United Nations War Crimes Commission, *Law Reports of Trials of War Criminals*, vol. X (1949), quoted in A.R. Thomas and James C. Duncan (eds) *Annotated Supplement to The Commander's Handbook on the Law of Naval Operations*, Vol. 73 *International Law Studies* (Newport, RI: Naval War College, 1999), p. 293 (note 6) [hereinafter *Annotated Supplement* (1999)] and in Marco Sassoli, '*Ius ad Bellum* and *Ius in Bello*', in Michael N. Schmitt and Jelena Pejic (eds) *International Law and Armed Conflict: Exploring the Faultlines* (Leiden: Martinus Nijhoff, 2007), p. 251. The same point is made in the 2004 British manual: 'It was formerly argued by some that necessity might permit a commander to ignore the laws of war when it was essential to do so to avoid defeat, to escape from extreme danger, or for the realization of the purpose of the war. The argument is now obsolete as the modern Law of Armed Conflict takes full account of military necessity. Necessity cannot be used to justify actions prohibited by law. The means to achieve military victory are not unlimited.' Ministry of Defence, *The Manual of the Law of Armed Conflict*, 2.3.

6. Jean-Marie Henckaerts and Louise Doswald-Beck, *Customary International Humanitarian Law*, Vol. I, *Rules* (Cambridge: Cambridge University Press for the ICRC, 2005), p. 3.

7. Henckaerts and Doswald-Beck, *Customary International Humanitarian Law*, I, p. 25.

8. 1977 Geneva Protocol I, Art. 48.

9. 1977 Geneva Protocol I, Art. 50 (1).

10. On the question whether terrorists should be considered a third category, see Adam Roberts, 'The Laws of War in the War on Terror', in Fred L. Borch and Paul S. Wilson (eds) *International Law and the War on Terror*, Vol. 79 *International Law Studies* (Newport, RI: Naval War College, 2003), 175, pp. 208–20.

11. 1977 Geneva Protocol I, Art. 57 (1). Identical language appears in Rule 15 of customary law – see Henckaerts and Doswald-Beck, *Customary International Humanitarian Law*, I, p. 51.

12. Henckaerts and Doswald-Beck, *Customary International Humanitarian Law*, I, p. 46. For identical language, see 1977 Geneva Protocol I, Art. 51 (5) (b), Art. 57 (2)(a)(iii) and (2)(b), and Art. 85 (3)(b), as well as the British Manual, 5.33.

13. Michael Walzer, *Just and Unjust Wars: A Moral Argument with Historical Illustrations*, 4th edn (New York: Basic, 2006), pp. 155–6.

14. Walzer, *Just and Unjust Wars*, p. 156.

15. Asa Kasher and [Maj. Gen.] Amos Yadlin, 'Military Ethics of Fighting Terror: An Israeli Perspective', *Journal of Military Ethics*, 4L1)(2005), p. 15. This proposal should be compared to actual positions taken by the Israeli High Court of Justice on the requirements of customary law, such as the 'targeted killing decision' reached the following year: *The Public Committee Against Torture* v. *The Government of Israel* (2006), HCJ 769/02 ('PCATI'), available in English from http://elyon1.court. gov.il/eng/home/index.html. I am grateful to an anonymous reviewer for

the preceding source. For a powerful general argument for equal protection for all civilians, see Hugo Slim, *Killing Civilians: Method, Madness and Morality in War* (London: Hurst, 2007), pp. 260–3.

16. Avishai Margalit and Michael Walzer, 'Israel: Civilians and Combatants', *The New York Review of Books*, 56(8) (14 May 2009). Margalit and Walzer are responding to a different article by Kasher and Yadlin, but it rests on all the same dubious assumptions.

17. Henckaerts and Doswald-Beck, *Customary International Humanitarian Law*, I, p. 29. Also see 1977 Geneva Protocol I, Art. 52 (2) for identical language.

18. Ministry of Defence, *The Manual of the Law of Armed Conflict*, 5.4, 5.4.1 and 5.4.5.

19. *Annotated Supplement* (1999), 8.1.1; United States, Department of Defense, *Military Commission Instruction* No. 2, 'Crimes and Elements for Trials by Military Commission' (30 April 2003), para. 5D. Although the US Congress has subsequently revised various elements of the instructions for military commissions in response to court rulings, this proposed definition of 'military objective' has remained unchanged through 2009. The 2008 edition of the *Operational Law Handbook* quotes the correct definition of 'military objective' but then undermines it by asserting: 'A decision as to classification of an object as a military objective and allocation of resources for its attack is dependent upon its value to an enemy nation's war fighting or *war sustaining* effort (including its ability to be converted to a more direct connection), *and not solely to its overt or present connection or use*', United States, Army, Judge Advocate General's Legal Center and School, International and Operational Law Department, *Operational Law Handbook* (Charlottesville, VA: 2008), p. 20 (emphasis added). Besides its invoking 'war sustaining' extended indefinitely into the future, this comment is notable for focusing on the enemy *nation*, not just its military forces. For fuller documentation, see the sources cited in note 22.

20. United States, Air Force, JP 3-60 (13 April 2007), *Joint Targeting*, Appendix E, 'Legal Considerations in Targeting', p. E-2 (emphasis added), available at http://www.dtic.mil/doctrine/jel/new_pubs/jp3_60.pdf, accessed 17 March 2009. It is true that only 'war-fighting capability', and not also war-sustaining capability, is mentioned explicitly, but the same sentence also speaks of objects that support 'or sustain' war-fighting capability.

21. 'Rule 2. Acts or threats of violence the primary purpose of which is to spread terror among the civilian population are prohibited', Henckaerts and Doswald-Beck, *Customary International Humanitarian Law*, I, p. 8. And it is morally objectionable to have terror as one's purpose at all.

22. This paragraph draws upon Henry Shue, 'Target-selection Norms, Torture Norms, and Growing US Permissiveness', in Sibylle Scheipers and Hew Strachan (eds) *The Changing Character of War* (Oxford: Oxford University Press, 2010), where ongoing US efforts to loosen the definition of 'military objective' are discussed more fully (on file with author). Also see Janina Dill, 'The Definition of a Legitimate Target in

US Air Warfare: A Normative Enquiry into the Effectiveness of International Law in Regulating Combat Operations', D.Phil., Oxon., in progress, ch. 2 (on file with author).

23. *1868 St Petersburg Declaration*, in Adam Roberts and Richard Guelff, R. (eds) *Documents on the Laws of War*, 3rd edn (Oxford: Oxford University Press, 2000), p. 55.

24. Robert A. Pape, *Bombing to Win: Air Power and Coercion in War* (Ithaca, NY: Cornell University Press, 1996), p. 68. For a largely confirmatory follow-up study, see Michael Horowitz and Dan Reiter, 'When Does Aerial Bombing Work? Quantitative Empirical Tests, 1917–1999', *Journal of Conflict Resolution*, 45(2) (April 2001), pp. 147–73. Pape's main finding in *Bombing to Win* was: 'The supposed causal chain – civilian hardship produces public anger which forms political opposition against the government – does not stand up. One reason it does not is that a key assumption behind this argument – that economic deprivation causes popular unrest – is false' (p. 24).

25. 'Relational' is more accurate than 'relativistic', which makes matters sound looser than they are; compare James Turner Johnson, *Just War Tradition and the Restraint of War: A Moral and Historical Inquiry* (Princeton, NJ: Princeton University Press, 1981), p. xxxiv: 'Obviously, the principle of proportionality implies relativistic thinking.'

26. Bothe, Partsch and Solf, *New Rules for Victims of Armed Conflicts*, p. 310 [Art. 51, 2.6.2.]; also see p. 363 [Art. 57, 2.4.4], which suggests the standard of the judgement of a 'reasonable and honest' commander.

27. This is the fallacy that leads people to think that they cannot decide whether to cut down an ancient forest in order to build a new airport runway without a cost–benefit analysis that (arbitrarily) monetizes all values on all sides in order to make them (artificially) commensurable.

28. See note 18, above.

29. Bothe, Partsch, and Solf, *New Rules for Victims of Armed Conflicts*, p. 365.

30. See Henry Shue and David Wippman, 'Limiting Attacks on Dual-Use Facilities Performing Indispensable Civilian Functions', *Cornell International Law Journal*, 35 (2002), p. 559; and Michael N. Schmitt, 'The Principle of Discrimination in 21st Century Warfare', *Yale Human Rights & Development Law Journal*, 2 (1999), 143, p. 168.

31. Henckaerts and Doswald-Beck, *Customary International Humanitarian Law*, I, p. 189. Also see 1977 Geneva Protocol I, Art. 54. and Ministry of Defence, *The Manual of the Law of Armed Conflict*, 5.27.

32. 1977 Geneva Protocol I, Art. 54 (2).

33. Ministry of Defence, *The Manual of the Law of Armed Conflict*, 5.27.2. The US takes the same position, see *Annotated Supplement* (1999), 8.1.2.

Chapter 8

The Ethical Challenges of a Complex Security Environment[1]

TED VAN BAARDA

Introduction

La Drang Valley in Vietnam is hallowed by the soldiers who died there in 1965. It was the scene of the first serious battle between US forces and the North Vietnamese Army. Historians regard it as one of the bloodiest battles of the Vietnam War with over 300 Americans and possibly over 3,000 Vietnamese killed in the space of a few days. Many of those who died came from a rural background and had limited knowledge of the political merits of the war. Both sides fought with perseverance and courage and it was only thanks to the superior firepower of the Americans, with their fixed-wing close air support, howitzers and Huey helicopters that the men of the 1st Cavalry Division (Airmobile), 7th Battalion, were not overrun by the Vietnamese, who vastly outnumbered them. What sets the Battle for la Drang Valley apart, from the ethical perspective, is the fact that the then commander, Lt Col. Harold Moore, returned to the valley nearly 30 years later, together with a handful of men from his former battalion.[2] During his 1993 trip he met a number of his former enemies, including the commanding officer of the attacking North Vietnamese regiments, Nguyen Huu An, who very nearly defeated him. The Vietnamese welcomed their American counterparts warmly. They exchanged memories and together they visited the battlefield, which the Vietnamese now called the 'Forest of the screaming souls' in commemoration of the many who died far too young. Throughout his account, Moore expresses his astonishment at the mutual feelings of friendship which develop. On another occasion, Moore met General Nguyen Vo Giap, who is one of history's chief architects of modern guerrilla warfare. When the moment of departure arrived, Moore, acting on impulse, took off his inexpensive wristwatch and gave it to Giap, telling that it is a token of appreciation.

[A] gift from one soldier to another. Giap held the watch in his both hands, looking at it with amazement, as tears gathered in his eyes and mine. Then he turned and clutched me to himself in a full embrace. It was my turn to be stunned as this former enemy – arguably one of the greatest military commanders of the twentieth century – held me like a son in his arms for a long moment.[3]

Patricio Perez was 19 years old in April 1982. Fresh from high school, he had been summoned for military service and found himself as a Private serving with A Company, 3rd Infantry Regiment of the Argentinean Army, following the takeover of the Malvinas (or Falkland Islands) from the British. Perez was positioned on the main island, East Falkland, just north of the capital. He had rejoiced in the islands returning to Argentinean control and was eager to defend them (although nobody seriously expected the British to attempt to re-take the islands by force). Within a few months he was defending his position from a British amphibious force.

Combat is an extraordinary experience ... What I felt at that moment was mostly hatred. I wanted revenge. I had forgotten fear by then, what sort of risk I was taking; the only thing I wanted to do, my obsession, was to avenge my fallen comrades. Whenever I saw one of my friends hit it was worse, it just made me want to continue fighting, it didn't matter for how long or at what cost ... I once heard a Vietnam veteran talk about the 'drunkenness of war'. He was quite right, and it is like being drunk and I enjoyed it at that moment.

By mid-June the Argentinean garrison had surrendered and the British had recaptured the islands, leaving Perez feeling humiliated and frustrated. He recounts his experience on board a British Royal Navy ship, HMS *Canberra*, returning him home as a prisoner of war (POW):

On the return home on *Canberra* I became friendly with one of the guards. Little by little we realised we had a lot in common and I tried to communicate with him in my broken English. We both loved music. I played the guitar and he played the piano. He was Welsh, his name was Baker, and we also realised we also loved rugby. He played for a Welsh club and I played in Buenos Aires. We used to sing along together in our cabin to the music from the BBC. We both realised the war was over now and now we could be friends, but was hard to think that I could have killed him.[4]

Similar stories have been quoted from the both the First and Second

World Wars; Köbben recalls the story that British First World War stretcher-bearers risked their lives, carrying a heavily wounded German back to their own lines while being under heavy German fire. The wounded German, who was conscious, was contacted by the leading stretcher-bearer after the Great War and they became friends.[5] It remains striking that in war, individuals, who have never met and who have no personal hatred towards each other, can change their opinion so rapidly that they become eager to slay each other in such a short space of time. One can be amazed, Köbben continues, by 'the human capacity to see others as things to be destroyed or damaged, once the society they are part of so defines them'.[6] Harold Moore and Patricio Perez are no exceptions. They appear to be ordinary men with no lust for murder. Yet, Perez describes his joy when he is killing British soldiers. Neither Moore nor Perez display emotions which are unusual; research demonstrates that feelings of gratification and pride can hold sway as soon as an enemy of particular prowess has been defeated.[7] Far less common, though not unique, is the change of heart demonstrated by both men once the battle was over. Despite the important differences in their respective cultural backgrounds, education and rank, the recollections of Patricio Perez and Harold Moore show how barriers of stereotyping and animosity can come down. Former enemies begin to see each other as human beings again. They can relive the commonality of humanness which was lost during the heat of battle. Moore writes that:

> there were those on our side who denied them their humanity, who spoke of our enemies as if they were robots who served an alien cause, Communism, only because some commissar had a gun pointed at the back of their heads. No thought was given to the possibility that they were fighting so hard because, like America's own revolutionaries, they had a burning desire to drive the foreigners out of their native land; that nationalism was a far more compelling reason for them to fight than Communism. They were good soldiers, implacable foes in battle, and now that the guns had fallen silent and peace had returned to their land they proved to be proud fathers, good husbands, loyal citizens, and, yes, good friends.[8]

The cases of Patricio Perez and of the stretcher-bearers differ from this slightly, since their sense of commonality with their respective enemies commences sooner rather than later. Perez was still serving – albeit as a POW – in the Argentinean armed forces. The lead stretcher-bearer, not wanting to compromise himself by rendering his own private address, instead wrote down the name and address of the German, though he wisely decided to wait until the war's end before contacting his counterpart. As we will see, cases of magnanimity are not limited to warfighting situations alone. They are also present in blue-helmeted peacekeeping operations as well as humanitarian operations, where an impartial stance needs to be taken.

Rare though these stories may be, the magnanimity of the individuals involved is striking, allowing them to rise above the confinements of their circumstances. Their situation is one of war and stark contrast: black and white, friend or foe, kill or be killed. Anything else can work to the advantage of the enemy who may not have the scruples to hesitate before firing. The purpose of fighting a war is to prevail over an enemy. In the process, that enemy becomes an object: at best he is anonymous, he is a target, to be removed or 'wasted'. The nuance is lost that within the enemy, shades of opinion usually exist, ranging from the extreme to the moderate, that 'the' enemy has a face and a family as well.[9] At its worst, the enemy becomes the focal point for degrading thoughts and corresponding language – Jerries, ragheads, gooks, cockroaches, etc. – and the enemy is dehumanized. This has, in turn, the important consequence that it becomes easier to kill:[10] killing a cockroach sounds far less vicious than killing a fellow country-boy who just happened to be drafted into the army of the enemy.[11] Comparatively few reports exist of soldiers who transcend the conceptual framework of friend – or foe. This chapter will, first, identify a number of factors which put pressure on the elementary considerations of humanity in conflict. They prevent soldiers as well as civilians from seeing the humanity of those who are on the proverbial other side of the frontline. In other words, they hinder those involved from rising above the confinements of their situation. Second, this essay will briefly discuss the question of how to maintain moral and intellectual integrity under such adverse circumstances.

Seven factors putting moral considerations of humanity under pressure

First, social conflict can be defined as a struggle between two or more competing value systems. This will often be accompanied with uncertainty as to which value system will prevail.[12] The existing value system, including its codes of conduct, laws, governmental systems, enforcement mechanisms, means of distributing wealth, jobs and status, etc., may find that it no longer has the support among a certain strata of society. It can no longer count on its legitimacy being accepted by, at least, a sizeable and influential minority. Well-known cases are the struggle for emancipation of African-Americans in the 1960s under the leadership of Martin Luther King, or the opposition among large sections of the European public to the international military presence in Iraq after 2003. Many argue the presence of large numbers of disillusioned youths (particularly those of foreign decent) in Europe's main cities as another example. Social conflict – whether it evolves peacefully or into armed conflict – is frequently focused on the distinction between perceptions of justice and injustice. If one happens to live in a society where social conflict exists, one may feel that one is caught in the middle. Though many prefer to pass over painful questions, social conflict may force such questions to the surface.[13] Those concerned may feel compelled to set priorities; it is a *defining moment* – they are forced to choose and they, for better or for worse, have to live with it.[14]

Second, not all differences of opinion need to result in conflict. As long as a difference of opinion remains at the conceptual level while retaining mutual respect, it can simply be a competition for the most convincing ideas and concepts. If managed properly, a difference of opinion may even prove to be culturally and intellectually interesting and stimulating. Differences of opinion become problematic as soon as their adherents disclaim the moral integrity of the other. It is one thing to disagree with someone's *ideas*, it is quite another to criticize someone for who he or she *is*. Being a liberal/socialist/humanist/Darwinist/creationist etc. is a matter of choice. Being black, Muslim, Jewish, British, French, male or female, etc., by-and-large, is not. A difference of opinion becomes painful, confusing and even discriminatory if one has to defend not only one's *conviction*, but also one's *being*. History is rife with examples where people have been defined on the basis of collective characteristics such as colour of the skin, ethnicity, gender or similar grounds, which one receives by virtue of being born. The struggle against national-socialism and apartheid demonstrate *inter alia* that it is an affront to man's dignity to be solely defined on the basis of collective characteristics, as an exemplar of a herd, with no individuality or sense of destiny of one's own.

Being a member of a democracy, in the true sense of the word, implies that one's own views will have to be held alongside those who hold differing and opposing views. A crucial aspect about having a democratic culture is the reciprocal willingness to engage, and to keep engaged in a dialogue which respects the identity of the other. In turn, this presupposes not only that one may call into question the ideas and concepts of the other, but that the other may do likewise. At the level of democratic dialogue this inevitably leads to the conclusion that, within the context of the democratic debate, concepts such as 'truth' and 'justice' can be relative (even if various schools of thought in theology and philosophy consider 'truth' and 'justice' *not* to be relative). We have our freedom to be loyal to our convictions, as much as our neighbour may be loyal to his. It is one thing to be personally committed to one's own truth, which may give a sense of purpose in life. It is quite another to impose one's truth on someone else. The essence of peaceful, democratic dialogue is that the force of arguments prevail over the argument of force.

In military history, many wars have been fought over ideology, and various ideologies exist which proclaim One Absolute Truth – fascism, communism or religious extremism. When one truth becomes accepted as valid, it is to the detriment of all others. In 1938 the President of the District Court of Amsterdam held his farewell address against the backdrop of the ascendance to power of Hitler in Germany. He warned against the quest for absolute truths and the corresponding absolute certainty in law and ethics:

> I know that it is customary to introduce as much as possible the conviction that the law is beyond doubt. This is surely understandable and within certain constraints, a positive effect emanates from legal certainty. However, it also harbours a great danger: that of intolerance. ... In certain countries, wars are fought ruthlessly and in others, significant tensions exist. Why those struggles? Why those tensions? A battle is always a battle concerning justice, both in peace and in war. Never is the conviction of each of the warring parties that they are in the right stronger than in time of war; he who does not share this opinion is by necessity a traitor in the eyes of his compatriots. However, tensions also arise through a difference in opinion concerning what is just, and the more dearly the convictions are held, the stronger the tension. He, who feels absolute certainty about what is just, as he sees it, cannot think otherwise and those who hold a different view are inferior, stupid or evil, regardless. Towards these people he may do anything, literally anything, and it is not only permissible, but also obligatory. Absolute certainty concerning justice has annihilated complete populations, religious sects, and social classes. He who has this certainty *cannot* be tolerant; he *may* not be it. Everyone who doubts the concept of justice that he believes in with such vigour, has to be an enemy who must be silenced at all cost. *Or* he ought to recognize, at least realize, that he himself can err and that, even if he doesn't err, there exists a higher consideration which demands respect for each honestly held view. Certainly, there is cause to warn: thou shalt not be too certain of thine law.[15]

The thoughts of the President of the District Court raise many fundamental philosophical issues which we cannot discuss in detail in the current chapter. His warning against a quest of absolute certainty suggests, however, that in a true democracy a measure of moral and legal uncertainty will have to remain. Absolute certainty is intolerant: it stifles open debate and runs counter to the concept of freedom that underpins democracy.

Third, during conflict a presumption of moral exclusion may collide with the concept of humanity or dignity, which has a universal pretence. Moral exclusion occurs, according to Opotow, 'when individuals or groups are perceived as outside the boundary in which moral values, rules and considerations of fairness apply. Those who are morally excluded are perceived as non-entities, expendable or undeserving.'[16] Moral exclusion is accompanied by the distinction between the *in-group* and the *out-group*, with the members of the *in-group* being eligible for preferential treatment – frequently at the expense of members of the *out-group*. The degree to which it occurs and its actual appearance may vary. For example, virtually all of us will feel a stronger sense of loyalty to members of our family than to a complete stranger. Within the military, the boundary between the *in-group* and the *out-group* may correspond with the membership of a certain regiment or branch of the armed services. The phenomenon of moral exclusion is ancient: Aristotle, known for his thoughts on the ethic of virtue, had no qualms in calling foreigners the *barbaros*, who knew of no civilization and who could therefore be enslaved. Severe cases therefore include large-scale violations of human rights and other mass atrocities. The *out-group* is frequently based on the collective characteristics which we have already addressed: blacks, Jews, etc. The phenomenon of moral exclusion is known under a variety of names: partisanship, nationalism and – in the field of ethics – as the nonuniversalist school of utilitarianism (see Chapter 1).

During armed conflict the appearance of moral exclusion that frequently holds sway is nationalistic: *my country, right or wrong*. In a world of stark contrast, of black-and-white conceptual frameworks, the distinction between morally good and morally wrong becomes virtually synonymous with the national, that is, partisan interest. Morally right is what supports the nation's ability to prevail, almost no matter of the cost. Morally wrong is anything contrary to this. The nationalistic premise is pertinent to issues of *jus ad bellum* (i.e., issues concerning the right to wage war in the first place), as well as *jus in bello* (i.e., rules that regulate conduct during armed conflict). While a violation of the *jus in bello* which has been committed by the enemy is likely to be portrayed as a typical example of the less-than-human nature of the enemy, propaganda may also suggest that a violation of the *jus in bello* committed by one's own side is merely a mishap, an accident, or an act which was at the least justified by the pressure of enemy advances. One author on international criminal law was able to quote with irony a Minister of Justice who stated that a war crime is by definition an act which is committed by the enemy.[17] The frightening result is that there develops a strong possibility that war crimes can be committed without them being recognized as such by the perpetrator. After all, it is apparently a good cause he is fighting for (at least, that is what the population is told by propaganda). The cause, the national interest, appears to justify the actions. The chances are that the nationalistic premise *in war, my country can do no wrong* will collide head-on with an international value system embodied in universally accepted Laws of Armed Conflict and core human rights. At the level of international law as well as the ethic of duty, texts including the Charter of the United Nations and the Universal Declaration on Human Rights explicitly refer to human dignity as a value which must be expected and protected, regardless of one's national allegiance.

Fourth, there is a possibility that the set of moral values generally upheld by the military may, under certain circumstances, become contradictory. Though partly an extension of the previous cause of moral confusion, a collision of values – a moral dilemma – does not necessarily emanate from moral exclusion. Thus, a soldier may feel loyalty to two or more time-honoured military values but, given the circumstances, cannot uphold both. The accompanying moral confusion may be considerable. Heart-wrenching moral dilemmas are not only for officers high up in the chain of command, they are just as much for NCOs and young lieutenants. In the well-known book *Bravo Two –Zero*, the account is given of a group of SAS soldiers operating on foot somewhere in the desert of Northern Iraq in the winter of 1990–1.[18] The group's mission was to prevent Scud missiles being launched at Riyadh and Tel Aviv – the fear at the time was that Iraq would arm them with chemical warheads. Due to bad luck, they were dropped by their helicopter at night at a location only 100 yards from an Iraqi anti-aircraft battery, manned by some 40 soldiers. While waiting for nightfall to leave their uncomfortable hiding position in a *wadi* – a dry river bed – they were accidentally discovered by a boy of about seven years old. They were presented with a profound moral dilemma. If they allowed the boy to run, he would almost certainly alert the Iraqi soldiers nearby; if they killed him, they would be executing a noncombatant. Given the stakes involved – considerations of force protection as well as the protection of hundreds, possibly thousands of civilians in Riyadh and Tel Aviv – the decision was a particularly tough one. Such decisions often lie on the shoulders of an NCO, not in a position to request instructions from their superiors.[19]

Crucially, international humanitarian law, as articulated in the four Geneva Conventions of 1949 and subsequent Additional Protocols, embody a tension between – in the main – the ethic of consequence and the ethic of duty. On the one hand, the concept of military necessity, which is found in some 30 provisions of the Geneva Conventions, relates to the necessity of a warring party to subdue the enemy as quickly and efficiently as possible, within the constraints of the law. Thus, and within important constraints, it emphasizes operational considerations.[20] The concept of military necessity is of key importance when collateral damage needs to be assessed: the value of the target to be destroyed has to be weighed against the number of innocent civilians being killed or of civilian objects being destroyed. Military necessity is, at its core, a case of moral exclusion since it serves as a focal point for defining the interests of one of the warring parties at the expense of the enemy or of innocent civilians. However, the elementary considerations of humanity embodied in the same conventions appear to have a root in the ethic of duty: deliberately killing the innocent is evil. It may not be done, regardless of the nationality of the civilian concerned. The ethic of duty has, as mentioned, a universal pretence. Seen from this perspective, it can be argued that if it is anticipated that civilian casualties may occur, the military should make an extra effort to avoid this from happening. Thus, the concepts of military necessity and humanitarian considerations can, and frequently do, collide. Resolving the remaining tension between military necessity and humanitarian protection has led to complicated and protracted legal debates on exactly where the balance lies. For example, authoritative sources such as Yoram Dinstein and the UK MOD have emphasized the operational and utilitarian aspects of the balance to be struck between military necessity and considerations of humanity. However, Henry Shue points out in Chapter 7 that military necessity is not a positive authorization allowing one to 'do whatever it takes (legal or illegal)' and that this is recognized in the British manual's negative formulation: 'The use of force which is not necessary is unlawful, since it involves wanton killing or destruction.' Necessity is also always subordinate to law and can never permit that which is itself illegal. The International Committee of the Red Cross has also emphasized the primacy of the universal nature of the ethic of duty.[21] In the Gali case (2003), the International Criminal Tribunal for the Former Yugoslavia has effectively come down in favour of the ICRC's position.[22]

Fifth, a cause for moral confusion on the battlefield concerns what Patricio Perez called the 'drunkenness of war' and which is more commonly known by Clausewitz's term 'fog of war'. The term refers to an ambiguity in situational awareness, a combination of factors that can severely impair the ability to make proper judgements. At the individual level, they include: stress, fatigue, memory of fallen comrades, the fear of appearing to be weak to one's peers, the joy of overcoming a strong opponent, etc. Fear may impel a soldier to 'fire at everything that moves'. The MHAT IV report, of the US Medical Health Advisory Team, discusses the relationship between mental health of US military personnel and their 'battlefield ethics' while on duty in Iraq. The study, virtually the first empirical study of its kind, notes a markedly negative relationship between anger and stress and the ability of US military personnel to behave in an ethically proper way in Iraq.[23]

The problem connected with this fog of war can be exacerbated by the very sharp difference between what might be morally acceptable on the battlefield, and what is morally right for a civilian in time of peace.[24] A civilian is forbidden to kill another human being except in cases of immediate self-defence; in armed conflict the killing of an enemy combatant may not only be considered necessary but also honourable.[25] People are brought up to respect human life; but once they put on a military uniform they are expected to comply with a different set of values.[26] Those who do not cope with this in the immediate situation may not fire their weapon, shoot deliberately into the ground, or even at their own foot – in an effort to avoid killing a fellow human being. Post-conflict ramifications such as PTSD and marital problems may take their toll months or even years after one has returned home.

Frequently, the problem is further exacerbated by a blurred distinction between civilian and military targets. This can come about when engaging 'dual purpose' targets, such as bridges, port facilities, electricity generators, etc. However, with increasing frequency, this blurring can be due to a disingenuous enemy who is deliberately seeking to 'draw the foul' (see Chapter 1). Being numerically the weaker side, guerrillas, to paraphrase the Chinese revolutionary Mao Tse Tung, have to vanish among the crowds as fish disappear in the sea. The result is that the moral dilemma and corresponding responsibility is shifted towards the regular soldier who engages in counter-insurgency warfare: *he* has to decide if a given target is civilian or military, and *he* has to live with the consequences. Unfortunately, military psychologists have grown used to nightmarish stories of regular soldiers returning to Britain, Holland or the US from Iraq and Afghanistan who are agonized about the fact that they might have made a mistake and inadvertently killed innocent civilians. Although a professional ethicist will undoubtedly argue that Mao Tse Tung has unduly shifted the moral responsibility for civilian deaths with his figure of speech about fish in the sea, this will do little to diminish the anguish of the regular soldier involved.

Sixth, the distinction between humanitarian operations carried out by humanitarian organizations, and military operations carried out by combatants, has become blurred. Since the end of the Cold War, UN peacekeeping missions have received mandates that overlap with those of humanitarian organizations. In turn, this has blurred the distinction between the core businesses of both.[27] One result is that moral dilemmas typical of humanitarian organizations have now been experienced by blue-helmeted peacekeepers. An issue of moral responsibility in humanitarian circles focuses not on the number of deaths which one has caused, but on the number of lives which one was unable to save. A strong humanitarian component in the mandate of the military may motivate soldiers to take actions which they would not have taken without that component in the mandate. The BBC drama *Warriors* provides a dramatized account of a British battalion of UNPROFOR, operating in central Bosnia at the beginning of the disintegration of the former Yugoslavia. While on patrol with two armoured personnel carriers (APCs), a platoon witnesses ethnic cleansing: militants are kicking in the doors of dwellings in a village where Muslims are known to live, throwing in a hand grenade or emptying their machine guns. One block further, the platoon finds about 20 women and children huddled frightened together, hiding in a cellar. After a brief hesitation, the lieutenant decides to evacuate the people in his APCs, thinking that they would meet a virtual certain death if they remain in the cellar. While they are loaded into the APCs, the lieutenant is countermanded on the radio by his superior at battalion HQ, who orders him to unload these people. In a heated conversation that evening at HQ, his superior told him that evacuating those people was not within UNPROFOR's mandate. The lieutenant frankly answered that he has been prevented from saving innocent lives. Dilemmas such as this one raise the ethically thorny question to what extent one is responsible for one's inactions which result in serious damage or death.[28]

Peacekeepers are expected to remain impartial, both as individuals and as a peacekeeping force as a whole. The United Nations, obviously desirous of maintaining open negotiating channels with all parties in a conflict, cannot afford to side or be seen to side with one of those parties. Reports from peacekeepers indicate that they sometimes witness horrendous crimes. Common decency and morality dictates that they prevent the crime on the spot, yet their orders may be not to interfere because of the posture of the peacekeeping force. This conflict has given rise to profound moral confusion, excruciating questions of conscience and, at times, the decision to disobey orders. However, it is one thing to remain impartial to the political issue which keeps the warring parties apart; it is another to do so in the face of gross human rights violations. To do so confuses impartiality with neutrality – one can remain impartial in a political sense by the peacekeeping force refusing to take a position on the issue which fuels the conflict, much like a good referee in a football game is not partisan. However, that does not mean that infringements of the rules or in this case, atrocities, must be ignored to preserve some sense of neutrality. A neutral referee would be wholly ineffective, as he would be trying to remain equidistant between the two sides, even if one of those sides was breaking the rules and deserved censure. If one drifts from impartiality to moral neutrality, the ethical distinction between morally good and morally bad loses its practical significance. Thus, when a choice is made in favour of moral neutrality, it becomes virtually impossible to take a stance against those parties to the conflict which commit war crimes.[29] Thus, one can somewhat provocatively argue that neither effective humanitarian assistance nor UN peacekeeping are neutral in the moral sense of the word. In fact, a choice must be made in favour of the dignity of the abused, the weak and the poor, who have nowhere left to go.

Particularly since the Anglo-American attack on Iraq of 2003, the blurring of the distinction with humanitarian operations has spilled over from peacekeeping into regular warfare as well as counter-insurgency operations. In his book *Plan of Attack*, Bob Woodward quotes an internal memo from the Pentagon explicitly stating that humanitarian assistance was an integral part of American military plans to defeat the forces of Saddam Hussein.[30] In turn, this spill-over raises the confusing issue of whether one can be a party to the conflict while offering impartial aid at the same time.

The chain of command and the corresponding military penal code raise a seventh issue of moral confusion. One of the aims of military penal codes is to enforce discipline and uphold the chain of command. A soldier who receives an order should be able to assume its legality, in which case, he is under a statutory obligation to carry it out. The military penal code ensures that an army will operate in battle as a coherent, unified, effective and responsible force. However, reasonable and justified this may be, there are consequences that are often underemphasized despite the legacy of Immanuel Kant (see Chapter 1). Kant argued the existence of a universal moral law which obliges us *never* to treat the humanity in the person of another (or, for that matter, in one's own person) merely as a means – a tool – but always at the same time as an end in itself. Kant's legacy thus demands that we demonstrate respect for individual rights.[31] To merely *use* someone is to treat them not as a responsible individual but simply as an instrument, thus denying them the possibility of consenting, bearing responsibility, or defining their own purpose. Morality, however, is part of our ability to consider a purpose to life itself and to make our actions meaningful. It is also part of our ability to consider whether events are just or unjust by being able to distance ourselves from those events to some extent. In that sense therefore, morality is part of freedom – the quality that makes us truly human. However, the military chain of command requires, almost by definition, a measure of *instrumentality*. On the battlefield soldiers do not have time to meet in working groups and discuss the range of military options available in a democratic way.

The context requires both hierarchy and speed of decision. Soldiers have to be both obedient and confident: obedient in the sense they should follow lawful orders; confident in the sense that they can trust the moral integrity of their commander who will not order them to do something that is wrong or throw their lives away needlessly. Total, unthinking obedience, however – in German *Kadaverdisziplin* – is extremely dangerous and highly undesirable.[32] The risks of pure instrumentality within military culture and in particular, the danger of a human acting like a machine, following external impulses without considering their real nature or their moral significance, are discussed by both Arendt and Isarin.[33] In doing so, they raise some profound issues. For example, if laws effectively codify the nationalistic premise of *my country right or wrong*, if there exists a statutory obligation to follow military orders, if the distinction between morally good and morally evil under the influence of propaganda becomes virtually synonymous with the distinction between friend or foe, then it becomes particularly difficult for a soldier to make their own, independent assessment of what is morally right. In discussing the Second World War, Isarin asks:

> What is the essence and function of the human ability to pass judgements? Are people able to distinguish good and evil when they are completely dependent on their own ability to judge, when their own judgement conflicts with that of virtually everyone else? ... Are individuals able to decide in a situation which is new to them what good and evil are, when their legal surroundings have become evil and murder has become the rule?[34]

There is certainly no reason for militaries in democratic countries to sit back and think that war crimes are only committed in far-away countries which lack a tradition of respect for human rights. Can 'ordinary people' with no previous record of violence or criminality commit war crimes? The disturbing answer, explored in more depth in Paolo Tripodi's chapter (Chapter 9), is quite simply: yes.[35] Amid the many thousands of stories of atrocities committed during armed conflict, the accounts of magnanimity and humanity like those of Harold Moore and his Vietnamese counterparts, Patricio Perez and his British captors on the Falkland Islands and of the British stretcher-bearers of the First World War are regretfully rare.

Moral integrity

This chapter has listed seven factors which can put moral considerations of humanity under pressure:

1. The struggle between two or more competing value systems, with no certainty yet on which values system is to prevail.
2. The distinction between challenging someone's *ideas*, which may even be stimulating, and challenging someone's *being*, which is likely to be threatening, discriminatory and dehumanizing.
3. The distinction between moral exclusion and human dignity, with the latter frequently perceived as universal in nature. Dignity is, hence, an inclusive concept, whereas nationalism is an exclusive concept.
4. Time-honoured military values may, under adverse circumstances, become contradictory. This collision of values leads to a moral dilemma, which, in turn, demands added moral decision-making skills, moral competence, of the on-scene commander. Even the *jus in bello* suffers from some contradictions, as we can find both utilitarian arguments and semi-deontological arguments codified in the Geneva Conventions.
5. The 'drunkenness of war,' commonly referred to as the 'fog of war', which includes psychologically important elements as stress, fatigue, uncertainty, fear and chaos.
6. The blurring of responsibilities between humanitarian organizations and the military, as soon as the military is mandated with humanitarian responsibilities.
7. The incongruence between the instrumental nature of the chain of command on the one hand, and the concept of humanity as defined by Immanuel Kant on the other.

If one attempts to summarize the various causes for the elements of pressure discussed above, a feeling of bewilderment may surface. We can easily agree with Colonel Hartle's statement that the hostile environment in which the military operates poses a severe threat to consistent moral behaviour.[36] This may present itself in a range of appearances. It can appear as forms of self-indulgence – such as looting, vanity, lust for revenge, or inappropriate sex. Or it can take the appearance of segmentation (compartmentalization), as when we divide our responsibilities between the professional and private, arguing that rescuing a refugee 'is not within our mandate', while making out a pay cheque to charity back home. They can take the appearance of callousness and cold-heartedness, seeing the fellow human being as an object of treatment rather than a subject of treatment – that is to say as a pawn, a target or an instrument, rather than as someone with dignity and a sense of purpose of their own.

We can also agree with General Krulak that in the era of the three block war, members of the military 'require unwavering maturity, judgment, and strength of character'.[37] Consequently, a high standard of moral integrity for officers and men is essential, along with the skill and the moral competence, to make the moral judgements necessary. Perhaps one answer to the question of Isarin quoted above can be found in the well-known radio-address of Theodor Adorno, 'Education after Auschwitz'. He argues that the premier demand on all education is that Auschwitz will not happen again. Thus, if education is to make any sense it ought to be aimed at critical self-reflection in order to avoid blind obedience and the loss of personal identity.[38] Moral education will, henceforth, not only have to focus on the applied ethics of moral competence, it will also have to focus on the development of one's own ethic, which, importantly, includes becoming aware of the inner *source* of one's values and intentions. It is the self, the I, which is performing intellectual and moral activity.[39] Of course, this refers to the neo-Aristotelian thought that man, throughout his life, has to strive for excellence – it is a case of *éducation permanente*. 'What kind of a person do I want to be, and what would such a person do? ...The answer will have to be determined in advance by personal traits the individual has acquired up to the moment of crisis,' Bonadonna has been quoted as arguing.[40]

Naturally, the perfect moral judgement will never be reached: it does not exist. Yet, it is still possible to educate our men and women in uniform to pass moral judgements which are both operationally and ethically wise, and where possible, magnanimous. The crucial question is, of course, how this can be done. Although scholarly literature still has to find agreement on the answer to this question,[41] many experts appear to agree on the ideal that a number of key values should be deeply engrained in each officer, NCO and soldier that he or she will abide by them under even the most adverse of circumstances.[42] This does not suggest, of course, that key values should be adhered to unreflectively: quite the contrary. Various philosophers, including Martha Nussbaum, have resisted the premise that a kind of 'moral arithmetic' exists: that it would be possible to decide each moral dilemma solely on the basis of concepts which are universalizable.[43] Thus, the ethic of duty states that morally just is that course of action which follows a universal moral law (categorical imperative) and that all other considerations are subordinate to it, while the ethic of consequence argues that morally just is that course of action which achieves the least amount of pain for the largest number of people. Moreover, 'moral arithmetic' in the sense of a clearly and universally accepted hierarchy of moral values does not exist. Since the days when Plato described Socrates' dialogues with Protagoras, numerous attempts have been made to arrive at an all-encompassing hierarchy of values; none has been beyond self-contradiction and none has met with general approval. Nussbaum and others argue that specific dilemmas may juxtapose values and considerations which are not necessarily comparable – as is the case, in our view, when a commander has to apply the principle of proportionality while balancing the value of a military target against the innocent loss of life.[44] When one demands that all moral decisions are always and completely related to universals, then one must accept that moral decisions become by their very nature impersonal – just as judges are, by analogy and in Montesquieu's view, no more than a mouthpiece of the law. Such a demand may be contrary to moral integrity itself and ignores the possibility and, indeed, necessity, of moral judgements having an element of uniqueness which is pertinent to the merits of a particular case. The demand can thus become insensitive, as it does not take into account the factual merits of the case, nor the moral emotions, ethical and moral competence of those involved in the decision-making process.[45] Moral decision-making, properly applied, holds a delicate balance between the application of generally valid moral principles, while also weighing unique merits. Indeed, one could move a step further and argue that the strength of the concepts of dignity and human rights is that they bridge the gap between the universal and the particular; between general principle and the true respect of each individual.[46] The real challenge for all of us would then be to live a meritorious life, where we undertake the balancing act between universals and particulars virtually on a daily basis.

Cox, La Caze and Levine argue that the complexity of moral integrity relates to us as fragile human beings.[47] They refer to Aristotle, who argued that we are neither god, nor animal. At best we are a bit of both. Cox *et al.* note that it is this interjacent position between 'divinity' and 'bestiality' that makes us human.[48] The symbolic duality of the god and the beast in us, including their respective endeavours to pull us into their sphere of influence, makes us fragile human beings. Thus, the upholding of integrity by an individual presumes a certain degree of inner tension and conflict: man is almost continuously in the middle of conflicting considerations, each pulling him in a different direction. Inner tension can be a mark of integrity, in a way that possessing absolute certainty may not be. Having an answer available to all the questions that arise in dilemmatic situations does not necessarily indicate the presence of integrity. On the contrary, it could be – as we have discussed – the indication of a lack of integrity, which is the case with the certitude and self-righteousness of the dogmatist and fanatic. 'Such inner tension is one measure of integration and is commensurate with the inner richness and magnitude of the values which it seeks to organize ... This Faustian sense of inward division and struggle may be the basis of integrity.'[49] Faust cries out: 'Two souls, oh my! Live in my breast,' in verse 1112 of Goethe's play. Faust refers to, on the one hand, his longing to grasp the sensual facets of life on Earth, be it financial wealth or the attractions of the opposite sex. On the other hand, there is his memory of 'higher', moral and spiritual notions.

According to Colby and Damon, a strong chance exists that an individual will uphold moral values if those values are not perceived as a thin, politically correct layer which has no deep root in personal convictions. Rather they must come about as a result of moral (self-) education, and therefore can become a part of one's sense of personal identity. 'When there is perceived unity between self and morality, judgement and conduct are directly and predictably linked and action choices are made with great certainty.'[50] Seen in this perspective, the value of selfless service – a typical military value – may receive an additional dimension: one may not only be dedicated to one's own nation, colleagues and compatriots, but also to humanity itself. 'The essential character of military ethic is based upon the conviction that there is something worth living for that is more important than one's own skin,' Toner argues.[51] This statement appears to be true, although its validity is by no means limited to men and women in uniform. We recall the nearly forgotten example of Lea Feldblum who lived in occupied France. She was entrusted with the care of 44 Jewish children aged between four and 17 – children whose parents were no longer able to care for them. She possessed excellently forged papers, so that the Gestapo did not discover her real identity. With the children she hid on a farm, in the village of Izieu, North of Lyon. In 1944, the Gestapo raided the farm and arrested the children; their own documents could be identified as forgeries. When the Gestapo was about to release Lea Feldblum on the misunderstanding that she was not Jewish, she revealed her true identity: she wished to stay with the children and comfort them on their travel to the gas chambers of Auschwitz. Unfortunately, none of the children survived, but by a miracle, she did. The Gestapo chief concerned, Klaus Barbie, was put on trial as recently as 1984 and Lea Feldblum was able to give testimony in court.[52] This is, of course, an extreme example. It is clear, however, that many UN peacekeepers have been prepared to risk their lives in 'someone else's war', simply to rescue civilians from a murderous death, thus demonstrating their dedication to the moral concept of humanity. The memoirs of Lt General Dallaire – Force Commander of United Nations Assistance Mission for Rwanda (UNAMIR) during the genocide of 1994 – demonstrate the point.[53] Although many peacekeepers did not rise to fame, they did display a measure of courage and magnanimity 'beyond the call of duty'.

The concept of moral integrity remains notoriously hard to describe. However, the previous argument would appear to confirm Calhoun's observation that integrity is not – or at least not merely – a virtue of the individual who wants to 'keep his hands clean' or wishes to 'stand on principle no matter what the consequence'. She suggests that integrity is a social virtue in which we take into account not only our own views, but also the views of fellow-deliberators, as well as, of course, the interests of others. She adds that this social virtue of 'standing for something' may help to explain 'why we care that persons have the courage of their convictions. The courageous provide spectacular displays of integrity by withstanding social incredulity, ostracism, contempt, and physical assault when most of us would be inclined to give in, compromise, or retreat into silence.'[54]

Someone with moral integrity is likely to resist the harsher forms of moral exclusion which we discussed above under the third impediment. Someone with moral integrity would perhaps refer to its opposite: moral inclusion. The latter refers to 'relationships in which the parties are approximately equal, the potential for reciprocity exists, and both parties are entitled to fair processes and some share of community resources'.[55] Armed conflict is, of course, a prime example of moral exclusion, usually with both parties disclaiming the membership of the other to any decent moral community. However, both the Law of Armed Conflict and military ethics contain clear elements of moral inclusion, such as the principle of the moral equality of soldiers, respect for the honour of the innocent, wounded and the captured, as well as the universal concept of humanity as enshrined in the prime principle of the Red Cross movement. It would appear that it is at this juncture that the experiences of Harold Moore, Patricio Perez and the British stretcher-bearers serve as examples.

'Integrity' is not by definition synonymous with norm-conforming behaviour or political correctness.[56] Martin Luther King's bus boycott was far from norm-conforming. Wakin expressed his concerns on military culture when he wrote that the military profession 'leaves us feeling that a man must give up his rationality, his very creativeness, the source of his dignity as a man, in order to play his role as a soldier'.[57] It is the point Adorno warned against. Wakin's words relate to the second impediment to rise above the confinements of a situation discussed above; certain elements of military drills may predetermine man and thus prejudice his potential to think freely at moments when a judgement call is needed. In his and our opinion it would be a misunderstanding to reject the distinction between integrity and norm-conformity on the basis of the argument that a plea in favour of an independent mind would by definition imply an erosion of military discipline. He emphasizes that 'it is the disciplined mind that is most truly free'.[58] It is important to note that Wakin focuses on mental discipline, self-discipline – that is to say discipline which one has to realize through personal education. Self-discipline is to be distinguished from discipline as it is traditionally understood in the military, which is discipline imposed by higher authority through a chain of command and the military penal code.[59] 'The unskilled is not free. The uninstructed is not free. The inexperienced is not free. Whether the field is carpentry, athletics, or space technology, only the skilled, the instructed, the experienced, and the disciplined have both power and freedom.'[60]

With these thoughts in mind, we return to the examples given at the beginning of this chapter, concerning Patricio Perez, Harold Moore and the British stretcher-bearers. They were able to live their lives to a point where they were morally able to transcend the world of black-and-white concepts, of a world divided between friend and foe, and where the distinction between good and evil had become virtually synonymous with the friend-or-foe distinction. They were able to see the humanity in the eyes of the other, on the opposite side of the frontline.[61] Scholarly literature suggests that if a person – civilian or soldier regardless – adopts moral values only because his or her peer group does, without having any considerations of his or her own as to *why* these moral values might or might not be the right ones, then his or her value system is not his or her own. Then his or her judgements and volitions will not be his or her own. On this count, at least, he or she will lack moral integrity. A poignant issue during armed conflict is, of course, that the friend-or-foe distinction has a collective nature, as defined by the political and military leadership. The use of violence – and thus the deployment of the armed forces – is virtually by definition instrumental – it is always in need of a justification, while peace, by comparison, needs no justification.[62] Precisely because violence is always in need of a justification, our men and women in uniform have to be able to maintain an independent mind in defiance of the erroneous claim that 'theirs is not to reason why'. While they are, on the one hand, subordinate to the chain of command, they should, on the other hand, retain their intellectual freedom, and thus their ability to decide when an act of magnanimity is in order. Bandura refers to the well-known case of the My Lai massacre. When Sgt Thompson recognized the atrocities being committed by his compatriots against Vietnamese women and children, he ordered the gunner of his Huey helicopter to aim at the Americans and to open fire if the massacre continued. Thompson realized he had a son of the same age as a two-year-old boy who, between a stack of corpses, clung to the dead body of his mother. Bandura recalls the case under the heading 'The Power of Humanisation'.[63] Where dehumanization of the enemy facilitates the commission of atrocities, the power of (re-)humanization serves as an antidote and a source for moral courage. Sgt Thompson, thus, transcended the confinements of his situation.

The values of integrity, dignity and magnanimity transcend distinctions between traditional war-fighting, counter-insurgency operations, peacekeeping or humanitarian operations. They are required regardless of the nature of the operation concerned. *Integritas* (the Latin root of integrity) literally means 'whole, pure, uncorrupted, untainted, undivided or untouched'. This is something that transcends the distinctions between the operations mentioned; they are core values for all men and women in uniform. It also transcends the friend-or-foe distinction. Fortunately, military history offers shining examples, as we have seen with Harold Moore, his Vietnamese counterparts, Patricio Perez and his British captors on the Falklands Islands. They demonstrate how to exercise this awesome responsibility.

Notes

1. An earlier version of the discussion in this chapter appeared in my article 'Military Ethics in Peacekeeping and War', published 1 March 2004 in the *Journal of Humanitarian Assistance*.
2. Lt Gen. Harold Moore (ret.) and Joseph L. Galloway, *We Are Soldiers Still: A Journey Back to the Battlefields of Vietnam* (New York: HarperCollins, 2008). Their book is a sequel to *We Were Soldiers Once ... and Young* (New York: Random House, 1992).
3. Moore and Galloway, *We Are Soldiers Still*, p. 32.
4. Patricio Perez, quoted by Michael Bilton and Peter Kosminsky, *Speaking Out: Untold Stories from the Falklands War* (London: André Deutsch, 1989), pp. 191–3.
5. A. J. F. Köbben, 'What is a Human Life Worth?', speech delivered on 10 December 1993 at Leiden University, *PIOOM Newsletter,* summer, 1994, 6(1); Alain Finkielkraut, *L'humanité perdue* (Paris: Le Seuil, 1996), pp. 31, 48–51.
6. Köbben, 'What is a Human Life Worth?'; A. Bandura, 'Moral Disengagement in the Perpetration of Inhumanities', *Personality and Social Psychology Review,* vol. 3, 1999; A. Bandura, 'Social Cognitive Theory of Moral Thought and Action', in William M. Kurtines and Jacob L. Gewirtz (eds) *Handbook of Moral Behavior and Development. Volume 1: Theory* (Hillsdale, NJ: Lawrence Erlbaum, 1991).
7. Joanna Bourke, *An Intimate History of Killing: Face to Face Killing in 20th Century Warfare* (New York: Basic, 1999); Chris Hedges, *War is a Force that Gives Us Meaning* (New York: PublicAffairs, 2002).
8. Moore and Galloway, *We Are Soldiers Still*, footnote 2, p. 145.
9. Lyse Doucet, BBC presenter and correspondent, criticized the Western media on this count. 'What's lacking in the coverage of the Afghans is the sense of the humanity of the Afghans ... You knew that the bombs were dropping in that direction and the guns pointing in that direction, but you never got a sense of how Afghans are as a people,' she said. When asked what was missing in the news coverage, she added: 'It may sound odd but the humanity of the Taliban, because the Taliban are a

wide, very diverse group of people.' Lyse Doucet, quoted in *The Telegraph*, 24 August 2008.

10. Bandura, 'Moral Disengagement'.

11. Perhaps the commonplace saying that the first casualty of war is truth, is not quite correct. According to Schuyt, the first casualty is language: our ability to speak about a situation in a factually objective manner and to distinguish factual observations from our normative judgments. As soon as man can only speak about his fellow man in derogatory language, then truth is lost as well. C. J. M. Schuyt, 'Maatschappelijke achtergronden van schendingen van rechten van de mens', *Internationale Spectator*, November 1981.

12. C. J. M. Schuyt, 'Recht, orde en burgerlijke ongehoorzaamheid', thesis, 1971, pp. 78–9; Ted Robert Gurr, *Why Men Rebel* (Princeton, NJ: Princeton University Press, 1970), pp. 42 and pp. 134ff.

13. Gunnar Myrdal, *An American Dilemma: The Negro Problem and Modern Democracy* (New York: Harper, 1944).

14. Joseph L. Badaracco, *Defining Moments: When Managers Must Choose Between Right and Right* (Boston: Harvard Business Press, 1997).

15. J. W. Huysinga, *Een afscheidsrede* (The Hague: Nederlands Juristenblad, 1938), pp. 472–7 (this author's translation; emphasis in the original).

16. Susan Opotow, 'Moral Exclusion and Injustice: An Introduction', *Journal of Social Issues*, 46(1), Spring 1990; Feitse Boerwinkel, *Inclusief denken. Een andere tijd vraagt een ander denken* (Amsterdam: Nederlands Uitgevers centrum, 1966).

17. C. F. Rüter, *Enkele aspecten van de strafrechtelijke reactie op oorlogsmisdrijven en misdrijven tegen de menselijkheid* (Amsterdam: AUP, 1973).

18. Andy McNab, *Bravo Two Zero* (New York: Bantam, 1993).

19. Within the space of hours, and perhaps even minutes, an NCO or even a corporal may have to be able to shift mentally from: (1) a full combat operation, to (2) a defensive peacekeeping posture and (3) to sensitive humanitarian posture. Gen. Krulak's analysis of the 'strategic corporal' in this 'three-block war' demonstrates the nearly super-human demand on the moral, psychological and leadership competences of military personnel of virtually all ranks. A failure on this point may not only lead to a failure of the mission – it may also have severe legal and political repercussions. Charles Krulak, 'The Strategic Corporal: Leadership in the Three Block War', *Marines Gazette*, January 1999. See also Chapter 1 in this volume.

20. The generally accepted definition of military necessity reads: 'Military necessity is an urgent need, admitting of no delay, for the taking by a commander of measures, which are indispensable for forcing as quickly as possible the complete surrender of the enemy by means of regulated violence, and which are not forbidden by the laws and customs of war.' This definition contains four elements: the urgent necessity of a commander to take immediate measures; which are indispensable for enforcing a complete submission of the enemy; by means of organized, regulated violence; in so far as that violence has not been prohibited by

the laws of war. See W. G. Downey, 'The Law of War and Military Necessity', *American Journal of International Law*, 47 (1953), p. 254; Y. Sandoz, C. Swinarski and B. Zimmerman (eds) *Commentary on the Additional Protocols of 8 June 1977 to the Geneva Conventions of 12 August 1949*, ICRC (Geneva: Martinus Nijhoff, 1987). pp. 392–6.

21. Yoram Dinstein, *The Conduct of Hostilities under the Law of Armed Conflict*, (2004), p. 121; UK Ministry of Defence, *The Manual of the Law of Armed Conflict* (Oxford: Oxford University Press, 2004), 2.6.3ff.; Sandoz, Swinarski and Zimmerman, *Commentary on the Additional Protocols*, pp. 625–6, section 1979. Sandoz, Swinarski and Zimmerman chose a semi-deontological position (rejected by the ICRC) when they stated: 'The idea has also been put forward that even ... very high, civilian losses ... may be justified if the military advantage at stake is of great importance.'

22. ICTY, *Prosecutor v. Gali*, Trial Chamber, case no. IT-01-98-29-T, of 5 December 2003, section 58. The Trial Chamber quoted section 1979 of the ICRC's official commentary approvingly in footnote 108 of its judgment when it stated: 'The basic obligation to spare civilians and civilian objects as much as possible must guide the attacking party when considering the proportionality of an attack.' For a detailed analysis of the move away from reciprocity as a basis for following laws in war towards a more universal ethic of duty regardless of others' adherence, see Th. A. van Baarda, 'Moral Ambiguities Underlying the Laws of Armed Conflict: A Perspective from Military Ethics', *Yearbook of International Humanitarian Law 2008*, vol. 11 (2010).

23. www.armymedicine.amry.mil/reports/mhat/mhat_iv/MHAT_IV_Report 17NOV06.pdf This study had a sequel in the MHAT V study, published in 2008.

24. That assumes of course that the soldier involved has known a time of peace. In armed conflicts which continue for two or more decades, a generation has grown up without having any experience of living under peaceful conditions.

25. Rüter, *Enkele aspecten*, pp. 23–4.

26. This graphically illustrates the fact that military ethics are *professional* ethical standards. As Col. Hartle has observed, a given society may have a number of professions which uphold moral values which are at variance with the values of the society to which they belong. Those professions include the police, medical doctors, etc. The members of these professions are legally permitted to commit certain acts which 'ordinary' individuals are not allowed to commit. See Anthony Hartle, *Moral Issues of Military Decision Making* (Kansas: University Press of Kansas, 1989). It can be argued, however, that the difference between military values during armed conflict and civilian values in time of peace is particularly large and thus confusing.

27. Larry Minear, *The Humanitarian Enterprise: Dilemmas and Discoveries* (Bloomfield, CT: Kumarian Press, 2002), pp. 99ff.; Th. A. van Baarda, 'The Involvement of the Security Council in Maintaining International Humanitarian Law', *Netherlands Quarterly of Human Rights*, 12(2),

1994. While a number of humanitarian organizations are prepared to cooperate on a limited basis with peacekeeping forces, the ICRC is more reluctant. Any association of a humanitarian organization with 'the' foreign military – perceived or real – can entail loss of neutrality and independence. This, it is claimed, can lead to an increased insecurity for humanitarian fieldworkers.

28. Bernard Williams, 'A Critique of Utilitarianism', in J. J. C. Smart and Bernard Williams, *Utilitarianism: For and Against* (Cambridge: Cambridge University Press, 1973). For a more modern, detailed discussion, see Deen K. Chatterjee (ed.) *The Ethics of Assistance: Morality and the Distant Needy* (Cambridge: Cambridge University Press, 2004).

29. Th. A. van Baarda and A. H. M. van Iersel, 'The Uneasy Relationship Between Conscience and Military Law: The Brahimi Report's Unresolved Dilemma', *International Peacekeeping*, 8(3), 2002.

30. Bob Woodward, *Plan of Attack* (London: Simon & Schuster, 2004), p. 55.

31. Onora O'Neill, 'Kantian Ethics', in Peter Singer (ed.) *A Companion to Ethics* (Oxford: Oxford University Press, 1991), pp. 178–9; Christopher McCrudden, 'Human Dignity and Judicial Interpretation of Human Rights', *European Journal of International Law*, 19(4), 2008; Paolo Carozza, 'Human Dignity and Judicial Interpretation of Human Rights: A Reply', *European Journal of International Law*, 19(5), 2008.

32. For example, see Hannah Arendt, *Eichmann in Jerusalem. A Report on the Banality of Evil* (London: Penguin, 1994), pp. 135–7; Mark J. Osiel, *Obeying Orders: Atrocity, Military Discipline and the Law of War* (New Brunswick, NJ: Transaction, 1999).

33. An extreme case is that of Franz Stangl, the camp commander of Treblinka. When asked how he could cope with the fact that he could be a loving husband and father while sending thousands of people to misery and death, he said: 'You know, outside of course doing my job properly, I wasn't really interested. You see, I had just got married. I had for the first time a home of my own. All I wanted was just to close the door of my home and be with my wife ... I was just a police officer doing a job.' A few weeks later, Stangl added: 'the only way I could live was to compartmentalise my thinking'. Franz Stangl, quoted by Gita Sereny, in *The German Trauma: Experiences and Reflections 1938–2000* (London: Allen Lane, 2000), pp. 99 and 120. For discussion of this limited interpretation of professionalism, see A. H. M. van Iersel and Th. A. van Baarda (eds) *Militaire Ethiek: Morele Dilemma's van Militairen in Theorie en Praktijk* (Budel: Damon, 2002), p. 25.

34. Jet Isarin, *Het kwaad en de gedachteloosheid: Een beschouwing over de holocaust* (Baarn: Ambo, 1994), p. 70 (this author's translation).

35. Stanley Milgram, *Obedience to Authority* (New York: HarperCollins, 1974); Philip Zimbardo, *The Lucifer Effect: Understanding How Good People Turn Evil* (New York: Random House, 2008); Hubert Michael Mader, '"Aber das machen meine Leute doch nicht!" Können auch "ganz normale" Menschen Kriegsverbrechen begehen?', in *Truppendienst*, 5, 2000, pp. 368–9; Christopher R. Browning, *Ordinary Men: Reserve Police Battalion 101 and the Final Solution in Poland*

(New York: Harper & Brothers, 1992); Major Jeffrey F. Addicott and Major William A. Hudson, 'The Twenty-fifth Anniversary of My Lai: A Time to Inculcate the Lessons', *Military Law Review*, vol. 139, 1993. See also the contribution by Paolo Tripodi in this book.

36. Hartle, *Moral Issues*.
37. Krulak, 'The Strategic Corporal: Leadership in the Three Block War'.
38. Theodor Adorno, 'Education after Auschwitz' (German original: 'Erziehung nach Auschwitz', Frankfurt, 1966).
39. C. Otto Scharmer, *Theory U: Leading from the Future as it Emerges* (San Fransisco, CA: Berrett-Koehler, 2009).
40. Reed Bonadonna, quoted by Osiel, *Obeying Orders*, p. 255.
41. Paul Robinson, Nigel de Lee and Don Carrick, *Ethics Education in the Military* (Farnham: Ashgate, 2009). Fortunately, a variety of good books on military ethics and the moral education of the military have been published since the end of the Cold War – a number of which stem from British soil.
42. See, for instance, Shannon French, *The Code of the Warrior: Exploring Warrior Values Past and Present* (Lanham, MD: Rowman & Littlefield, 2003); Edwin Micewski, 'On the Philosophical Framework for the Ethical Debate in Professional Military Education', in E. R. Micewski and H. Annen (eds) *Military Ethics in Professional Military Education – Revisited* (Bern: Peter Lang, 2005); James Toner, *Morals Under the Gun* (Kentucky: University Press of Kentucky, 2000).
43. Martha Nussbaum, *Non-Relative Virtues: An Aristotelian Approach* (Helsinki: World Institute for Developmental Economics Research, 1987); Martha Nussbaum, *Wat liefde weet: Emoties en moreel oordelen* (Amsterdam: Boom/Parrèsia, 1998).
44. Alison McIntyre, 'Doing Away with Double Effect', *Ethics*, p. 111, January 2001; van Baarda, *Moral Ambiguities*, 3.1.
45. Michael Stocker, 'The Schizophrenia of Modern Ethical Theories', *Journal of Philosophy*, 1976; Elisabeth Anscombe, 'Modern Moral Philosophy', *Philosophy*, 33, 1958; Lawrence Hinman, *Ethics: A Pluralistic Approach to Moral Theory* (Belmont, CA: Wadsworth, 2007).
46. McCrudden, 'Human Dignity'; Carozza, 'Human Dignity'.
47. Damian Cox, Marguerite La Caze and Michael Levine, *Integrity and the Fragile Self* (Farnham: Ashgate, 2003).
48. See also McCrudden, 'Human Dignity'.
49. James Gutmann, cited in Cox *et al.*, *Integrity*, p. XIII.
50. Anne Colby and William Damon, 'The Uniting of Self and Morality in the Development of Extraordinary Moral Commitment', in Gil G. Noam and Thomas E. Wren (eds) *The Moral Self* (Cambridge, MA: MIT Press, 1993), p. 150.
51. Toner, *Morals Under the Gun*, p. 19.
52. Francine du Plessic Gray, 'Bearing Witness', *The New York Times*, 30 August 1987.
53. Lt Gen. Roméo Dallaire, *Shake Hands with the Devil: The Failure of Humanity in Rwanda* (London: Arrow, 2003). Gen. Dallaire asks: 'In

the face of such crimes against humanity, should commanders continue to put their troops' lives at risk from a sense of moral duty, or should they cease operation, protect themselves, and withdraw because they no longer have a mandate? ... Thirteen days later, as both the civil war and the slaughter gained momentum, I received the order to withdraw all forces. I refused outright' (pp. 36–7).

54. Chesire Calhoun, 'Standing for Something', *Journal of Philosophy*, XCII(5), 1995, p. 259.

55. Opotow, 'Moral Exclusion', who argues at p. 4 that moral inclusion contains a number of cluster of attitudes: '(1) believing that considerations of fairness apply to another, (2) willingness to allocate a share of community resources to another, and (3) willingness to make sacrifices to foster another's well-being'.

56. Daryl Koehn, 'Integrity as a Business Asset', *Journal of Business Ethics*, 58, 2005, p.126.

57. Brig. Gen. Malham Wakin, *Integrity First: Reflections of a Military Philosopher* (Lanham, MD: Lexington, 2000), p. 61. Osiel, *Obeying Orders*, p. 168, observes that soldiers are frequently trained, if not conditioned, during military training on the basis of the philosophies of Skinner and Pavlov.

58. Wakin, *Integrity First*, pp. 65–8. Compare: Isaiah Berlin, 'Two Concepts of Liberty', in *Four Essays on Liberty* (Oxford: Oxford University Press, 1969).

59. Interestingly, self-discipline is one of the values mentioned in the Code of Conduct of the Dutch Military Police.

60. Wakin, *Integrity First*, p. 66.

61. For a remarkable and well-documented account, see Pumla Gobodo-Madikizela, who was a member of South Africa's Truth and Reconciliation Commission under the chairmanship of Bishop Desmond Tutu. In her *A Human Being Died That Night: A South African Story of Forgiveness* (Boston, MA: Houghton Mifflin Harcourt, 2003) she recounts her conversations with Eugene de Kock – one of the most notorious police officers under the apartheid regime. De Kock is being described as someone who undergoes a complicated moral odyssey in prison, who begs for, and receives, forgiveness from the widows of some of his victims.

62. Hannah Arendt, *On Violence* (San Diego, CA: Harcourt, 1970).

63. Bandura, 'Moral Disengagement', p. 203; Lawrence P. Rockwood, 'The Lesson Avoided: The Official Legacy of the My Lai Massacre', in Th. A. van Baarda and D. E. M. Verweij (eds) *The Moral Dimension of Asymmetrical Warfare: Counter-Terrorism, Democratic Values and Military Ethics* (Leiden: Brill, 2008).

Chapter 9

Understanding Atrocities: What Commanders Can Do to Prevent Them

PAOLO TRIPODI

Introduction

In a BBC documentary on torture, William Menold comes across as a thoughtful man probably in his 70s. Nearly 50 years ago, in 1961, he had just left combat duties in the US Army as he decided to volunteer to participate in Stanley Milgram's famous experiment on obedience.[1] Menold remembers that experience very vividly and in particular 'the fact that you are acting crazy ... I really think I kind of lost track of right and wrong, and good and evil.'[2] He was one of several individuals who complied with 'orders' and inflicted what he believed to be severe harm on another individual through the deliberate administration of electric shocks.[3]

Menold's experience is echoed by a few among those who participated in the infamous events at Abu Ghraib. Two documentaries, *Taxi to the Dark Side* and *Ghosts of Abu Ghraib*, have provided an interesting perspective on the individuals who were directly involved in abusing and torturing prisoners. For a US Army Military Intelligence Sergeant, Abu Ghraib was 'the most surreal place you can imagine. Like a combination of *Apocalypse Now* meets *The Shining*. Except that this is real and you are in the middle of it.' Adapting to such an environment was also a sort of surreal experience as noted by a US Army Military Police specialist. She said that 'something in your brain clicks, everything you see is normal, you go crazy if you do not adapt to what you are seeing'. Another noted the pressure of complying with superiors' orders. She said that 'I do not think it was my place to question anything ... If you question your orders, you and your battle buddy are going to die.' Finally, there were individuals who lived the experience of Abu Ghraib as if they were not the same individuals they used to be. A US Army Military Police Sergeant realized that 'at that moment I was a different person, at that moment I was not the person that is sitting here talking to you right now. At that moment I was somebody else.'[4] Many of the individuals who were involved in these events did not expect to be able to perform acts that would be considered immoral and illegal. Many simply did not know that, placed in a stressful, extraordinary situation, they might perform acts that are harmful to other human beings. Most people who have committed atrocities in combat often are not obvious perpetrators; more likely they are ordinary people. As noted by Steven Baum 'most evil is the product of rather ordinary people caught up in unusual circumstances'.[5]

This chapter offers an explanation for atrocities and the role played by situational forces. It will also analyze how the military system interacts in a positive or negative way with those situational forces. There are cases, as will be illustrated, in which the dispositional explanation offers a better way to understand an atrocity; however, even in these cases the situation, though maybe to a smaller extent, still plays an important role. The objective of this chapter is to raise awareness among commanders about the 'hidden' risks that exist in any deployment. Nothing could be more dangerous than for commanders to convince themselves that they and their troops are somehow immune or invincible to situational forces. Arthur Miller warned us that 'people typically do not acknowledge the pervasive, often subtle, influences of situational pressures, preferring instead to account for behavior ... in terms of dispositional characteristic of the actor'.[6] Zimbardo, Maslach and Haney have shown rather compelling evidence that 'any deed that any human being has ever done, however horrible, is possible for any of us to do – under the right or wrong situational pressures.'[7]

Situational forces are extremely powerful. Therefore relying on the strength of character of the individual, or a group of individuals, to resist such forces may simply not be enough. It should be noted that, while engaging in an explanation of individual's behaviour that focuses heavily on the situation, this chapter does not intend to condone such behaviour: this is not the objective of the analysis. In the author's view the individuals remain fully responsible for the acts they commit. It is of great value, however, for us to understand what made individuals behave in an unusual way when they were placed in such situations.

The chapter takes an interdisciplinary approach, exploring what scholars from several fields, from social psychology to genocide studies, have discovered about atrocities and human behaviour. It then applies such findings to the field of military studies with the main focus being on the role of military leaders.

Categories of atrocity

Professor Robert Zajonc rightly noted that 'Massacres are extremely complex phenomena that will not be explained by a single factor, and no one discipline alone can explain the hundreds of massacres all over the globe. The task is novel and quite difficult.'[8] Indeed, in order to provide such a complex explanation it is beneficial to attempt a categorization of military atrocities. I believe that military atrocities can be broken down into three categories. The first one, very likely the most important because of the magnitude of victims, is about atrocities perpetrated systematically by a state's military, paramilitary, or militia against the state enemy. Inside this first category, one extreme is represented by the Nazi military and the Waffen SS, with their commitment during the Second World War to exterminate an entire ethnic group. On the other extreme might be, for example, the Chilean armed forces during the Pinochet dictatorship. The enemy of the state were taken to be all those opposed to General Augusto Pinochet and in particular those who supported the socialist government of Salvador Allende. The Chilean armed forces' objective was not the extermination of the enemy, like in Nazi Germany, but rather to suppress the enemy. In Chile, 3,000 political opponents and innocent civilians were killed in the dark days of September and October 1973 at the National Stadium and in the cities touched by the infamous *Caravana de la Muerte*. Many were killed in the following years; over nearly two decades, thousands were tortured. In such circumstances, potential enemies might fear that an active opposition to the regime will be met with the strongest response. However, should they decide to give up their political opposition, it is unlikely that they will be eliminated.

Under these conditions atrocities are condoned and, indeed, even promoted by those in power. Perpetrators will find a strong motivation in the belief that they are just participants of a political project that requires the type of atrocities that they will support. Christopher Browning, the author of *Ordinary Men*, defines this first category as 'atrocity by policy'. He explains that 'these were not the spontaneous explosions or cruel revenge of brutalized men, but the methodically executed policies of government'. Perpetrators of 'atrocity by policy' 'act not out of frenzy, bitterness and frustration, but with calculation'.[9]

Browning's explanation also helps to define the second category of atrocities – those committed by people who, indeed, act out of anger, revenge and rage. Those individuals would find it difficult to justify what they did in those circumstances and, very likely, under normal conditions would not have done the things that they did. Glenn Gray, a Second World War veteran who had seen combat in many battles, noted that, 'If the war taught me anything at all, it convinced me that people are not what they seem or even *think themselves to be* [my emphasis]. Nothing is more tempting than to yield oneself, when fear comes, to the dominance of necessity and to act irresponsibly at the behest of another.'[10] The category of actions that can take place in this context is referred to as 'atrocity by situation' and it will be addressed in greater detail in the next section of this chapter.

The third category is 'atrocity by crime'. Explanation for these atrocities can be more dispositional than situational. For instance, some individuals might be inclined to commit a crime because of the situation they are in, yet these are individuals who have the potential to commit a crime no matter what the situation. The situation they are in, the deployment far away from home, the anonymity associated with it and the position of power they might enjoy, may make it easy for them to commit a crime. Even in the case of 'atrocity by crime', commanders still have a great deal of responsibility. A good knowledge of their troops and concern for discipline are normally the best tools commanders should possess. However, if potential criminals are allowed to wear the uniform, the screening process is at fault. 'Atrocities by crime' probably cause a smaller number of victims, yet they may have a similar level of brutality of 'atrocity by policy' for those involved.

In order to illustrate this last category, two cases among many are particularly telling. The first one involves a US Army Staff Sergeant with the 82nd Airborne Division, who in 2000, while on deployment in Vitna, Kosovo, 'raped, sodomized and killed Merita Shabiu',[11] an 11-year-old girl. The army investigation ordered by the then chief of the Army, Gen Shinseki, found that a combination of the wrong mindset and poor discipline were among the factors that allowed the sergeant and several other members of the unit to adopt immoral behaviour in dealing with locals.[12] Dana Priest covered the incident in her book *The Mission*. She noted that the sergeant 'regaled his soldiers with stories of his Mafia connections and how he had raped two young sisters in Haiti ... [he] and some members of his weapons squad became the neighborhood bullies.'[13]

The second case involves five soldiers with the 101st Airborne Division, who since September 2005, were deployed in the Iraqi city of Mahmudiyah, just a short, but then very dangerous, 15 minutes' drive from the Green Zone in Baghdad. The Regiment was engaged in major counter-insurgency operations, several members of the unit were killed and danger was everywhere. Among the many tasks performed by the five soldiers, they also manned a checkpoint from which they acquired some knowledge of the local population. They had noticed an attractive 15-year-old girl, Abeer Qassim Hamza, who often passed by the checkpoint. On 12 March 2006, the soldiers went into the house where the young girl's family lived. They killed her family and then raped and killed her.[14] The Iraqi doctor, who inspected the bodies, told prosecutors that he was ill after witnessing the crime scene.[15]

Clearly in 'atrocities by crime', several factors must be considered. Primarily the personality of the individuals involved is going to be important, but the role of the military institution itself can enable such situations. In these cases, the institution in which the men served underestimated or simply ignored reports of alcohol and drug abuse and petty crimes. Yet, what was even more important in these cases was the inability of the military leadership to maintain a good level of discipline and morality in the unit.

Atrocity by situation

Robert Secher was an extremely dedicated Marine Corps Captain, with tour of duties in Afghanistan and Iraq. He was 33 when, in October 2006, while out on patrol in Hit, Iraq, he was shot and killed by a sniper. Secher's father made the email correspondence he had with his son public. It is significant what the Marine officer wrote to his dad after the events of Haditha and Hamdania, two incidents in which Marines, under significantly different circumstances, yet in an overall similar situation, were responsible for the killing of civilians. Secher wrote:

> Of course you've heard about two different sets of Marines being charged with murder ... I feel bad for those guys ... And anyone who calls those young Marines killers should think twice. War puts perfectly ordinary young men in situations that can't be judged by laws. They are the situations of survival. The dirty little secrets of war, no one would want to know the horrible things that the 'greatest generation' did to German and Japanese soldiers and civilians.

Secher ended the message in an emotional fashion, explaining that in order to be successful in war 'you have to be brutal' like 'Sherman and raze Georgia as you march to the sea'. However, as he had time to reflect more on the events and what he had written, particularly in the last part of his message, he wrote again to his father:

> I don't want you to think that I condone the actions of the Marines, and if in fact they are guilty they should be held responsible and punished. My point is that this is the reality of what war is. This is what war does to normal young men ... People are so quick to only criticize the Marines and to demonize these young men. I pity them.

Their lives are ruined, ruined by their actions which are judged by men who have never been in those situations.[16]

Captain Secher had a great appreciation of how a situation of war impacts on ordinary individuals. He understood how atrocity by situation might happen.

At this point, before proceeding to a deeper analysis of atrocity by situation, it is important to define exactly what is meant by the term 'situation'. According to Christina Maslach, 'The concept of *situation* has not always been clearly defined, but it usually refers to the immediate setting in which the individual is located. As such, it includes the physical environment, other people, social norms or constraints, and other types of physical or social stimuli.'[17] Clearly, in Maslach's definition, individuals interact with the environment and are actively engaged with the group, so they contribute to the determination of social norms, the perception of morality and what is right and wrong. Thus, it would be inaccurate to perceive the individual as completely passive; I would argue that, indeed, the individual is actively engaged with the situation, yet it is mostly the situation that influences the individual rather than vice versa. The more powerful the situation the higher the chances will be that the individual might succumb to inappropriate behaviour.

Philip Zimbardo provided a strong exploration of individuals' behaviour in stressful situations. His study demonstrated that human nature can be significantly transformed when people are within a powerful social setting.[18] In *The Lucifer Effect* he wrote:

> good people can be induced, seduced, and initiated into behaving in evil ways. They can also be led to act in irrational, stupid, self-destructive, antisocial, and mindless ways when they are immersed in 'total situations' that impact human nature in ways that challenge our sense of the stability and consistency of individual personality, of character and of morality.[19]

Soldiers deploy in complex, stressful, often extreme situations. The issue is that the military believes the best preparation for soldiers and their leaders focuses on the individual, while very little is done to explain how the individual might interact with a particularly demanding and stressful situation. Leaders at all levels, especially those operating at the tactical level, should acquire a solid awareness of how situational forces might impact on them and those serving under them. As will be explained below, small unit leaders who are aware of the power of situational forces are those who can best use the tools available to them to deal with those forces. According to Zimbardo, we like to believe in the 'unchanging goodness of people, in their power to resist external pressures'. However, having a strong confidence in individuals' invulnerability to situational forces is to 'set ourselves up for a fall by not being sufficiently vigilant to situational forces'. Therefore 'we are best able to avoid, prevent, challenge, and change such negative situational forces only by recognizing their potential power to 'infect us' *as it has others who were similarly situated* [my emphasis]'.[20]

If we look at the experience of soldiers who have participated in atrocities, we can learn much about their behaviour. It is important that in undertaking such an analysis we focus on what their experience was like, how they perceived their situation, and what they did, rather than focusing on the fact that they might attempt to morally justify their actions. If we truly want to learn more, we should look at their experience in a cold, detached fashion and should not be distracted by our own moral judgements.

Some examples can help us understand the impact of the situation on individual soldiers. Varnado Simpson was not even 20 years old when he found himself committing horrible atrocities on 16 March 1968 in My Lai. His testimony is helpful to understand how the situation took over this young man. That day, the men of Charlie Company arrived in the small Vietnamese village with a strong desire to take revenge on anyone who they believed might have been responsible for the violent death of some of the most popular members of the company over a period of a few months. The 'assault' was to be their first combat operation, engaging an enemy they had been hunting for weeks; finally they would be able to fire at this constantly vanishing, deadly ghost. Simpson and the rest of the company were ready. Fred Widmer, the Company radio operator, remembers 'My understanding was we were going in, we were going to get into one hell of a fight and we were going to kick some ass when we got down. And there wasn't going to be anybody left. It didn't turn out that way.' They were ready for a fight. Simpson was among those who brutally killed dozen of civilians. He remembered the defining moment of that day. It was when he was ordered to kill a running woman who, apparently, was carrying a weapon. He did. He soon realized that he had killed a woman and the child she was carrying.

> She was running with her back from a tree line ... but she was carrying something, I didn't know if it was a weapon or what, but it was a woman, you know, I knew it was a woman. I didn't want to shoot a woman, you know, but I was given an order to shoot, so I'm thinking that she had a weapon running. So when I shot and I turned over it was a baby, you know. Shot about four times, three

or four times, and the bullets just went through and shot the baby too, you know. And I turned over and I saw the baby's face with the half gone, you know, and I just blinked, I just went.

Simpson blamed the training he received, 'the programming to kill ... I just start killing.' After that, 'It wasn't hard to kill, it wasn't hard to find anyone to kill.'[21]

Simpson spent his remaining life dealing with a persistent remorse for what he did. He tried to kill himself a few times and finally succeeded in 1997. Among the men of Charlie Company, many stood back, some were strong enough to resist the situation and pressure to harm defenceless human beings. Many defied the orders to kill. One of the GIs in Lt Calley's platoon, Harry Stanley, had no intention of killing innocent civilians. He disobeyed a direct order to kill:

Lieutenant Calley ordered certain people to shoot these people and I was one of them. And I refused and he told me that he was going to have me a court-martial when we got back to base camp, and I told him what was on my mind at the time. Ordering me to shoot down innocent people, that's not an order, that's, that's craziness with me, you know, and so I don't feel like I have to obey that ... And if you want to court-martial me, you do that. If you can get away with it. I feel like it was, it was horrible, you know, just a terrible thing to be going on, and American boys doing this, you know. And I feel like I'm a red-blooded American boy just like any of the rest of the guys that was there, you know. And to see that, I'm talking about black or white, you know, black or white guys doing this, you know, it didn't make any difference. I'm saying it just seemed like a horrible thing. I'm talking we all came from the same place ... and I know they all had to have the same values that I had somewhere along the line. If ... they didn't get it in school, they had to get it in a religion or church or some place ... If you didn't go to school, you could pick it up from a stranger, it's just simple. But then to go and do something like this, it's, it's immoral to me. That's just the way I feel about it.[22]

Stanley and others never wanted to become perpetrators of an atrocity, yet they remained bystanders and not, like many would expect, rescuers. None of the members of Charlie Company actually tried to stop the slaughter.[23]

Joanna Bourke, in *An Intimate History of Killing*, notes that an attempt to explain the atrocity in My Lai 'in terms of the personal characteristic of the men of C Company has only limited usefulness'. She noted that there was little to distinguish Charlie Company from other American military units in Vietnam at the time.[24] General Peers provides an interesting reflection on both Charlie and Bravo companies, noting that in Vietnam 'as in all units ... there was a sprinkling of toughs – in Task Force Barker they were almost gangsters. In the absence of effective leadership by junior officers and NCOs [as was the case for Charlie and Bravo companies in My Lai] some of the lower-ranking enlisted men probably followed along with these hoodlums.' Thus, the investigators explored whether such a line of analysis would have explained the massacre, 'we thought that perhaps the units had included an unusual number of men of inferior quality'. However, when they asked the deputy chief of staff for personnel to provide an analysis of the men of Charlie Company, the result was a fact sheet that indicated that the members of the unit were average when compared with other Army units. A similar outcome was provided for the men of Bravo Company. General Peers found that: 'It seems to follow that if these men were average American soldiers, and if other units with the same kind of men did not commit atrocities of this order, there must have been other overriding causes.'[25]

Atrocity by situation and the military system

Joanna Bourke noted that:

> widespread military and civilian complacency about the killing of non-combatants, failure in military leadership, fear of punishment, ignorance of the rules of engagement, a belief in the value of unquestioned obedience, guerrilla tactics, and racism can all be blamed for the incidence of atrocities and the grotesque behaviour indulged in by combatants. Taken together, it was surprising that more atrocities did not occur, yet presented singly, each factor failed to 'explain' why a particular atrocity took place.[26]

Bourke's list was recently echoed in a US Marine Corps (USMC) publication, *War Crimes*, in which a special emphasis was placed on the role of leadership and training to 'eliminate' some of the factors that historically led to the commission of war crimes. It lists such factors as:

> high friendly losses; high turnover rate in the chain of command; dehumanization of the enemy or use of derogatory names or epithets; poorly trained and inexperienced troops; poor small unit discipline standard; the lack of a clearly defined enemy; unclear orders; [and] high frustration level among the troops.[27]

Bourke is right to argue that no single factor will provide an adequate explanation for an 'atrocity by situation'. Normally, several factors will have to align in an extremely dangerous way; it is very likely that when this alignment happens atrocities might occur. Commanders therefore need to understand, appreciate and have a full awareness of the situation; they must also understand how valuable or deleterious the military system can be when dealing with situational forces. Philip Zimbardo provided an enlightening explanation of how situation and system come together to allow evil to happen. In his view, 'System Power involves authorization or institutionalized permission to behave in prescribed ways or to forbid and punish actions that are contrary to them.'[28] Indeed, there are key components of the system that can actually interact with the situation and might increase chances that atrocities will happen.

Rarely can commanders make a significant impact on the situation, yet they can shape the system so individuals who are part of it are better prepared to deal with situational forces. In a counter-insurgency environment commanders cannot remove soldiers' fears of being a constant target, and the fear that they might be killed at any time, yet they can provide soldiers with the type of preparation that would make them more resilient in such an environment. Soldiers still operate in the same situation, yet the impact that the pressure will have on a well-prepared individual will be significantly smaller than it might have been on an individual who has developed a mindset that is inadequate for such an environment. Commanders must learn about the positive and negative sides of the military system. It is clear that one of the key components of that military system is training and education. In a counter-insurgency environment, soldiers who have been trained for conventional war-fighting, characterized by a simple friend–enemy dichotomy, will find it extremely difficult to adopt appropriate courses of action. Soldiers who have the wrong mindset will find operating in a counter-insurgency environment extremely frustrating, for example, blurring the line that separates combatants from noncombatants. A US Army Sergeant who was deployed in Iraq maintained that 'The frustration that resulted from our inability to get back at those who were attacking us led to tactics that seemed designed simply to punish the local population that was supporting them.'[29] One of the Marines involved in the Hamdania incident, stated that 'This is a war and the other side, they don't have rules.'[30] It is evident that in such an environment training and education play a key role in preventing atrocities.

The 1940 USMC *Small Wars Manual* provided much wisdom to Marines who operated in such difficult situations. The manual, written by Marines who had experienced decades of deployment in what we would label today as counter-insurgency, but also peacekeeping, peace enforcing and humanitarian intervention, is enlightening. The manual explains that:

> The application of the principles of psychology in small wars is quite different from their normal application in major warfare or even in troop leadership. The aim is not to develop a belligerent spirit in our men, but rather one of caution and steadiness. Instead of employing force, one strives to accomplish the purpose by diplomacy.[31]

Such an approach is reinforced in many passages and some key concepts are spelled out with greater clarity. The manual suggests that 'In major warfare, hatred of the enemy is developed among troops to arouse courage. In small wars, tolerance, sympathy, and *kindness* [my emphasis] should be the keynote of our relationship with the mass of the people.'[32] Such an approach is inspired by common sense, wisdom and a clear understanding of what the best mindset is in order to succeed with the local population, rather then focusing exclusively on defeating the local insurgents. Leaders who apply such principles prepare their troops with the best mindset to avoid atrocities in situations in which the very presence of civilians might be perceived as hostile.

Commanders must also be able to use cohesion to their best advantage, while at the same time acknowledging that such a fundamental component of the military system might also have a negative side. In this respect, Glenn Gray posed an important question: 'Why can men do together without conscience things that would torment them unendurably if done singly?'[33] Indeed, troops will do their best to help their brothers in arms and might feel justified in doing something immoral and/or illegal in the process. Commanders need to understand when cohesion is turning into such a factor with the potential to lead to atrocities. In My Lai the small unit 'group' thinking affected the positive values of the GIs and slowly replaced them with what can be considered anti-values.

One of the most important components of the military system is obedience. This is a particularly complex element for commanders. Military organizations strongly rely on a rather strict response from individuals to superior orders. Only illegal, unlawful orders should not and must not be executed, any other order should be carried out. In this process, however, the military system places significant stress on the execution of legal orders, rather than on disobedience of illegal ones. The average soldier trusts that his or her commander will never give them an illegal order. In addition, the execution of an immoral order might be perceived as a perfectly appropriate thing to do if the order comes from a respected commander. Thus, a commander should give small unit leaders more autonomy, but also more moral responsibilities for the actions small unit leaders perform, even in the execution of superior orders.

Finally, discipline and a close monitoring of troops are components of the military system that help commanders establish the appropriate command climate. It is known that, in many situations, commanders might be tempted to relax unit discipline and, indeed, for limited periods of time and under very unique circumstances, it may well be beneficial to the unit. However, when such an approach becomes a consolidated practice, commanders might be indirectly encouraging unwanted behaviour from their troops. In the famous Stanford Prison Experiment in which several students participated in research conducted by Zimbardo on how they would behave as inmates and guards, Zimbardo, playing the role of the warden, decided not to enforce any discipline. It took the 'guards' just two days to understand that they would not be disciplined; they quickly began abusing the 'inmates'. In just six days 'Stanford Prison' was, in the words of Christina Maslach, a mad house. Commanders must be aware of the potentially negative consequences of relaxed discipline. 'Atrocities by situation' will not happen because a single element of the military system is corrupted; rather several of them must align in an extremely dangerous way. Yet, the one element that stands above all the others and has the power to influence all the others in a positive (but also negative) way is leadership. As long as commanders retain the ability to understand how to best use the system, they will always be the element that does not align, thus 'atrocities by situation' will not happen because the system will not surrender to the situational forces.

Atrocity by situation: the leader's trap

Leaders play a major role in preventing or enabling the three types of atrocities. In 'atrocities by policy' often they are perpetrators themselves as they have accepted to be active participants in an overall evil project. In 'atrocities by crime' leaders have played a shallow, passive role, much like Zimbardo's in the Stanford Prison Experiment. In 'atrocities by situation' leaders might have acted in good faith, but at some point they gave in to the situational forces and sent negative input to the system or, potentially just as dangerous, did not send any input at all. General Peers was adamant in the identification of leadership as the main failure that led to My Lai. He stated that 'In analyzing the entire episode, we found that the principle breakdown was in leadership. Failures occurred at every level within the chain of command, from individual non commissioned-officer squad leaders to the command group of the division.'[34] At the end of the investigation, General Peers provided General Westmoreland with a series of suggestions for leaders who operate in counter-insurgency environments. The entire list is extremely interesting, but the conclusion of this chapter will focus on just two of the points. General Peers emphasized that:

> in the heat of battle, some officers and men tend to lose sight of the more fundamental issues upon which the war is being waged. Along with the attitudinal changes ... this can lead to winning a battle or two but losing a war. It is an inherent and paramount responsibility of the commander to insure that his officers and men understand ... and put into practice the principles of discriminate and tightly controlled application of firepower; genuine and practical concern for private property ... humane treatment and care of refugees, non-combatants and wounded (whether friendly or enemy).

General Peers fully appreciated the power of situational forces and stressed the role of commanders to be proactive in tempering the impact of such forces. He also suggested a practical approach, as he warned that a commander must be:

> constantly alert to changes in the attitude and temperament of his men and the units to which they belong. Ground combat in a counterinsurgency environment may develop frustration and bitterness which manifest themselves in acts quite apart from that which would *normally* [my emphasis] be expected ... commanders must be quick to spot such changes and to take appropriate corrective action. Any indications of an attitudinal change from one of physical toughness in combat to senseless brutality requires immediate remedial action by the commander concerned.[35]

In a deployment in a stressful situation, the proper functioning of the military system is critical to prevent 'atrocity by situation'. The recommendations by General Peers demonstrate that the system will respond in a positive way as long as leaders send the right input to its key components: training/education; cohesion; discipline; unit's values and culture; and obedience, just to cite an important few. The challenge for commanders, however, is to avoid what can be called the leader's trap, which is the commander's ability to remain aloof from the situation and to resist the power of situational forces. This is possibly the greatest challenge, as commanders live the situation like anyone else in the unit; they experience the fear, grief, rage and frustration, with the same level of intensity. Thus, it is of great importance for commanders to retain a high level of awareness and, as much as possible, the ability to look at the system they are in charge of in a cold, objective way. As he makes clear in his own account of the Stanford Prison Experiment, Zimbardo himself provides us with a particularly useful illustration of what happens when this is not the case.[36] Despite the fact that he created the 'situation' in which the students played the role of guards and inmates, in just a few days, he lost his ability to view his experiment in an objective way. In doing so, he failed to use his leadership position to prevent what become an unacceptable situation for an experiment. The degrading conditions at 'Stanford Prison' were not recognized by Zimbardo and the team of scholars who worked on the experiment: they simply viewed as normal what clearly was not. It was a researcher who was not associated with, nor part of the daily activities related to the experiment, who realized how abusive the behaviour of the guards had become in the space of a few days and persuaded Zimbardo to bring it to an end even though he initially could not understand why she was upset.

Larry Colburn, a member of the crew serving under CWO Hugh Thompson, the helicopter pilot who stopped the killing of several civilians at My Lai, did not hesitate to provide Thompson with the support he needed to confront some of the GIs involved in the atrocity. Over multiple occasions he expressed firm condemnation for what happened that day in March 1968. Without sympathy, but with a clear sense of objectivity, he noted that those soldiers on the ground had gone through all the negative experiences of a counter-insurgency situation. Over a period of a few months, they had lost the ability to act as normal individuals. Their leaders became part of this deadly dynamic and, if they had ever had it, they lost the ability to send positive input to the system they were supposed to control.

Notes

1. Stanley Milgram's experiment findings were published in several articles and in *Obedience to Authority* (New York: HarperCollins, 1974). Recently, a new version of the Milgram experiment was replicated by Jerry Burger. Despite several limitations to the original experiment, Burger found that 'my partial replication of Milgram's procedure suggests that average Americans react to this laboratory situation today much the way they did 45 years ago'; in addition he suggested that the 'same situational factors that affected obedience in Milgram's participants still operate today'. Jerry M. Burger, 'Replicating Milgram: Would People Still Obey Today?' *American Psychologist*, 64, 2009, pp. 1–11.
2. *We Have Ways of Making You Talk*, broadcast on 5 April 2005 on BBC 2.

3.   Unbeknown to the person in the role of 'teacher', who believed they were administering the increasingly severe electric shocks, there were no electric shocks and the 'learner' was simply acting to see how far the 'teacher' would go.

4.   *Ghosts of Abu Ghraib*, HBO, 2007, directed by Rory Kennedy.

5.   Steven Baum, *The Psychology of Genocide Perpetrators, Bystanders, and Rescuers* (New York: Cambridge University Press, 2008), p. 170.

6.   Arthur Miller *et al.*, 'Explaining the Holocaust: Does Social Psychology Exonerate the Perpetrators?', in Leonard Newman and Ralph Erber (eds) *Understanding Genocide: The Social Psychology of the Holocaust* (Oxford: Oxford University Press, 2002), p. 302.

7.   Philip Zimbardo, Christina Maslach and Craig Haney, 'Reflections on the Stanford Prison Experiment: Genesis, Transformations, Consequences', in Thomas Blass (ed.) *Obedience to Authority* (Mahwah, NJ: Lawrence Erlbaum Associates, 2000), p. 206.

8.   Robert Zajonc, 'The Zoomorphism of Human Collective Violence', in Newman and Erber, *Understanding Genocide*, p. 235.

9.   Christopher Browning, *Ordinary Men: Reserve Police Battalion 101 and the Final Solution in Poland* (New York: HarperCollins, 1998), p. 161.

10. J. Glenn Gray, *The Warriors: Reflections on Men in Battle* (Lincoln: University of Nebraska Press, 1998), p. 169.

11. Martina Vandenberg, 'Peacekeeping, Alphabet Soup, and Violence against Women in the Balkans', in D. Mazurana, A. Raven-Roberts and J. Parpart (eds) *Gender, Conflict and Peacekeeping* (Lanham, MD: Rowman & Littlefield, 2005), p. 150.

12. On 1 August 2000 a court-martial sentenced the Staff Sergeant to life in prison without the possibility of parole, in addition to a dishonorable discharge and reduction to the lowest enlisted grade. For details see http://news.bbc.co.uk/2/hi/americas/862063.stm, http://www.nytimes.com/2000/12/02/world/army-orders-peacekeepers-to-sessions-on-rights.html and http://www.armfor.uscourts.gov/opinions/2004Term/03-0520.pdf.

13. Dana Priest, *The Mission: Waging War and Keeping Peace with America's Military* (London and New York: W. W. Norton & Co., 2004), p. 335.

14. Of the five soldiers responsible for the killing in Mahmudiyah, four were charged with premeditated murder, conspiracy and rape. They were sentenced by the Military Tribunal at Fort Campbell to 100 years' confinement, 110 years' confinement, 90 years' confinement and 27 months' confinement. For more details about the sentencing see http://www.msnbc.msn.com/id/17247852/, http://www.guardian.co.uk/world/2006/nov/17/iraq.usa1 and http://www.nytimes.com/ 2007/08/05/us/05abuse.html. A fifth member of the unit was sentenced in May 2009 by a jury in Kentucky to life in prison. For more details see http://www.nytimes.com/2009/05/22/us/22soldier.html?scp=1&sq=steven%20green&st=cse and http://www.guardian.co.uk/world/2009/may/08/steven-dale-green-guilty.

15. Troops 'took turns' to rape Iraqi, BBC, Monday, 7 August 2006, avail-

able at http://news.bbc.co.uk/2/hi/middle_east/5253160.stm, accessed on 16 February 2009.

16. Captain Robert Secher, quoted in Dan Ephron and Christian Caryl, 'A Centurion's E-mails', *Newsweek*, 6 November 2006.

17. Christina Maslach, Richard T. Santee and Cheryl Wade, 'Individuation, Gender Role, and Dissent: Personality Mediators of Situational Forces', *Journal of Personality and Social Psychology*, 53(6), December 1987, p. 1088.

18. Philip Zimbardo, *The Lucifer Effect: Understanding How Good People Turn Evil* (London and New York: Random House, 2008), p. 210.

19. Zimbardo, *The Lucifer Effect*, p. 211.

20. Zimbardo, *The Lucifer Effect*, p. 211.

21. 'Remember My Lai', *Frontline*, broadcast 23 May 1989. Transcripts available at http://www.pbs.org/wgbh/pages/frontline/programs/transcripts/714.html.

22. 'Remember My Lai', *Frontline*, broadcast 23 May 1989. Transcripts available at http://www.pbs.org/wgbh/pages/frontline/programs/transcripts/714.html.

23. Early in October 2009 former Lieutenant Calley, in a rare address about the My Lai massacre, said: 'Not a day goes by that I do not feel remorse for what happened that day in My Lai ... I feel remorse for the Vietnamese who were killed, for their families, for the American soldiers who were involved and their families. I am very sorry.' Calley's remarks are available at http://www.miamiherald.com/opinion/other-views/story/1266964.html.

24. Joanna Bourke, *An Intimate History of Killing* (New York: Basic, 2000), p. 194.

25. Lt Gen. W. R. Peers, *The My Lai Inquiry* (New York: W. W. Norton & Co., 1979), p. 231.

26. Bourke, *An Intimate History of Killing*, p. 194.

27. *War Crimes*, MCRP 4-11.8B, (Quantico: US Marine Corps, 2005), pp. 7–8.

28. Zimbardo, *The Lucifer Effect*, p. 226.

29. Quoted in Chris Hedges, 'The Death Mask of War', *Adbusters, Journal of the Mental Environment*, 15(4), July/August 2007, p. 72.

30. Thomas Watkins, 'Released Marine Details Deadly Day', *Associated Press*, 10 August 2007.

31. *Small Wars Manual*, US Marine Corps, Washington, 1940, Chapter 1, Paragraph 11, p. 18.

32. *Small Wars Manual*, US Marine Corps, Washington, 1940, Chapter 1, Paragraph 16, p. 32.

33. Gray, *The Warriors*, p. 168.

34. Peers, *The My Lai Inquiry*, p. xi.

35. Peers, *The My Lai Inquiry*, p. 248.

36. Zimbardo, *The Lucifer Effect*, pp. 168ff.

Chapter 10

To Whom Does a Military Medical Commander Owe a Moral Duty?

DUNCAN BLAIR

I swear by Apollo Physician and Asclepius and Hygeia and Panacea and all the gods and goddesses, making them my witnesses, that I will fulfil according to my ability and judgement this oath and this covenant.[1]

The original Hippocratic Oath sets out, in a clear and didactic manner, the duties of a physician. In the arena of military medicine, however, conflicting elements of duty can set up friction between the medical and military precepts of the medical officer. There are also natural tensions that arise from the differing roles and requirements of a mission commander and a mission's medical officer. This is further compounded by the dual roles of the medical officer both as a doctor and as an officer. An exploration of these issues follows below, prompted by further quotes from the Hippocratic Oath.

Introduction

This chapter explores the inherent conflict engendered in the dual roles of being both a doctor and officer within a Western armed force (while the author naturally brings a United Kingdom perspective, the issues discussed are pertinent to all Western armed forces and Western medicine as a whole). As a direct result of this duality, some moral and ethical dilemmas will inevitably present themselves. It is recognized that the majority of practitioners of Western medical art undertake similar oaths to that mooted by Hippocrates 2,400 years ago and this chapter will not stray into issues of alternative or complementary medical practices, which may not fall under the same aegis as Western medicine.

The chapter will first examine the duties of the doctor; how one's professional practice is set against a legal and social framework. In the UK, this is circumscribed by the Medical Act, the General Medical Council (GMC) and the specific Royal Colleges of which one may be a Member or Fellow.[2] Similar practices exist across Western medicine. For example, South Africa, New Zealand and Singapore all have a central regulator similar to the GMC, while in Australia, Germany and the United States, each state has its own regulatory board for doctors. Such medical regulating bodies are in regular contact to ensure best practice is maintained across national borders. These duties place certain limitations on the way one may interact with patients and, more pertinent within this context, how this interaction may be limited within the wider military environment. Understanding what information may be divulged, what may not and under what specific circumstances, is key in maintaining the doctor–patient relationship, as well as keeping the higher levels of command informed about the wellbeing of their personnel. The chapter will then move on to explore the similar duties of the officer or, more specifically, the military commander, setting out the duty of care which is owed to personnel under their command. Again, as one would expect, a legal framework exists, but not so much within civil law as within military law.[3] Such military frameworks also place a burden upon the doctor by setting up a friction between military requirements and norms and those of the medical profession, with both coming from differing background precepts of duty, right and wrong.

Having set out the two competing concepts of the duties of the doctor and those of the officer, this chapter will draw out specific moral dilemmas that may come to light within the role of the medical officer performing the task of medical commander on a deployed operation. Examples regarding the treatment of combatants, noncombatants and medical confidentiality will be used to refine and explore the competing demands placed upon the medical officer. It will not attempt to delve too deeply into other ethical dilemmas, preferring to concentrate on specific practical ethical issues encountered within regular military medical practice. The chapter will argue that the medical officer has to steer a precarious course between the Scylla of medical legality and the Charybdis of military duty, guided by the Circe of their personal ethical construct (closer to medical legality, or Scylla, if the Odyssey metaphor holds true).[4] Invariably they will have had to reflect upon their own ethical code and work out where it fits within their societal and communal outlook.[5] Only by being aware of the presence of these Titans can a doctor expect to be able to execute their duties to men, cadre and command in a balanced and ethical manner.

The chapter concludes that medical commanders owe a prime moral duty to their patients but that it may be abrogated, in certain circumstances, to the wider needs of the military community served. They must be ever mindful of this duality, reflect upon it and act in accordance with the imperatives of law, tempered by their own ethical construct. Specific arguments may be applied in a court of law and knowledge of these will enhance the outcome for both patient and command; but the moral duty still falls to the doctor and their interpretation of the situation.

The duties of a doctor

The Medical Act of 1983 (amended in 2002) sets out the regulations for the registration of medical practitioners in the United Kingdom, establishing and defining the requirements of the General Medical Council (GMC). The GMC is charged with the regulation of medical practitioners' registration and disciplinary matters; it has the right to remove doctors from its register of practitioners, permanently or temporarily (thus stopping them working), as well as the ability to place limitations upon their practice. The GMC takes this responsibility seriously and has laid down guidance on how doctors should conduct both themselves and their practice in the pamphlet 'Good Medical Practice'.[6]

Though the Hippocratic Oath, or one of its modern equivalents, is sworn at degree convocations for doctors, it does not have the legally binding power that the GMC holds and enacts through the Medical Act of 1983. It is, however, worth mentioning that modern Western medicine is still referred to as 'Hippocratic medicine' and though debates still continue as to its relevance in modern medical practice, the majority view remains that it still holds true.[7] As the GMC has the ultimate power to permit or deny a doctor's ability to practise medicine, its pronouncements must be heeded and complied with. The Royal Colleges (such as those of the General Practitioners, Surgeons, Physicians, etc.) also set specific standards to which they expect their Members and Fellows to adhere. The Royal College of General Practitioners (RCGP) set out in 2002 what it felt were the elements of 'Good Medical Practice for General Practitioners'; committing to paper that which it feels essential for best clinical practice, giving evidence of the 'Excellent GP' or the 'Unacceptable GP'.[8] As can be seen from the above, a firm and didactic regime exists to regulate, constrain and guide the actions of medical practitioners to ensure their compliance with the laws of the land (through the Medical Act 1983) and the requirements of the GMC and the Royal Colleges. Again, the medical regulating bodies of other countries run the licensing of medical practitioners in similar ways. These duties have to be considered in each and every patient interaction undertaken by any medical practitioner and their actions should be guided by the precepts evinced in their personal, vocational and legal ethical constructs:

> *What I may see or hear in the course of the treatment or even outside of the treatment in regard to the life of men, which on no account must be spread abroad, I will keep to myself holding such things shameful to be spoken about.*[9]

One of the greatest burdens placed upon a medical practitioner is that of patient confidentiality, mentioned in the Hippocratic Oath and being a long-standing duty that enables the full disclosure of patient information to the practitioner to enable accurate and timely diagnosis. This is almost akin to that confidentiality afforded the sacerdotal function under the 'seal of the confessional'. Confidentiality is viewed as so important by the GMC that it is given a unique publication, 'Confidentiality: Protecting and Providing Information', which details the GMC's requirements for patient confidentiality. These are further expanded in 'Good Medical Practice'[10] and 'Good Medical Practice for General Practitioners',[11] to fully inform the practitioner of the legal framework and ethical view from the regulating authorities.

Confidentiality is pivotal in the doctor–patient relationship. The process of diagnosis is underpinned by the taking of a history from the patient. This provides the key to the greater part of the diagnosis and the understanding of the patient as a person.[12] If a patient could not trust that their disclosed personal history would not be spread abroad without explicit consent then they would be extremely unlikely to confide all the necessary information to the doctor and therefore might hold back important details that could make the entire diagnostic process harder for both patient and doctor. This could irreparably degrade the key diagnostic process and irretrievably compromise the all-important doctor–patient relationship.

The issue of patient confidentiality is the most frequent cause of clashes between the medical practitioner and the military commander; both have a responsibility for the welfare of the personnel under them but what information to disclose, at what point in time is invariably a difficult question to render in 'black and white' terms. The specific point at which disclosure to higher command may be justified can be neither predicted nor easily defined, but, 'you must always be prepared to justify your decisions in accordance with this guidance'.[13] In essence, the only readily justifiable reason for divulging information is given within Personnel, Legal and General Orders (PLAGOs) 106: 'The right to the protection of confidential information is modified in the case of service personnel. In that context, confidential information may be disclosed on a "need to know" basis only for the purpose of maintaining the security, health, safety, discipline and welfare of the unit as a whole.'[14] It is important to note that, in the above quote, the right to protection is only 'modified'; it is not removed, thus reinforcing the importance of confidentiality even within the armed forces.

It should be clear from the above that this sort of situation cannot be laid out in an easy formula and that each time such a situation arises the context and information must be carefully considered before action is taken. However, given the time constraints under which military commanders are usually working, such reflection may not be easily achieved and the pressure for a quick decision may compromise the duty of care owed to a patient's confidential information. The medical officer should, though, be aware of this conflict of interests and be equally conscious of their duties as an officer as well as those as a medical practitioner.

The duties of an officer

Common to all levels of command from independent sub-units upwards is the responsibility of the commander for: 'The command, training, safety, security, discipline, education, health, welfare, morale and general efficiency of the troops under their command'.[15] This quotation summarizes the somewhat extensive duty owed by the commander to their troops, no matter at what level or rank they are working. This does not apply solely to the Army but similar duties are also laid down in the analogous regulations for the Royal Navy and Royal Air Force, and parallel regulations also apply across other Western forces. The basic precept of this charge is that the officer takes responsibility for their own actions and those of their subordinates, while achieving what is asked of them by higher command. Taken to the logical conclusion, in the UK this would involve service on behalf of 'Queen and country' to achieve the requirements and orders passed on by Her Majesty's Government through the Ministry of Defence.

In striking similarity with the medical profession, military professionals 'subordinate their personal interests to the requirements of their professional functions'.[16] Admittedly, the above quote is about American military professionals but this has a resonance with the approach of any professional armed force; the idea being that service to the country overrides and supersedes any personal interests, while recognizing that they have to: 'adhere to the laws of war and the regulations of their service in performing their professional functions'.[17]

Can it be said that doctors have to subordinate their personal interests to that of the patient? One could contend that, in the era of the patient-centred approach to healthcare, this is very true. A doctor is not supposed to tell a patient what to do; instead, they are charged with enabling the patient to make an educated and informed choice from a range of options. Very rarely does a modern doctor work without the express and informed consent of a patient, except in some emergency situations – and even then factors may intervene that modify the doctor's own autonomy (e.g. a rapidly exsanguinating Jehovah's Witness in urgent need of, and refusing, a life-saving blood transfusion).

There is some argument that the 'profession of arms' does not really exist as a profession. One aspect of the definition of a professional is one whose activities may only really be judged, or understood, by others from within the same profession. It can be claimed, with a modicum of authority, that the professional soldier's actions can only be recognized and understood by similar members of the armed forces. Even the *Oxford English Dictionary* recognizes this status within definition 7.a. of 'profession'.[18] The element of self-sacrifice is what is pivotal here and the officer is also required to take on the ultimate responsibility to protect the lives and best interests of those under their command. To best achieve this the commander needs to understand how they function and how best to get them to do what is required of them; this is the most basic tenet of leadership. Equally, they have to know if they are placing undue strain upon them, be it physical or emotional. This is where the medical profession comes into its own and can provide the apposite guidance to ensure a healthy (both mentally and physically) group of subordinates.

This does, however, engender an area of friction; that which the medical officer knows, that which he is permitted to reveal and that which the commanding officer wishes to know may be at variance. This is why the problems of confidentiality were mentioned above. While the medical officer should have insight, as an officer, into why the commander wishes to know what they do, the commander does not necessarily, unless a medic him or herself, have the same understanding of the medical officer's position. This, as may well be imagined, has the potential to further exacerbate the perceived intransigence of any medical officer unwilling to compromise patient confidentiality. The onus falls upon the medical officer to educate and gain the full understanding of the two partners within their medical remit, that of the patient and the commander, so that they may fulfil the 'ideas, concerns and expectations' of both parties without compromising their personal ethical construct.[19]

The other area of friction that may compromise the professional relationship between medical officer and commander is the well-intentioned but ill-advised order to carry out a procedure or process that compromises medical ethics, be it subverting triage to the detriment of wounded enemy combatants or suggesting access to medicines against professional advice. A further example would be the threat to withdraw or withhold treatment in order to glean information from an enemy combatant: the information may well prove to be vital and lifesaving in the short term but such action would be contrary to numerous legal and ethical constructs. The commander may not be fully aware of the mass of medical law that focuses a medical officer's actions so, once again, the education process falls to the medical officer. This should not be seen as onerous but rather as a chance to enter into a dialogue with the commander to glean their intentions and what they wish to know while allowing an insight into the medical officer's constrained framework of ethics. The following section looks at some of the moral and ethical problems that may be encountered in operational medicine and within the remit of the medical commander.

Moral duty and the medical officer

The medical commander (Comd Med in military jargon) in any deployed role is tasked with keeping the commander informed of medical issues that may affect the efficiency and operational capability of the unit. Comd Med has to maintain a watching brief over all the subordinate medical units ranging from the Role 3 field hospitals to the regimental aid posts and dressing stations nearer the frontline. This role may not even have a clinical requirement for the Comd Med, who instead may be required to use their medical knowledge to collate and order information ensuring that both the overall commander and the medics under their own particular command get what they need at the right time.

> *I will apply dietetic measures for the benefit of the sick according to my ability and judgement; I will keep them from harm and injustice.*[20]

Comd Med will also be responsible for making sure the field hospitals are run appropriately by their own commanding officers and that appropriate levels of clinical governance and clinical care are provided for all entitled personnel. Technically this is limited to the provision of care for military personnel of friendly or enemy forces. Within deployed United Kingdom forces, no formal specific provision is made for the care of noncombatants and children. However, it will inevitably occur that as soon as the tents with the Red Cross appear, the local populace will seek aid. It is at this point that the Comd Med has to give clear guidance and support to their medical teams as well as letting higher command know what they intend to do and how they intend to achieve it. With a potential new drain on the logistic train, is it reasonable to start treating persons for whom one may not have the right kit or training? For example, while paediatric equipment is now a part of UK medical stores, the provision of paediatric training is still being addressed. If one does start treating the local populace, what happens if supplies run out or the field hospital is moved in the course of operations? Having set a precedent for treatment how does one continue to fulfil a newly generated moral imperative? Equally, is it correct to bring a level of medicine to the local populace that cannot or will not be sustained after withdrawal from the theatre? What does one do if faced with such a disparity with an injured person from the host nation. Does one provide treatment that would only delay death because follow-on treatment would not be available to keep the person alive? This would be evident in the case of, for example, a spinal injury, usually survivable in the developed world but less so in developing countries that do not have the facilities to rehabilitate or even prevent bed sores. This sort of ethical dilemma arises at a more distant, political, level and reflects external constraints that have a direct impact on what care is provided, to whom, by whom and with what intent.

The moral duty starts from the basic principle of the physician, espoused within the Hippocratic Oath, to do no harm. The medical commander has to interpret this imperative within the context of the operation and facilities available. The tools used to guide such a duty will come from their personal ethical construct, be it purely deontological or teleological or one of the many other modalities of dissecting ethical problems.[21] This brings a personal, and hence unpredictable, element into the duty required of Comd Med and how it is expanded to their higher command elements as well as the units under their command. Again this is a potential cause of friction, not only between Comd Med and higher command but also the other medical practitioners within the medical units – each will have their own ethical construct which may be impaired or constrained by that of Comd Med.

> *Whatever houses I may visit, I will come for the benefit of the sick, remaining free of all intentional injustice, of all mischief and in particular of sexual relations with both female and male persons, be they free or slaves.*[22]

To make this matter easier, though, there are certain elements of medical care that are clearly enshrined within law, especially the

Geneva Conventions (which, having been ratified by a majority of nations, give a robust legal framework for these medical ethics).[23] Most of these centre on the provision of treatment in an unbiased and open manner to all who are in need thereof.[24] This will encompass noncombatants and enemy combatants as well as those of friendly forces. It would appear clear-cut what one should do and how it should be done; however, what happens if higher command starts to interfere in the treatment process by demanding that enemy casualties are treated later than friendly casualties? Or what if the treatment is delayed or used as a coercive tool to glean information from an enemy prisoner, requiring them to give information before being treating for their injury? The Geneva Conventions are very clear on this but examined from a consequentialist point of view, there may be an argument that lives can be saved if vital information is gleaned. Looked at from the deontological viewpoints of beneficence and nonmaleficence then the prisoner has an absolute right to treatment in the appropriate course of time relative to their injuries.[25] The prime duty, according to the Geneva Conventions and the Hippocratic Oath, would be the treatment and wellbeing of the patient. However, there may be the perception of a conflicting duty to do the right thing for the wider unit and all the personnel under higher command.[26] At this point the personal ethical construct and conflict between deontology and consequentialism come to the fore. It is unlikely that any responsible, or reasonable, commander would ask such things of their Comd Med, but consideration must be given to such an eventuality to allow reflection on how one would deal with such a situation and the potential outcomes.

To put the elements of friction generated by the above conflict of interest in a different, but more common, context: when is it reasonable to divulge privileged information to command without the patient's explicit consent?[27] Both the GMC and the RCGP emphasize the primacy of patient confidentiality: 'patients have a right to expect that information about them will be held in confidence by their doctors',[28] and 'the context of your work within a defined community means that confidentiality is of exceptional importance'.[29] Against these rather didactic invocations lies the ability to divulge information in specific and heavily constrained circumstances: 'disclosures without consent to employers, insurance companies, or any third party, can be justified only in exceptional circumstances, for example when they are necessary to protect others from risk of death or serious harm'.[30]

However, there is a wider community at play to which the medical officer owes a moral duty and the reference in the duties of a doctor section above detailed that confidentiality may be compromised in the interests of the 'unit as a whole'. This is further enabled by the recognition that, 'the responsibility of the Commanding Officer and the mutual involvement of the Medical Officer for the health and welfare of the unit as a whole may necessitate the disclosure of restricted medical matters concerning individual cases'.[31] With the GMC and RCGP agreeing that information can be released in terms of the 'public good', it could be argued that the community of a ship, field hospital or operational deployment is a specific and uniquely vulnerable subset of 'the public'. As such it would be reasonable to defend such an action from the points of view of the PLAGOs and the Queen's Regulations for the Royal Navy, as well as the GMC and RCGP guidelines. Having achieved some degree of legal backing for such an action, does it fulfil the requirements of one's personal ethical construct? This is more likely as the deontological approach can now be mitigated by the application of the legal 'get out clause' of serving a community in need and permitting a, now, legitimate action. Consequentialist-motivated medical officers would continue to be comfortable in the greater ends of the whole deployed community being served.

To give an example of a potential area of friction between command and the medical officer it may be worth considering the following scenario. A commander has instigated a plan for counselling of victims and casualties of major incidents, against the advice of the medical officer and evidence of best medical practice. It is known that such immediate counselling, if not peer-led and appropriately guided, does not decrease the risk of post-traumatic stress disorder but may actually increase the risk of it occurring. If a major incident occurs, what is the medical officer to do? On one hand there is an imperative to obey the orders of the commander but on the other hand the medical officer has to work for the benefit of the patients in order to protect them from harm. If the commander were to discover that the plan was not being followed and issued a direct order to enact the plan, could this be viewed as a lawful order? The medical officer would then have to consider carefully their potential courses of action: is the order obeyed or are the patients protected from potential harm? The patients are not in a position to exercise their own ideas, concerns and expectations and are, effectively, disenfranchised from the decision-making process. Clearly, disobeying a direct order is a serious issue and any such action may need to be justified in front of a court-martial. Is the protection of patients from harm sufficient? In this example would the action be supported by higher elements of the medical command chain? With regard to the patients' autonomy in such a situation, the decision would be out of their hands, so the test would be: did the medical officer act with 'beneficence-in-trust'[32] on their behalf?

From the example above it is clear that what may appear to be a simple situation may well develop into a wider and more complex scenario with disturbing rapidity, forcing the military medical officer to reflect and consider carefully the full impact of actions taken. How to prepare for this is inherently very difficult; the societal and personal context of the physician as well as their ethical construct will have a bearing upon the outcome for a problem. The Swann Scenario is an exercise designed to expose and explore the ethical difficulties that may arise during military operations. The basic situation is a 'Cold War' scenario in Europe where a casualty station is about to be overrun by enemy troops, who have been reported to be killing those wounded and dying that they encounter. The commander has ordered the medical officer to withdraw from the casualty station as soon as practicable. The significant problem that needs to be considered is what to do with the many unmoveable casualties at the casualty station. When this was presented at a medical conference it gleaned three basic courses of action. These were: to leave the patients behind and hope for the best; to stay in place to tend the wounded soldiers no matter what occurred; or to administer opiate overdoses to grant a swift and painless death to those deemed unsalvageable; but it was noted that, 'there are almost infinite variations of these'.[33]

> *I will neither give a deadly drug to anybody if asked for it, nor will I make a suggestion to this effect.*[34]

Valid concerns lie around the 'doctrine of double effect',[35] classically described as the giving of a drug to ease suffering that, as a side effect, may shorten the life of the recipient (one of the suggested solutions to the Swann Scenario above). This scenario is played out on a regular basis in primary and secondary care units across the world and the physician has to weigh each instance upon the needs of the patient and their personal ethical construct. Palliative care is vital to enable the autonomy of a patient to choose the location and manner of their death; the use of opiate analgesics can help achieve this but may hasten their demise. Is this an illegal act of killing or is it the art of removing suffering with a foreseeable, but welcome, side effect? In the military scenario, what if the patient has no hope of survival and, in the course of pain, is moaning and likely to give away the position of the patrol – what is the rationale behind the excessive use of opiates at that stage? Again reflection on the ethical construct of the physician is imperative to ensure that they comply with their own personal ethical construct and fulfil the moral duty to their charges – both the patient and the unit with which they are serving.

A further area of concern exists in the use of medical personnel within the interrogation process. This may be relatively innocent through the provision of medical care to detainees, but may become more 'ethically ambiguous' when straying into 'active' interrogation techniques. Physicians and psychiatrists have been involved in such activities over the years, despite the clear incompatibility and illegitimacy of any partnership between medicine and torture. The World Medical Association's Declaration of Tokyo[36] has been used by medical societies around the world as a model for their own codes.[37] To cite just one example, in 1982 the UN agreed the Principles of Medical Ethics Relevant to the Role of Health Personnel, Particularly Physicians, in the Protection of Prisoners and Detainees against Torture and Other Cruel, Inhuman or Degrading Treatment or Punishment:

> Principle 2: It is a gross contravention of medical ethics, as well as an offence under applicable international instruments, for health personnel, particularly physicians, to engage, actively or passively, in acts which constitute participation in, complicity in, incitement to or attempts to commit torture or other cruel or degrading treatment or punishment

The absolute, deontological prohibition under all circumstances of such action (or inaction) could not be any clearer than:

> Principle 6: There may be no derogation from the foregoing principles on any ground whatsoever, including public emergency.

In accordance with this clear statement of principle, the American Psychiatric Association have adopted the position that involvement in interrogation or even using background knowledge of medicine, psychiatry or of the patient's medical history is prohibited.[38] However, it is worth noting that the US military have in the past taken a different view, claiming that psychiatrists may be involved in providing advice regarding interrogation and that this input is, 'warranted by compelling national security interests'.[39] This worrying difference of opinion arises directly from the tensions between an alleged military imperative and the medical ethical imperative. Despite medical codes of ethics, domestic and international legal agreements and the personal ethical construct of the healers involved, unfortunately, medical complicity in prisoner abuses are well documented.[40]

Inferences

It can be seen from the scenarios and situations discussed above that the medical and military professions have areas of friction which are difficult to overcome. There is a perception that the role of a physician (to preserve life) is diametrically opposed to that of the military (to use violence for national, political, goals).[41] At a superficial level this may be true, but deeper examination of the issues could lead one to conclude that the basic ethos behind the physician and the military is to transcend one's will to achieve the best for the patient or the state. Both are charged with the wellbeing of a group of personnel and may be forced to take decisions regarding their health and wellbeing; equally the execution of their duties may take away their own autonomy and place another's at the core of their actions, be it the state or the patient.[42]

> The doctor is in a professional force field, with loyalties to both the organisation and the patient. His knowledge of the organisation and the responsibilities involved can also help the patient in terms of enabling individual needs ... Even though he cannot please everyone all the time, the doctor's actions must be characterised by transparency and reliability.[43]

The above quote gives a clear view that the physician really has only one recourse once a moral dilemma has been recognized; that being their own personal ethical construct that will guide their thought processes when evaluating the situation. Be it grounded in deontological, consequentialist, teleological or one of the other evolutions of the basic ethical principles, they must be able to apply it consistently and clearly, within the specific context of the situation in which they find themselves.[44] This will invariably be a product of all the personal and societal influences that have come to play during upbringing and education and will mature further with reflective practice and consideration of similar problems.

It is virtually impossible to imprint, or expect, one physician to have precisely the same ethical construct as one of their colleagues; the variance of human nature militates against this. Consideration of ethical issues, dilemmas and coping strategies, however, will have been given during the lengthy training process which should habituate the physician to the process of reflecting upon moral dilemmas. It may well have been tested as part of the under-graduate or post-graduate curriculum.[45] There is no current modality of testing this within military medicine, even though the legal basis for decision-making does form an element of the medical officer training for all forces; this could well form a future part of the curriculum.

Given the medical predilection for simple acronyms to help with basic treatment, a prime example being the ABCDE of emergency medicine (Airway, Breathing, Circulation, Disability, Exposure) a similar one has been created for medical ethics:[46]

A   Autonomy
B   Beneficence
C   Confidentiality
D   Do no harm
E   Equality (justice)

The relative importance of the patient's autonomy is evident as is the continuing need for confidentiality. The fact that such aides-memoire exist shows the extant need for training and reflection in matters pertinent to medical ethics. As stated above, no one physician has the right or wrong answer, every situation will also have its own context that may alter the perspective and so no two outcomes will be the same. It still behoves the physician to be able to reflect upon and rationalize the decisions made in relation to their own ethical construct, guided by previous education and experience.

Conclusion

The essential, and primary, moral duty is to the patient; no matter to which ethnic, religious or combat grouping they may belong. The patient's autonomy may be compromised by the authoritarian constraints of the military but it is still beholden upon the physician to apply a firm moral code to ensure the best possible outcome for the patient. There is a further moral duty to higher command that needs to be discharged with the primary duty in the back of the mind at all times and an awareness of the legal and ethical constructs that may permit marginal compromises to ensure the health not only of the individual but the wider community within which that individual is placed. The key factor that may define the situation and guide outcomes will be the context in which the situation is unfolding; as each scenario develops it is likely to hold both common and unique elements when placed alongside another scenario, as will be the case for the physician and patient concerned. This will inevitably lead to a unique outcome for each context against a background of uniform guiding principles.

The art of being a reflective medical practitioner directly aligns with the art of being a good military medical officer. The background knowledge and ethical approach inculcated at medical school and during further professional training should provide a robust framework on which to base well-founded, coherent and transparent decisions regarding most situations. The ability to reflect upon and review events and tease out the common strands that may form a future framework for decision-making is essential in this respect. The inherent uncertainty of medical practice reinforces the need for periods of reflection to define what has or has not worked to the benefit of the patient. Ethical dilemmas form an integral part of this process and the lack of 'correct' answers clarifies the importance of context within any proffered solution. Numerous tools now exist to guide physicians through this process and it is becoming enshrined in the annual appraisal undertaken by all clinicians, both military and civilian in the UK.

The recognition that a specific situation is beyond one's experience should prompt a search for alternative views and advice from medical or military mentors, possibly those involved within the appraisal process. Being aware of the dichotomy inherent in the professional roles of the physician and the soldier is the first step on the road to dilemma dissection. Equally, being aware of the impact of the personal ethical construct is pivotal in maintaining an ethical and moral approach that is above reproof. The basic tenet still exists; 'To depart from the morality of medicine is to repudiate what it is to be a physician.'[47]

If I fulfil this oath and do not violate it, may it be granted to me to enjoy life and art, being honoured with fame among all men for all time to come; if I transgress it and swear falsely, may the opposite of all this be my lot.[48]

Notes

1. Edmund D. Pellegrino, 'The Moral Foundations of the Patient–Physician Relationship: The Essence of Medical Ethics', in Thomas E. Beam and Linette R. Sparacino (eds) *Military Medical Ethics*, 2 vols (Virginia: Office of the Surgeon General, United States Army, 2003), p. 6.
2. *The Medical Act 1983 (Amended) Order 2002* (London: The Stationery Office, 2002).
3. In the UK, this is under the auspices of the Queen's Regulations for the Royal Navy, Army and Air Force. See: Ministry of Defence, *The Queen's Regulations for the Royal Navy* (London: The Stationery Office, 1997); Ministry of Defence, *The Queen's Regulations for the Army (Including Amendment 27)* (London: The Stationery Office, 1996); Ministry of Defence, *The Queen's Regulations for the Royal Air Force* (London: The Stationery Office, 1999).
4. This personal ethical construct is analogous with the idea of a 'moral compass' which serves to guide a person's decisions.
5. This may even have been tested at some point during their examinations for their respective professional bodies.

6. General Medical Council, 'Good Medical Practice' (London: General Medical Council, 13 December 2006), inside front cover.
7. Edward Roddy and Elin Jones, 'Hippocratic Ideals are Alive and Well in the 21st Century', *British Medical Journal*, 325, p. 496.
8. Royal College of General Practitioners, 'Good Medical Practice for General Practitioners' (London: Royal College of General Practitioners, September 2002), p. xi.
9. Pellegrino, 'The Moral Foundations', p. 6.
10. General Medical Council (2006), 'Good Medical Practice', Para. 37.
11. Royal College of General Practitioners, 'Good Medical Practice for General Practitioners' (2002), pp. 23–4.
12. John Macleod (ed.), *Clinical Examination: A Textbook for Students and Doctors by Teachers of the Edinburgh Medical School*, 6th edn (Edinburgh: Churchill Livingston, 1983), p. 1.
13. General Medical Council, *Confidentiality: Protecting and Providing Information* (London: General Medical Council, September 2000), Para. 2.
14. Personnel, Legal and General Orders, 106, *Medical Information – Disclosure to the Command*, Para. 2.
15. Queen's Regulations for the Army 1996, Chapter 3, Para. 3.001.
16. Anthony E. Hartle, 'The Profession of Arms and the Officer Corps', in Beam and Sparacino, *Military Medical Ethics*, p. 147.
17. Hartle, 'The Profession of Arms and the Officer Corps'.
18. *Oxford English Dictionary* Online, *Profession*, from http://dictionary.oed.com/cgi/entry/50189444?query_type=word&query_word=profession&first=1&max_to_show=10&single=1&sort_type=alpha, accessed 31 December 2008.
19. Neighbour Roger, *The Inner Consultation: How to Develop an Effective and Intuitive Consulting Style*, 2nd edn (Oxford: Radcliffe, 2004), p. 55.
20. Pellegrino, 'The Moral Foundations', p. 6.
21. David C. Thomasma, 'Theories of Medical Ethics: The Philosophical Structure', in Beam and Sparacino, *Military Medical Ethics*, pp. 28–9.
22. Pellegrino, 'The Moral Foundations', p. 6.
23. For example: Geneva Convention III Relative to the Treatment of Prisoners of War, 1949, http://www.icrc.org/ihl.nsf/7c4d08d9b287a4214 1256739003e63bb/6fef854a3517b75ac125641e004a9e68;   Geneva Convention IV Relative to the Protection of Civilian Persons in Time of War, 1949, http://www.icrc.org/ihl.nsf/385ec082b509e76c41256739003e 636d/6756482d86146898c125641e004aa3c5, accessed 1 March 2010.
24. Lewis C. Vollmar, 'Military Medicine in War: The Geneva Conventions Today', in Beam and Sparacino, *Military Medical Ethics*, p. 767.
25. Thomas E. Beam, 'Medical Ethics on the Battlefield: The Crucible of Military Medical Ethics', in Beam and Sparacino, *Military Medical Ethics*, Case Study 13–5, pp. 397–8.
26. For an explanation of deontological and consequentialist approaches, see Chapter 1.
27. Edmund G. Howe, 'Mixed Agency in Military Medicine: Ethical Roles in Conflict', in Beam and Sparacino, *Military Medical Ethics*, pp. 344–6.

28. General Medical Council (2006), 'Good Medical Practice', p. 20.

29. Royal College of General Practitioners, 'Good Medical Practice for General Practitioners' (2002), p. 23.

30. General Medical Council (2000), *Confidentiality*, Para. 35.

31. Ministry of Defence, *The Queen's Regulations for the Royal Navy* (London: The Stationery Office, 1997), Chapter 15, Section 1501.

32. Thomasma, 'Theories of Medical Ethics', pp. 40–1.

33. Beam, 'Medical Ethics on the Battlefield', p. 388.

34. Pellegrino, 'The Moral Foundations', p. 6.

35. Victor W. Sidel and Barry S. Levy, 'Physician-Soldier: A Moral Dilemma?', in Beam and Sparacino, *Military Medical Ethics*, p. 300. See also Chapter 4 in this work.

36. World Medical Association's Declaration of Tokyo, 1975 (formally, the 'Guidelines for Medical Doctors Concerning Torture and Other Cruel, Inhuman or Degrading Treatment or Punishment in Relation to Detention and Imprisonment'), available at http://www.wma.net/en/ 30publications/10policies/c18/index.html, accessed 25 March 2010.

37. For example: the International Council of Nurses' position on *Nurses' Role in the Care of Prisoners and Detainees*, available at http://www.icn.ch/psdetainees.htm; The American Medical Association, *E-2.067 Torture*, available at www.ama-assn.org/meetings/public/ annual05/10a05.doc, accessed 1 March 2010.

38. American Psychiatric Association, 'Psychiatric Participation in Interrogation of Detainees: Position Statement', May 2006, Available at http://archive.psych.org/edu/other_res/lib_archives/archives/200601.pdf, accessed 1 March 2010. See also Jonathan H. Marks and M. Gregg Bloche, 'The Ethics of Interrogation: The US Military's Ongoing Use of Psychiatrists', *New England Journal of Medicine*, 359(11), pp. 1090–2.

39. Department of the Army, 'Behavioural Science Consultation Policy', OTSG/MEDCOM Policy Memo 06-029, dated 20 October 2006.

40. For a detailed account, based on public testimony and government documents, of the role played by members of the medical profession in the torture and abuse of prisoners and detainees, see Steven H. Miles, *Oath Betrayed: America's Torture Doctors*, 2nd edn (Berkeley: University of California Press, 2009). While this work focuses specifically upon US medics, such practices are regrettably rather more widespread.

41. William Madden and Brian S. Carter, 'Physician-Soldier: A Moral Profession', in Beam and Sparacino, *Military Medical Ethics*, p. 289.

42. Madden and Carter, 'Physician-Soldier', pp. 279–80.

43. A. J. van Leusden, and M. J. J. Hoejenbos, 'Moral Dilemmas in Military Health Care', in Th. A. van Baarda and D. E. M. Verweij (eds) *Military Ethics: The Danish Approach, A Practical Guide* (Leiden and Boston: Martinus Nijhoff, 2006), p. 210.

44. Thomasma, 'Theories of Medical Ethics', pp. 28–35.

45. For example, it is currently a feature of the membership examination for the RCGP: Royal College of General Practitioners, *Clinical Skills Assessment*, available at http://www.rcgp-curriculum.org.uk/mrcgp/ csa.aspx, accessed 4 March 2010.

46. Medical Educator, *The ABCDE of Medical Ethics for Medical Students*, available at http://www.medicaleducator.co.uk, accessed 31 December 2008.
47. Pellegrino, 'The Moral Foundations', p. 17.
48. Pellegrino, 'The Moral Foundations', p. 6.

Chapter 11

The Ethics of Nuclear Deterrence

SRINATH RAGHAVAN

Introduction

This chapter considers the ethical questions surrounding the strategy and policy of nuclear deterrence. Nuclear weapons are rightly held to have constituted a revolution in warfare, challenging and altering the traditional notions of strategy and war-fighting. But equally they have called into question our settled views about ethics and the use of armed force. The development of ethical thinking about nuclear weapons has run parallel to, and has been influenced by, strategic thinking about these weapons. The chapter therefore begins by examining the strategic idea of deterrence, particularly in the nuclear context. It then goes on to discuss the three main ethical approaches to nuclear deterrence: consequentialism, Just War theory and deontology (see Chapters 1 and 4). The chapter suggests that nuclear deterrence presents a real moral quandary; for our commonly held approaches fail to give us clear-cut ethical guidance in this realm of conflict. Extricating ourselves from this bind is unlikely to be an easy task.

Deterrence: nuclear and conventional

The word deterrence derives from the Latin root 'terre', which means to frighten. As its origins suggest, deterrence is based on the idea of scaring away one's opponents. As a strategy, deterrence is the use of threats to dissuade an adversary from initiating an undesirable act. The term deterrence, however, is also used to describe a relationship between two actors. In this sense, it refers to a situation where, owing to the use of a strategy of deterrence by both sides, neither party is likely to initiate actions that are inimical to the other side. The distinction between deterrence as a strategy and as a relationship needs to be borne in mind, although the term will be used in both senses in this chapter.

Thus defined, deterrence is a strategic concept that clearly has been relevant throughout history. Yet systematic thinking about deterrence only commenced in the years following the Second World War. The advent of the nuclear age, and more specifically the acquisition of nuclear weapons by the Soviet Union as well as the United States, pushed the issue of deterrence to the forefront of strategic debates. The devastating effect of nuclear weapons on Hiroshima and Nagasaki left few doubts about the catastrophic consequences that would ensue from their use in any future conflict. The subsequent testing and introduction of thermonuclear weapons buttressed and accentuated these concerns. In consequence, scholars like Bernard Brodie realized that in the nuclear age total war could not be waged: it could only be threatened to avert the possibility of war. But this begged a question that dogged the study of nuclear deterrence: if a nuclear war could not be fought, how could it credibly be threatened?[1]

While strategists grappled with the credibility of the threat of nuclear retaliation, philosophers and theologians began to engage with the ethical dimension of strategic deterrence. The arguments advanced by them were many and varied; but they all rested on an understanding that nuclear weapons at once represented a revolution in modern warfare and a momentous ethical challenge. For one thing, the destructiveness of these weapons was unprecedented in the history of warfare. The use of nuclear weapons could not just inflict a high-level of damage but could put an end to social order itself. For another, the use of nuclear weapons would have long-term effects, harming several generations to come. These effects would be both at the individual and environmental levels. It is impossible, therefore, to deny that a nuclear war (even in defence against aggression) would be an unrelieved material and moral catastrophe.

What about nuclear deterrence? Is it immoral to threaten nuclear retaliation in order to prevent an adversary from using nuclear weapons or from embarking on a war? Before we examine how different ethical traditions have approached this issue, it is worth considering how nuclear deterrence is different from conventional deterrence, and why it poses special ethical challenges. At one level, the answer appears to lie in a judgement of what would happen if deterrence failed. The awesome destructive capacities of nuclear weapons means that the failure of deterrence would have consequences of an altogether different order from those following a failure of conventional deterrence.

At another level, however, it is possible to discern a more fundamental difference between nuclear and conventional deterrence. A strategy of deterrence could assume two forms: punishment and denial. The former aims to deter an adversary by using the threat of retaliatory damage. The latter aims to deter by promising to prevent the adversary from taking the undesirable step. Conventional deterrence relies largely (if not exclusively) on denial. By erecting robust defences and by keeping our armed forces in readiness, we threaten to foil any offensive move by the opponent. Conventional deterrence cannot rely solely on a strategy of punishment, for it is not possible to threaten destruction of the opponent's society without defeating its armed forces in battle. Of course, it is possible to inflict some damage on the opponent's society by resorting to strategic bombing or conventional missile strikes. But, as the experience of the Second World War suggests, these capabilities alone are unlikely to deter an adversary. Nuclear deterrence by contrast cannot be based on a strategy of denial, for no credible system of defence against nuclear weapons has yet been designed. Conversely, with nuclear weapons it is possible to punish the adversary's society without having to engage his armed forces. Here lies the crucial distinction between conventional and nuclear deterrence. Unlike the former, nuclear deterrence directly threatens the destruction of the opponent's society.

Nuclear weapons could, of course, be used in the battlefield in response to an act of conventional military aggression. Even so, the collateral damage will be exceptionally high. It is unlikely that the defending country would want to use them on its own territory; for the harm to its own populace would be impossible to mitigate. This would create incentives to use battlefield nuclear weapons against the enemy's forces inside his territory. This in turn would result in colossal damage to the enemy's population. In short, using nuclear weapons for purely military purposes seems a difficult proposition.

Taken together, then, it is the *threat of unacceptable damage* to the adversary's *population* (as opposed to his armed forces) that marks out nuclear deterrence as a strategy that poses entirely different moral challenges.

The consequentialist case

The consequentialist approach evaluates a choice or an action solely in terms of the overall value or disvalue of its consequences for all of humankind (see Chapter 1). The approach is by definition comparative. From the consequentialist perspective, the moral action is the one that has the most value or the least disvalue among the available alternative courses. Such an approach can evidently be congenial to decision-makers. Yet it differs from realism or other forms of prudential reasoning. The underlying principle of realism is the pursuit of national interests – a quest in which the use of force is deemed acceptable. The consequentialist approach, however, requires us to consider the consequences as they affect all human beings and not any particular subset of them. Thus considerations of national interest or other narrow forms of self-interest may or may not accord with consequentialism. In the issue of nuclear deterrence, there is a good case to be made for convergence between strategic and consequentialist reasoning.

For a start, we may note that the consequentialist approach seems particularly appropriate in discussing nuclear deterrence. Choices in this area of strategy affect the possibility of a nuclear war and its devastating aftermath. The sheer magnitude of the consequences makes it difficult both to ignore them and to take seriously a moral standpoint that does not reckon with them.[2] To be sure, this is unlikely to be the case with a conventional conflict in the absence of nuclear weapons. But in discussing nuclear weapons policy, the consequences invariably force themselves upon our moral reasoning.

The consequentialist understanding of nuclear deterrence follows a three-pronged approach.[3] It begins by examining whether the strategic logic of nuclear deterrence comports with a moral perspective. It goes on to consider the alternatives to deterrence: unilateral or mutual disarmament. Finally, it weighs the overall value and disvalue of each of these options.

The strategic argument holds that nuclear deterrence is most effective in precluding war and defending national sovereignty. The efficacy of deterrence stems from the fact that mutual deterrence threatens unacceptable societal damage to both the antagonists. This condition of mutual vulnerability imposes a great degree of caution on both sides, and so obviates the possibility of a major war or a nuclear exchange. The consequentialist argument considers the implications of this logic for humanity as a whole. Nuclear deterrence makes it clear to both parties that there is no prudential advantage to be gained by resorting to war. Deterrence, in other words, has a stabilizing effect on a conflictual relationship.[4] The consequences of this stability are of great value to all of humankind, and not just to the states engaged in deterrence. Indeed, the entire world benefits from the fact that the chances of a war between two nuclear-armed nations are low – not least because of the potential for widespread damage in a nuclear exchange.

What about the alternatives to deterrence? Take the policy of unilateral disarmament. Consider what might happen if state A decides to renounce its nuclear arsenal while its rival state B continues to hold on to its own nuclear weapons. B is likely to take advantage of its preeminent military position in order to advance its own interests at the expense of A (while we do not need to impute particularly malign intentions to B, a realist view in particular would suggest that any state is likely to seek further its own interests vis à vis its rivals, especially in the absence of a countervailing power. To be sure, states might not always behave in this fashion, but in an anarchic international system there are strong incentives to do so).[5] In so doing, B need not resort only to war, but might use the coercive threat of its nuclear weapons to obtain concessions, or even simply rely on the unspoken latent threat that exists simply through the possession of the weapon to 'enhance' its diplomatic position when making demands. In the absence of a nuclear arsenal A will have few means to resist such nuclear blackmail. The defensive measures which are otherwise morally permissible are unlikely to have any effect in these circumstances.

Furthermore, if A and B are great powers, the former's decision to forsake nuclear weapons will have ripple effects in the international system. A's allies and other third parties who benefited from 'extended deterrence' will begin to feel vulnerable and will likely begin to consider developing their own nuclear deterrent. Those countries that already have nuclear weapons might seek to enhance their arsenal. These measures could prove destabilizing.[6] Moves by third parties to acquire nuclear weapons might be viewed with concern by B, who might then consider pre-emptive use of force to stop them from achieving this capability. The resulting conflicts might well turn nuclear. If such a crisis broke, A might reconsider its decision to abandon nuclear weapons and seek to rearm. This in turn could pre-cipitate a conflict between the two great powers with all its potentiality of escalating to nuclear use.

While these points are generic, they were entirely plausible in a context such as the Cold War. Even today, these dynamics could work in a situation involving the United States and China. If the former abandons its nuclear arsenal, Japan – despite its status as the sole victim of nuclear weapons and its pacifistic strategic culture – will be compelled to consider the option of acquiring its own deterrent vis à vis China. This is, of course, not inevitable and will depend on the state of Japan's relations with China. But short of a situation where Japan manages to build a relationship with China that entirely over-comes the legacy of past rivalry (like France and Germany have managed to do in this respect), the need to secure some form of deter-rence will necessitate a serious review of its approach to nuclear weapons. China, in turn, will view this development with grave concern. In short, if one of two great powers relinquishes its nuclear deterrent, the resultant instability could have serious negative conse-quences for the international order as a whole.[7]

What would be the consequences of mutual abandonment of nuclear weapons by the two rivals? Nuclear weapons deter not just the actual use of this capability, but also a major conflict that might escalate to the nuclear level. If deterrence is drained out of a relationship, major war might yet again become a prudential and rational instrument for either side to pursue its interests. Moreover, the forsaking of a nuclear capability is not the same as forgetting how to build one.[8] If a major war looms, the two sides might well scramble to reconstitute their nuclear arsenal. This might happen for purely defensive reasons, as each side believes that the other would try to do so. In these circumstances, the incentives and pressure for a pre-emptive strike is likely to be high. Mutual disarmament might ironically pave the way for an actual nuclear war. Unless accompanied by a normalization of the relationship between the rivals, forswearing nuclear weapons is likely to have deleterious consequences. If peace seems a distant prospect, it appears best to hold on to the nuclear deterrent.

On balance, the consequentialist approach would appear to suggest that nuclear deterrence is the morally appropriate course of action. On the positive side, the threat of mutually assured destruction promotes stability and peace, with all the beneficent consequences for humankind. On the negative side, the alternatives are much worse. Disarmament – whether unilateral or multilateral – could lead to war and could destabilize the international system. In either scenario, the possibility of a nuclear war is much stronger than in the case of deterrence. Anthony Kenny sums up the consequentialist position well:

> the conditional willingness to engage in massacre which is an essential element of the policy is a slight and almost metaphysical evil to weigh in the balance against the good of preserving peace. The moral blemish which this may taint us with in the eyes of the fastidious is at best something to put on the debit side, along with the financial costs of the weapons system, against the massive credit of maintaining our independence and our security from nuclear attack.[9]

Be that as it may, the consequentialist case for nuclear deterrence is hardly unproblematic – particularly when viewed from deontological or Just War traditions. But even in its own terms, the consequentialist position can be assailed for both its approach and its central argument. Given the limitations of our capacity to predict, and the epistemic uncertainty that has already been raised in Chapter 1 in these regards, a consequentialist argument must be considered in terms of likely rather than certain outcomes. Moreover, the assessment of the likelihoods must be rather rough. In consequence, an approach that relies on the expected value of possible outcomes can scarcely warrant a high degree of confidence. The likelihood of events such as aggression or nuclear war may not be determined with sufficient accuracy to make judgements about the efficacy of a nuclear deterrent. As Gregory Kavka observes, 'reliable quantitative utility and probability estimates' about the choice between retaining and abandoning nuclear weapons are simply not available.[10]

Furthermore, the central component of the consequentialist position – that nuclear deterrence promotes stability – can be contested. From an abstract perspective, the logic of nuclear deterrence is convoluted. If a nuclear war is an unmitigated catastrophe, then how can we credibly threaten to wage one? Nuclear strategists might respond that deterrence works better in practice than in theory. But even this is not clear. For one thing, the possibility of accidents or inadvertent escalation cannot be gainsaid. Even during the Cold War these posed significant dangers at various times.[11] The calculation and weighing of consequences, therefore, is not as straightforward as the consequentialist position suggests.

Just War and nuclear deterrence

The Just War Tradition provides the most familiar and perhaps the most widely held argument against nuclear deterrence. As we have seen in earlier chapters, the theory is divided into two logically distinct but related categories: *jus ad bellum* and *jus in bello*, or justice of war and justice in war. It is the latter that suggests most strongly that the practice of nuclear deterrence is immoral. *Jus in bello* comprises of two criteria: proportionality and discrimination. The criterion of proportionality states that the amount of force used should be proportional to the good that it intends to achieve. Put differently, for the use of force to be justified, the good that it is intended to achieve must outweigh any harm that might also be caused. The criterion of discrimination requires force to be used in such a fashion that the distinction between combatants and noncombatants is respected. Deliberate killing of noncombatants is expressly proscribed.

Nuclear deterrence stumbles against both these criteria, but especially the principle of discrimination. As Michael Walzer observes, 'Anyone committed to the distinction between combatants and noncombatants is bound to be appalled by the spectre of destruction evoked, and deliberately evoked, in deterrence theory.'[12] The problem, as Walzer suggests, is the intent to commit genocidal violence that underlies the strategy of nuclear deterrence. The theologian Paul Ramsey concurred with this view: 'Whatever is wrong to do is wrong to threaten, if the latter means "mean to do" ... if counter-population warfare is murder, then counter-population deterrent threats are murderous.'

Writing in the early 1960s, Ramsey believed that there was a way by which nuclear deterrence might comply with the Just War requirement. Interestingly, his views were shaped by the strategic debates of the period. At this point, nuclear strategists were arguing that it was possible to use nuclear weapons in a 'counter-force' role as opposed to a 'counter-value or counter-population' strategy. In a counter-force strategy, nuclear weapons would be used to target the enemy's military installation, logistic chain and fielded forces. The strategic assumption was that this would enable controlled escalation and prevent all-out nuclear war. Picking up these arguments, Ramsay claimed that 'collateral civilian damage' would be an 'indirect effect' of 'a plan and action of war which would be licit or permitted by the traditional rules of civilized conduct in war'.[13]

Ramsay's argument, however, is flawed on several counts. First, it is extremely unlikely that any such strategy could avoid escalation. The opponents' government would be unable to distinguish between a 'limited nuclear attack' and an all-out strike. Friction, as Clausewitz observed, permeates the activity of war. The use of nuclear weapons would greatly exacerbate this problem. Each side would have a considerable problem discerning the other's intentions: sources of information and links to them are highly vulnerable to disruption by nuclear weapons; furthermore, they cannot be sure that the information that they possess is either timely or accurate. It is unreasonable, therefore, to assume that a nuclear war could be kept within bounds. Second, the civilian damage wrought by even a limited counter-force attack would be immense. This is true both of strikes against military installations and armed forces. The number of people killed in the war would exceed the limits set by the principle of proportionality. Third, the term 'collateral damage' obfuscates the role played by civilian casualties in a counter-force strategy. As Ramsay writes at one point, 'The collateral civilian damage ... may itself be quite sufficient ... to preserve the rules and tacit agreements limiting conflict in a nuclear age.' The threat of civilian damage thus plays a central role in this form of deterrence. We are back to the key problem with nuclear deterrence from a Just War perspective.

The Just War Tradition usually positions itself on a middle path between military necessity and absolute moral standards. Hence, most Just War theorists recognize that although threat of nuclear retaliation is immoral, it may be the only way to maintain a precarious and imperfect peace. Walzer, for instance, invokes the notion of 'supreme emergency' to describe nuclear deterrence. Faced with an 'unusual and horrifying' threat to our entire way of life, we may respond in a manner that goes against the principles of justice. In an age of nuclear weapons, he suggests, supreme emergency has become a permanent condition. 'We threaten evil in order not to do it, and the doing of it would be so terrible that the threat seems in comparison to be morally defensible.'[14]

This, of course, brings the Just War approach uncomfortably close to a consequentialist position. Walzer, the foremost Just War theorist of our times, is too honest to deny this. As he concedes: 'Nuclear weapons explode the theory of just war. They are the first of mankind's technological innovations that are simply not encompassable within the familiar moral world.'

The deontological argument

Deontology is a tradition of ethical reasoning that focuses on the character of the intentions and motivations driving an action rather than the nature of the outcomes of those actions. Arguments from a deontological standpoint approach the question of deterrence in several ways. Many of them begin much as the Just War arguments do – by emphasizing the importance of distinguishing between combatants and noncombatants. They also agree with the Just War theorists that it is immoral to threaten something that would be immoral to actually do. But unlike the Just War theorists, they are reluctant to accord overriding importance to claims of necessity. Deontology, after all, is concerned with the intrinsic moral character of action.

Some deontologists contend, moreover, that the manner in which Just War theorists frame the issue is problematic. By focusing excessively (not to say exclusively) on the intention to retaliate, Just War theorists do not consider the strategy in its own right and in its entirety. Instead, they examine it only in relation to another action to which it might lead – the use of nuclear weapons.[15] Douglas Lackey, for instance, argues that a deontological analysis should evaluate the activity of threat-making and not merely its conditional intent. Making nuclear threats involves the creation of a risk that large numbers of civilians would perish in one's retaliation. To be sure, deterrence creates this risk owing to its conditional intention to strike. Yet it is the creation of risk, and not the intention to carry it out, that should be the central focus of moral enquiry. As Lackey argues:

> when we consider the act that is risked, we are considering ... an act that someone intends to perform if and only if, worst come to worst. But when we consider the act of risk creation, we consider a directly intended act, an act which is a means to a desired end.

By so doing, we can obtain a better insight into the moral character of the strategy.[16]

Steven Lee pursues this line of enquiry by considering nuclear deterrence as a form of hostage holding.[17] Hostages are individuals who are threatened with harm to influence the behaviour of another party. The people threatened are thus third parties in the situation. Hostages are deemed innocent because they are neither the agents of the behaviour that the hostage takers seek to influence, nor are they responsible for that behaviour. Moreover, hostages are usually held captive against their wishes. Hostage holding is morally unacceptable because '(1) it imposes a risk of harm (2) upon innocent persons (3) without their consent'. Taken as a whole, hostage holding is a clear example of treating persons as mere means and not as ends – an action that goes against fundamental principle of deontological perspectives such as Immanuel Kant's 'categorical imperative' (see Chapter 8 for further discussion on this).

Nuclear deterrence exhibits all the features that make hostage holding morally incorrect. In a condition of mutual vulnerability, each side threatens the other to kill large numbers of its civilians and to destroy its society. The bulk of the people being threatened are third parties to the transaction and do not control the behaviour that each side seeks to control. Nor yet do they consent to being threatened in this fashion. If the threat has to be perceived as credible, it cannot be a bluff. Since there is no assurance that the threat will succeed, retaliation might well occur. The risk is magnified by the possibility of accidental or inadvertent nuclear exchange. Owing to the creation of the risk together with the innocent and nonconsenting status of the people, nuclear deterrence violates the principle of not treating human beings as mere means.

It bears emphasizing that the use of third-party threats differentiates the moral status of nuclear deterrence from second-party threats, which may, at times, be morally permissible. For example, it may be acceptable to threaten a child with a beating if he or she does not rush out of a burning building. It is also important to underscore the point that the analogy of hostage holding does not break down simply because the persons threatened do not lose their liberty. Keeping hostages under custody is merely the most effective way to place them under threat. Nuclear weapons can do so without this requirement. In short, when viewed from this deontological perspective, nuclear deterrence is morally prohibited.

Some scholars have attempted to reduce the gap between nuclear deterrence and deontology. The best case has been advanced by Jeff McMahan,[18] who suggests introducing a deontological principle that 'it is wrong, other things being equal, to risk doing that which it would be wrong to do'. McMahan goes on to argue that it need not be absolutely forbidden to risk doing something it would be wrong to do.

Hence it is not necessary that the pursuit of a policy of deterrence must be as wrong as the actual use of nuclear weapons. The problem with such a weaker or nonabsolutist deontological position is that the final moral judgement turns on 'how great the risk is'. This, of course, brings us back to a consideration of consequences. It is not surprising that most deontologists steer clear of such a position and insist that nuclear deterrence is morally proscribed.

A moral conflict

Ordinary moral reasoning is pluralistic. When confronted with situations demanding a moral choice, individuals have moral intuitions of various kinds – in particular, consequentialist and deontological. When these intuitions conflict, they also recognize the need to weigh them against each other in their deliberations. The overall moral status of an action is usually determined by whether the consequentialist or deontological or another approach to assessing that action carries greater weight (see Chapter 1). For instance, there is a deontological requirement to keep one's promises, and yet there are times when keeping one's promise is not morally preferable in terms of its consequences. When there is a conflict, one weighs the differing assessments in order to decide the morally appropriate course of action. A moral conflict is resolved when it is clear, despite the conflict, what should be done morally.

Nuclear deterrence poses a moral conflict. The consequentialist argument shows deterrence to be morally necessary; the Just War argument gives way to the consequentialist position; and the deontological argument holds deterrence to be morally unacceptable. This moral conflict, however, does not seem susceptible of resolution: an intuitive weighing of conflicting prescriptions of these approaches does not yield a clear answer. The sheer magnitude of the stakes (both consequentialist and deontological) makes it difficult to achieve moral certitude. Nuclear deterrence thus presents a genuine moral quandary.

Is there a way out of this moral bind? On the one hand, it could be resolved by a technological breakthrough that makes defensive systems feasible. If nuclear defences become a reality, it may be possible to give up the morally vexatious strategy of deterrence. As of this writing, such a move seems unlikely. It is realized that even with the development of ballistic missile defences, it will not be possible to shift to purely defensive strategy.

On the other hand, we might be able to extricate ourselves by adopting policies that create and reinforce a normative consensus against nuclear weapons. One of the most striking things about nuclear weapons is that they have never been used since 1945. Scholars have argued that this is not simply because of deterrence but because of the creation of a 'taboo' against nuclear weapons.[19] A taboo requires us to eschew a certain course of action because it is seen as simply unacceptable: there is no question of calculating the costs and benefits of the action. A taboo structures our moral intuition definitively against a choice or action. The argument for the existence of a nuclear taboo rests on the fact that nuclear weapons were not used even in cases where the opponent did not possess them and where their use might have resulted in decisive victory: Korea and Vietnam, for instance. The fear of drawing universal condemnation played a critical role in these nonchoices.

Reinforcing such a taboo is unlikely to be easy in the second nuclear age.[20] Most countries that are now pursuing and acquiring nuclear weapons are certain of their relevance and importance in international politics. The task before the nuclear-haves is not merely to prevent the have-nots from getting these weapons, but to convince them that nuclear weapons are not really the currency of power in international politics. This task of delegitimation can only succeed if the nuclear powers are themselves prepared to forsake excessive reliance on nuclear weapons. Deep cuts to existing nuclear arsenals coupled with a serious effort to address the sources of insecurity for those countries that seek to acquire nuclear weapons will be a useful first step in this direction. In addition, we could think of a global convention on 'no first use' of nuclear weapons. In time, as the number of weapons is significantly reduced and as the taboo against their use is widely accepted, it may even be possible to think of abandoning them. For now, genuinely moving in this direction and being seen to move in this direction by the broader international society seems to be the most morally appropriate policy towards nuclear weapons:

> Deterrence is a slippery conceptual slope. It is not stable, nor is it static. Its wiles cannot be contained. It is both master and slave. It seduces the scientist yet bends to his creation. It serves the ends of evil as well as those of noble intent. It holds guilty the innocent as well as the culpable. It gives easy semantic cover to nuclear weapons, masking the horrors of employment with siren veils of infallibility. At best, it is a gamble no mortal should pretend to make. At worst, it invokes death on a scale rivaling the power of the creator.[21]

Notes

1.   There is an immense pool of works on nuclear strategy. For a magisterial

intellectual history, see, Lawrence Freedman, *The Evolution of Nuclear Strategy*, 3rd edn (London: Palgrave, 2003).

2. Russell Hardin, 'Deterrence and Moral Theory', in Kenneth Kipnis and Diana Meyers (eds) *Political Realism and International Morality* (Boulder, CO: Westview, 1987), pp. 35–6.

3. The following discussion draws on Steven Lee, *Morality, Prudence and Nuclear Weapons* (Cambridge: Cambridge University Press, 1996), pp. 36–42.

4. For the classic statement, see Kenneth Waltz, 'More May be Better', in Scott Sagan and Kenneth Waltz, *The Spread of Nuclear Weapons: A Debate Renewed* (New York: W. W. Norton & Co., 1977), pp. 3–45.

5. While the huge political changes that were taking place domestically and internationally (particularly in the USSR) at the time complicates the issue somewhat, the decision by the Republic of South Africa to abandon a successful nuclear weapons programme and dismantle its warheads (completed by June 1991) provides a fascinating case study of a state that did choose to follow this path. W. Stumpf, 'The Birth and Death of the South African Nuclear Weapons Programme', 50 Years After Hiroshima Conference, Castiglioncello, Italy, 1995. For full text, see Federation of American Scientists, available at http://www.fas.org/nuke/guide/rsa/nuke/stumpf.htm.

6. This effect, where each side is seeking to enhance its own security, but each rational individual step aimed at achieving this actually results in growing instability due to the perceptions of the other side, is referred to as the 'Security Dilemma'. See Robert Jervis, *Perception and Misperception in International Politics* (Princeton, NJ: Princeton University Press, 1976).

7. Cf. Russell Hardin, 'Risking Armageddon', in Avner Cohen and Steven Lee (eds) *Nuclear Weapons and the Future of Humanity* (Totowa, NJ: Rowman and Allenheld, 1986), pp. 206–7.

8. For an alternative view, see Jacqueline Cabasso and Andrew Lichterman, 'Section Three: Comment and Critical Questions', *Securing our Survival (SOS): The Case for a Nuclear Weapons Convention*, p. 117, available at http://www.icanw.org/files/SoS/SoS_section3.pdf: 'Abolition of nuclear weapons is seen as impossible because the knowledge needed to make nuclear weapons cannot be "disinvented". At the same time, many of the same people argue that we must keep and constantly modernize a huge complex of nuclear weapons research and testing facilities, because the knowledge needed to maintain an adequate deterrent is so fragile that it requires enormous effort to retain it.'

9. Anthony Kenny, 'Better Dead than Red', in Nigel Blake and Kay Pole (eds) *Objections to Nuclear Defence* (London: Routledge & Kegan Paul, 1984), pp. 13–27.

10. Gregory Kavka, *Moral Paradoxes of Nuclear Deterrence* (Cambridge: Cambridge University Press, 1987), p. 59. Also see John Finnis, Joseph M. Boyle and Germain Grisez, *Nuclear Deterrence, Morality and Realism* (Oxford: Clarendon Press, 1987).

11. Scott Sagan, *The Limits of Safety: Organizations, Accidents and Nuclear Weapons* (Princeton, NJ: Princeton University Press, 1993).

12. Michael Walzer, *Just and Unjust Wars: A Moral Argument with Historical Illustrations*, 3rd edn (New York: Basic, 2000), p. 270.

13. Paul Ramsey, *The Just War: Force and Political Responsibility* (New York: Scribner, 1968), pp. 251–2.

14. Walzer, *Just and Unjust Wars*, p. 274.

15. Jeff McMahan, 'Deterrence and Deontology', *Ethics* 95 (April 1985), pp. 517–36.

16. Douglas Lackey, 'Immoral Risks: A Deontological Critique of Nuclear Deterrence', in Ellen Frankel Paul *et al.* (eds) *Nuclear Rights, Nuclear Wrongs* (Oxford: Blackwell, 1985), p. 159.

17. Lee, *Morality, Prudence and Nuclear Weapons*, pp. 45ff.

18. McMahan, 'Deterrence and Deontology'.

19. Nina Tannenwald, *Nuclear Taboo: The United States and the Non-Use of Nuclear Weapons since 1945* (Cambridge: Cambridge University Press, 2008).

20. For an exploration of some of the ways that the West may be able to bolster deterrent effects by concentrating upon reinforcing norms, amongst other things, see John Stone, 'Al Qaeda, Deterrence, and Weapons of Mass Destruction', *Studies in Conflict and Terrorism* 32: 9 (2009), pp. 763-775.

21. General Lee Butler (retired Commander-in-Chief of United States Strategic Air Command), 'The Risks of Nuclear Deterrence: From Superpowers to Rogue Leaders', speech to the National Press Club, Washington DC, 2 February 1998, available at http://www.brookings.edu/projects/archive/nucweapons/deter.aspx.

Chapter 12

The Ethical and Legal Challenges of Operational Command

PETER WALL

Implications for fielded forces

This chapter will attempt to convey some thinking about the practical application of the Just War Tradition and the legal and ethical principles covered in the earlier chapters. Following a few thoughts on the issues that have prominence at the outset of any new campaign, it will be worth detailing the characteristics of the modern operating environment, before looking in detail at some practical aspects of the legal and ethical conduct of operations.

Let me begin by stressing the importance of legitimacy. Legitimacy and perceptions of legitimacy could not be more critical to contemporary operations, especially those that are considered in some way discretionary rather than linked to vital self-interest or direct self-defence. Legitimacy is judged by a number of audiences, each essential to the sustainability of an operation in political, moral and resource terms. These include: our own forces and the wider military coalition; the indigenous population; our domestic population which 'own' us in terms of political and financial support; and the international community, particularly through the United Nations. The implications of the judgements of the different audiences can be wide-ranging, but common to them all is their impact on shaping perceptions of military and political success, and turning the former into the latter.

Jus ad bellum: going to war

The majority of conflicts UK forces have been engaged in throughout our history have aroused relatively few concerns about their legitimacy and legality. The invasion of Iraq in 2003 was our most recent 'major combat operation' (to use modern doctrinal parlance) and it is certainly an exception to that norm. The stated rationale – to denude Iraq of its weapons of mass destruction (WMDs) – was less in question at the outset of the operation than it turned out to be subsequently, when no WMD could be found in Iraq. The lessons have been very public and there is no doubt that any state embarking on a major military endeavour in the near future will take even more care than before to ensure that their *casus belli* (the *jus ad bellum* criteria set out in Chapter 4) is scrutinized and thought through very thoroughly.

What are the responsibilities of the chain of command to ensure that forces are committed to such operations justly, in accordance with the 'tests' found in the Just War Tradition? As Chapter 4 notes, the degree to which a soldier has a duty to offer advice or even question the direction of their political masters is not a new debate. Is it the responsibility of a state's defence department to ensure that *all* of its people forming up for an operation are sufficiently informed to make their own personal judgements about the validity of the operation? Or is this impractical, because the demands of operational preparation preclude a detailed awareness of the debate that might be ensuing back home, or indeed because operational security prevents the sharing of intelligence to the lower tactical level? I tend strongly to the latter view (incidentally, a view shared by Vitoria and Suarez).[1] It is for the high command to ensure that intended operations satisfy *jus ad bellum* and to be prepared to reassure the members of the force that this is the case through the chain of command. To do otherwise would be a distraction of the highest order – potentially leading to chaos, especially with the media involved – at a time when the coherence of the force and its preparations for combat are at a premium for mission success.

This will place an onus on the most senior commanders to satisfy themselves about the legal basis for operations on behalf of the Forces they represent. At what level should this occur? Should senior theatre commanders be distracted from operational priorities by such issues? Preferably not, in my view, though they will wish to galvanize their force behind the mission, and legitimacy will be an important aspect of doing so. The decision to go to war is the ultimate grand strategic issue for ministers. It is for the Ministry of Defence in both its department of state and military strategic headquarters roles to be fully satisfied, with the Chief of the Defence Staff and other Chiefs of Staff assuming that mantle both collectively and on behalf of their respective services.

Challenges of the contemporary environment

The contemporary battlespace has tended to involve operations *among* and *about* civil society, including local politics. This is in contrast to many of the operations in the twentieth century, such as the Falklands conflict or the first Gulf War, which were predominantly between opposing military forces, where strategic decisions were achieved primarily through the military defeat of the enemy. Civilians are, of course, affected by such force-on-force operations, and there are important legal issues concerning noncombatants in the battlespace, which link directly to legitimacy. But the civil population was not the centre of gravity of the campaign in the same way that they would be, for example, in a counter-insurgency campaign.

Just War logic applies equally to both of these categories of conflict, but our most recent experience, and that of our potential opponents, is derived from 'conflicts among the people', and we seem set for the nature of future operations to be an extrapolation of these current campaigns for some time into the future. The term 'hybrid warfare' has been coined by some to explain this future context. Many readers will be all-too familiar with this genre: a mixture of counter-terrorism, counter-insurgency and stabilization activities, which in the Cold War era we might have regarded as 'low-intensity' operations. As many serving soldiers will testify, on the ground at the tactical level they feel anything but *low* in intensity. Recently we have seen ferocious ideologically motivated insurgents, attempting to galvanize and intimidate societies in support of their cause, by any means at their disposal. Iraq and Afghanistan have seen some of the hardest fighting anywhere for many decades: modern counter-insurgency involves high-intensity operations at the tactical level. The backdrop for these operations involves complex and unfamiliar cultural terrain, extreme physical environments, adaptive and tenacious opponents and embryonic indigenous forces. These are highly complex campaigns by any measure. Superimpose the dynamics of a diverse multinational military coalition, other international community actors and provocative domestic media and the challenge can seem daunting at every level.

In keeping with other forms of conflict, the demands of this environment are at their most acute at the lower tactical level. The stresses on young officers and soldiers are often extreme and they test battlefield discipline and behaviour to the limit. The intense 360-degree threat, the tendency towards ruthless, often nihilist, tactics and the asymmetries of moral and legal behaviour between the opponents and ourselves are but some of the aggravating factors. Ethical discipline is at a premium, and junior leadership has to deliver the safeguards to ensure that it is upheld – at all costs.

Improvised explosive devices (IEDs) – typically against vehicles and dismounted soldiers, and becoming increasingly sophisticated in terms of remote initiation and penetrative effect – and suicide tactics, in particular, test the skills and judgement of ground troops in a highly intimidating way. The utter disregard for civilian casualties, including women and children, and no hint of comprehension of the laws of war are the insurgents' currency. When a deliberate attack in, say, Basra, London or Kabul kills coalition and local soldiers or police and kills or maims local civilians, including children, what moral or legal tenets are they applying? Military advantage? Proportionality? Discrimination? Minimum force?

The demands of this type of environment on the proper behaviour of our soldiers are extremely exacting. And so they have to be; the implications of breaches of ethical conduct are corrosive to our own force, to the coalition effort and to campaign success in the wider sense. They will have particular resonance back home as we have seen. They have the potential to do far more damage than might be caused by an occasional tactical setback, even including localized defeat. They impede reconciliation at the culmination of campaigns and they tend to weigh on the scales of legitimacy in future campaigns, because they impact on the reputation of a nation's military as a whole. The My Lai massacre and certain examples from Iraq are cases in point.

Jus in bello: **beyond the line of departure: fighting the battle**

I propose to consider the legitimate conduct of operations in keeping with our standard hierarchy of military thinking – strategic to operational to tactical – albeit these do get blurred and compressed in the execution.

The key grand strategic issue – that of committal to conflict – is the business of *jus ad bellum* described above. Sustaining broad support for an operation by promoting its continued legitimacy is the next key strategic issue. This applies not only to the military but also to those other government departments and agencies that play such a critical role in counter-insurgency and stabilization operations – especially the intelligence community, the UK's Department for International Development and the Foreign and Commonwealth Office – or their coalition equivalents. Such legitimacy is a key driver in sustaining public support. The dynamics of this in the UK cannot be isolated from those in other nations and there is a real challenge in ensuring coherence of political leadership, where strategic patience is easily undermined by short-term pressures. There is a clear requirement to sustain a credible and accurate narrative to inform public opinion. This depends on the basis for operations and can be easily undermined by poor military conduct on the battlefield: civilian casualties, disproportionate enemy casualties and any hint of maltreatment or handling of detainees have an innate bearing on the acceptability of operations to the public audience and hence the ability of the government to sustain public support. Any perception of flaws in the justification for an operation or its conduct can be far reaching. Public inquiries often result – an entirely natural quest for truth and clarity, but an additional burden and a potential distraction for those charged with delivering campaign success.

At the operational level I will focus on two specific issues: targeting and detainee handling.

Targeting: a national responsibility

Commanders in the battlespace hold the direct responsibility for ensuring that the ethical and legal behaviour of their formations and units is proper. This is more straightforward, but still never easy, even in a purely national context where no coalition or indigenous forces are involved. However, such situations are now rare. It is much more likely that the UK will be part of a coalition, probably US-led or NATO, often with other national contingents under UK operational command (this is certainly not a situation that is unique to the UK and many smaller states have found this to be the norm for decades). Coalitions are an amalgam of national forces in legal terms, and each nation will be bound by its own laws and policies. This poses the challenge of coordinating coherent operations when a range of national targeting policies and rules of engagement apply. Recent conflicts have had the added dimension of close cooperation if not partnership with local forces, with their own code of practice and behaviour, usually in a separate command chain.

Chapters 5 and 6 in particular have focused on the types of international, as well as national legal and policy guidelines pertinent to the operational and tactical levels. Deliberate targeting, where the law governing direct self-defence does not apply, is a specialist field. The key guidance for this will be found in a targeting directive which is comprised of both legal and policy judgements, usually signed off by Ministers. It will prescribe targets sets – who may be attacked – and stipulate the tolerance of civilian casualties through 'collateral damage' estimates for those targets that pass the standard tests of necessity, distinction, military advantage and proportionality to ascertain whether the target is legitimate. In many campaigns, especially counter insurgency campaigns for example, there may be no tolerance of civilian casualties, irrespective of the short-term military advantage.

Some targets may be 'time sensitive'. They may be fleeting, for some tactical reason, there may be an imperative to incapacitate them quickly – for example in the case of a chemical, biological, radiological or nuclear (CBRN) threat, or our own aircraft having only a limited time over the target. This calls for a rapid judgement by the targeting board, which must be convened at short notice, and the potential for some demanding decisions to be taken in haste. Thus the board must be extremely well prepared and should have 'war-gamed' all the anticipated scenarios in the run-up to conflict. This calls for dedicated professionals with a high degree of understanding and training.

Handling of detainees

This is an area in which illegal behaviour stemming from poor regulation and oversight has had a deeply debilitating impact on the credibility of military forces and specific campaigns all too regularly throughout history. Recent campaigns have, sadly, been no exception. There are too many cases in which individual training and collective systems for the regulation and oversight for the handling of detainees have been found wanting. This is, of course, morally unacceptable, but it also has a long-term detrimental effect on the credibility and legitimacy of our armed forces and the campaigns upon which they are engaged. As a consequence we must treat these issues very seriously throughout the chain of command.

How should this key risk be handled? Establishing systems that formalise the handling of detainees from the point of apprehension is paramount. General education and training for all is critical and must be enforced, regardless of other priorities. Nominated experts in formal detainee handling are required at all command nodes, from battlegroup upwards – perhaps lower down in specific situations. No time can be wasted in getting detainees from their point of detention to formal custody to ensure that they are expertly handled. Regular retraining of soldiers during protracted operations and long tours of duty is also important, under the supervision of qualified instructors and their chain of command.

International organizations such as the International Committee of the Red Cross (ICRC) take a keen and credible interest in the conduct of operations and particularly their impact on civil populations, prisoners of war and detainees. There is proven benefit in inviting ICRC inspectors to visit detention facilities to see conditions and procedures for themselves and to offer advice on best practice, within the constraints of the situation.

The critical role of tactical commanders

The tactical level of operations, especially in the close battle, is where the ethical behaviour of an armed force is most severely tested. The nature of fighting ensures that there will be few noncombatant spectators and commentators. A military force must be confident in overseeing and controlling its own behaviour. The onus is on commanders at all levels and their soldiers to make their own decisions, drawing heavily on ethical foundations, their military understanding of legal and policy matters, and their judgement on the application of the military force that it is their *privilege* to apply.

In counter-insurgency campaigns commanders, especially those at formation level, will take a longer-term view of the military benefit of certain styles of operation or actions. It may be legitimate and, therefore, superficially attractive to mount an operation to detain a high-value target, for example, but the disruption to the civil community, the risks of destabilizing a civil population and the potential impact of collateral damage and civilian casualties need to be gauged against longer-term benefit, rather than purely in the context of a short-term action. Here it is the style and tone of operations that will be set by the commander at formation and battalion level – and perhaps at lower levels where the situations demands it. Consent for the continued presence of military forces, especially foreign ones, will be a function of the civilian population's tolerance of their conduct. This will, in turn, relate to the local political context in which tactical operations are set. 'Courageous restraint' is an essential element of operations among populations, and it can have a distinct influence on the course of a campaign. Acts of gallantry involving the withholding of force are especially noteworthy in setting the tone for a campaign.

Alternate measures to the use of force should be considered. This is often an ethical and moral issue as well as a tactical one. There is a professional onus on a military commander to have a sufficiently developed understanding of a complex situation, including the 'human terrain', to be able to think laterally about a problem and to deliver innovative solutions. In current operations these complexities do, if anything, penetrate further down the chain of command than hitherto. There is no doubt that these are weighty responsibilities for relatively junior commanders. The authority vested in officers and soldiers to employ lethal force brings these distinct and awkward provisos. And so it should in light of the chaos that ensues when military force is inappropriately used, even when it is strictly legal.

Delegation and trust

As noted in Chapter 1, the British Army's command doctrine – known as Mission Command – seeks to energize the initiative and imagination of subordinates by empowering them to carry out their mission in the way that is most pertinent given the situation they find themselves in. This is usually done by defining their tasks (ends), affording them sufficient resources (means) and permitting them the freedom to decide how to do the job (ways). This is now an established and emulated practice, which has, in the round, been extremely successful. It has more than reversed the impact of decades of defensive thinking during the Cold War – when the need for conformity to higher-level plans necessarily constrained initiative at lower levels. However, there is a caveat: the autonomy afforded to a subordinate should be tuned to his or her capability and experience. And there is no more significant issue than oversight of the application of military force and battlefield ethics. There is less scope for error here than in any other command function and delegation must depend on high levels of trust and competence.

Grey areas

The theories articulated in this publication are relatively clearly defined, and they can be interpreted with reasonable clarity by commanders, with advice from legal and policy experts. In practice, as an operation unfolds, issues that are less clear-cut, and for which specific rules and guidance may not be readily available, will inevitably arise. For example how should a junior commander partner an indigenous force with very different habits and ethical behaviour from his own? What level of force may be applied to prevent damage to property? How should we handle the passage of intelligence, or detainees, to indigenous forces with a different approach to human rights from our own? If situations are perceived as posing serious dilemmas, they must be referred up the chain of command for clear and timely direction to be issued. Junior commanders will find themselves making awkward judgements until such direction is available, not least because in most of these cases the constraints may seem counter-intuitive, not least in the heat of the moment. For example, as Chapter 6 makes clear, physical measures to exploit the shock of capture are unacceptable, and in UK policy, force may not be used purely to protect property.

Conclusion

Ethical behaviour is a moral requirement for all operations, and in campaigns *amongst and about people* it is crucial to campaign success. Commanders at all levels play differing but critical roles in ensuring operations are appropriately designed and conducted. The onus on all commanders in contemporary operations in understanding the parameters, setting the example and ensuring ethical conduct is clear, and is unlikely to diminish.

Note

1. See Chapter 4. See also George R Lucas, 'Advice and Dissent: The Uniformed Perspective', in D. Whetham and D. Carrick (eds) *Journal of Military Ethics: 'Saying No: Selective Conscientious Objection'*, special edition 8(2), June 2009.

Index

THE
PROFITABLE
PROFESSIONAL

The 10 key ingredients for building a highly profitable business coaching, consulting or advisory business

Profit

Kelly Clifford

'**The Profitable Professional** contains a wide range of practical advice for anyone running a professional services business. For those newly starting out there is excellent advice on getting things right from the start, for those already in business the succinct format with check lists and summaries is a good reminder to keep coming back to again and again. In my experience, an ongoing battle when you lead a business in which you are also a skilled practitioner is the tendency to be drawn into working in the business rather than on it. **The Profitable Professional** is a great reminder to keep assessing your progress and it offers vital guidance about how to achieve better results.'

Jefferson Lynch – *Client Director, Red Olive Analytics*

'I only wish this book had been around a few years ago...it would have made my other businesses a lot easier to run! Kelly has an incredible focus; particularly with regard to the numbers and in this book he manages to demystify the formula for financial success, by breaking it down into an organised and effective framework. **The Profitable Professional** is the sort of book that I will read and encourage others to read many times to keep the business on track'.

T Bovingdon – *Managing Director, Mind Your Own Business*

'**The Profitable Professional** is written from the heart, a really honest view of running a professional business. Business books often leave me 'cold' after reading them as they don't feel like they have been written by a human. Kelly has been brave enough to share some real insights into the problems that we all face and offers a solution that savvy accountants and business advisers can use. Selfishly, I don't want my peers to read this book! It is full of practical and valuable exercises to follow that will help move you from where you are now with your business to a much better place – wherever you decide to be.'

Martin Horton – *Managing Director, Rivington Accounts*

The Profitable Professional

First published in 2017 by

Profit in Focus Ltd
Registered Office:
Chaparral, Yoxall Rd, Kings Bromley, Staff, United Kingdom, DE13 7JJ
hello@profitinfocus.com
www.profitinfocus.com

Printed and bound by Ingram Spark/Lightning Source in the UK and USA.
Copy edit sourced through Write Business Results Ltd.
Interior design by Diana Russell

ISBN 978-1-68418-863-5

Dedication

For Jon and your unwavering support in all that I do. I dedicate this also to my family, friends and loved ones for helping to shape the person that I've become. A special mention is for my two little nieces Krysta and Alyssa, who are just starting out on the adventure that is life. I love you all.

In loving memory of my Nanna (Grace – aged 93), Nan (Freda – aged 94) and most unexpected and heart-breaking of all my cousin (Ben – aged 38). I have sadly lost all three in the past year. You may be gone but you will always be in my heart and will never be forgotten.

Lastly, I dedicate this book to all those people that have chosen to live their life on their terms and have been bold enough to do something about it in building their own business.

Contents

Introduction

Welcome to *The Profitable Professional.* This book is about giving business coaches, consultants, advisors and more forward thinking business accountants the tools, insights and resources needed for building a highly profitable business coaching, consulting or advisory business. The 12-step blueprint I provide you at the end of the book for implementing the 10 key ingredients we will cover will help you to dramatically fast track your success and increase the profits of your business in a more sustainable and enduring way.

I have written this book to help business coaches, consultants, advisors and business accountants who have been in their business for a while and are passionate about what they do but who have become frustrated that they have nothing really to show financially for all the effort they have put in.

If you are one of these business owners you may see your staff earning more than you do without the associated stress and wonder why you put yourself through it all. You might lack time to focus on growing the business and its profits as you are so involved in the 'doing' part of the business. You may also be overwhelmed by the sheer volume and 'noise' of what you think you should be doing versus what actually needs to be done, so you might tend to avoid dealing with it by making excuses and hoping that by ignoring it, it will go away or somehow magically take care of itself.

You may have reached a stage where you have had enough and want to change things, and have a real desire to take your business to the next level. Perhaps you don't know where to start, but you recognise that you need the right help to move the business forward.

Equally this book is useful for people just starting out in building a business coaching, consulting or advisory business to get off on the right foot and to learn from the mistakes that others may have made before them.

I am a qualified accountant with over fifteen years' experience and the author of the book *Profit Rocket: The five key focus areas to sky rocket your profit*. I'm not a typical accountant though, as tax returns and compliance type stuff quite frankly bore the life out of me! What I like to focus on is the much 'sexier' profit improvement and business growth driver stuff. The stuff that, in my view, actually makes a tangible difference.

Up until 2011 I was working in a corporate role as a finance director for company in London. I resigned by choice to embark on my entrepreneurial journey, because fundamentally what I see all the time is that too many people go into business thinking that being good at what they do or good at the doing part of their business is enough, and for a variety of reasons tend to overlook one of the most important aspects of running a business – the numbers! So I set out with the intention of making numbers simpler for small businesses and helping them to better understand the underlying business growth drivers that are key for their ultimate success.

Working as a finance director and working with many hundreds of business owners over that time has given me some unique insights that many business coaches, consultants, advisors and accountants face in building their businesses. I have learned much of what I will share with you in this book the hard way through being in the 'business trenches', trying to determine what works and what doesn't. I know what it is like as I have been where you are, and in many respects still am, as being in business is a continuous journey of discovery and learning. It comes with its up and downs, victories and challenges.

What my first hand experiences have enabled me to do is distil the 10 key ingredients that are vital for building a highly profitable business coaching, consulting or advisory business that works for you as opposed to you working for it. It is these insights that I share with you in this book, which builds on the very powerful 'Profit Accelerator Formula' that I introduced in Profit Rocket. I take you on a journey through the book, share parts of my story and provide other relevant case studies to help make the content come alive and feel real for you.

The case studies shared are based on UK businesses so the currency used is £GBP, but the principles we cover are in fact currency neutral. Exactly the same results are achievable no matter where you are in the world; the currency used is interchangeable in that regard. So if you are based in the US then read everything as $USD, if in Europe then €EUR, etc. Whatever your home country currency, replace £ with that.

It is my hope that you will see this book as a resource you can come back to time and time again for new ideas and inspiration. I am excited for you right now and look forward to sharing the journey with you through this book and beyond, if you feel you need further support.

I hope what you are about to read will give you the insights, blueprint and confidence to transform the way you approach building your business to ensure a highly profitable and rewarding outcome for both you and your family!

CHAPTER 1

Key ingredient #1

Doing things differently

'Doing things differently' is the first key ingredient for building a highly profitable business coaching, consulting or advisory business. After all, if you do what everyone else is doing then you are likely to be making the same mistakes or facing the same challenges that they are.

In many respects, adopting a 'time for money' model that many coaches, consultants and advisors tend to do presents a number of opportunities but also some key challenges. In my view, whilst usually being the easiest commodity to 'sell', a time-based offering is often the hardest business model to scale in a profitable and time efficient way.

So in this chapter, let's take a bit of a reality check upfront about the key challenges that many in this industry face and identify what, if any, you are facing in your own business. It may be an uncomfortable exercise, but being aware of the key challenges that apply to you will help bring them into your consciousness so that you can start to tackle them.

You may actually draw comfort from the knowledge that you aren't the only one facing these challenges, and that they are in fact very common across your peers. Becoming clear on what your challenges are now and writing them down in the exercise at the end of the chapter means that they will be at the forefront of your mind as you read and digest the contents of this book. Hopefully solutions and suggestions for tackling these challenges will present themselves to you as you read on.

Later in the chapter I'm going to introduce you to three case studies, the progress of which we'll follow right through the book so that you can understand their challenges, the approaches used to deal with them and ultimately the benefits and results that were realised in ensuring the 10 key ingredients that we will cover were in place as the foundation for growing their business.

But first, the common challenges many business coaches, consultants, advisors or accountants (which I will collectively refer to as 'business support professionals' moving forward in the book) that I come across typically face are some or all of the following five.

'Working in' vs. 'working on' conundrum

For some it's a constant struggle between client delivery and maintaining the future client pipeline. It feels a bit like catch 22 – you need to get the client work done because it produces revenue but you are very conscious that it means that no time is being spent on business development and ensuring that your future client work pipeline is full.

This often causes peaks and troughs in your revenue and earnings, which can only be described as feeling like FEAST versus FAMINE. When you have a client project, your revenue is strong and the going feels good, but as soon as that client work project stops you may be plunged into the depths of despair when there is nothing to replace it, and so you madly ramp back up your business development activities to replace the lost revenue. Sound familiar?

The challenge here is that without a strong future pipeline that you are actively nurturing in place, it is very difficult to turn that new client or project tap on and off at your discretion. This is usually purely down the time lag that exists between when you first engage with a potential client to when they actually agree to start working with you, sign your contract and pay your invoice.

Lack of consistent 'qualified' lead generation

Most business support professionals I speak with have no consistent way of generating leads. They are taking a patchwork quilt ad hoc type approach to lead generation. They have no system or process for it. They aren't embracing social selling and digital asset techniques when more and more business is being originated online.

Many are stuck approaching lead generation in the way they have always done it, whether this is face to face networking or otherwise, and getting the same or less results – good or bad. Many of the old ways of working are no longer as effective as they once were. Don't misunderstand my point here; nothing can ever replace the power of building enduring relationships in business. However, to build a profitable and enduring business you need to make best use of your time and have a constant flow of 'qualified' prospects to speak with.

The curse of unqualified FREE consultations

Spending too much time on non-chargeable work. We've all had it happen, or at least I know I have many times in the early days. People come to you asking for a "quick favour" or can I "pick your brains" on something. Then the email from them comes through with a long list of questions, at which your heart sinks.

90% of business support professionals that I come across tend to meet with most or all people that approach their business without any pre-qualification as to whether they are a potential client or not. Their first response is to say let's get a meeting in the diary or meet for coffee because they are just so excited to have a 'lead'.

Now I'm not saying that you shouldn't offer FREE consultations if that is important for you, but what I am saying is that you should be pre-qualifying who you are doing this for more. If there is little or no likelihood that the connections can add any 'value' for your business, then you need to question whether that is the best use of your time. Time is money, after all, and any time spent on FREE consulting is less time that you could be spending on chargeable work from which your revenue is derived. The opportunity cost for FREE consultations can be huge in a typical business.

CHAPTER 2

Key ingredient #2

A business by design
not by default

In business there are lots of twists and turns to navigate and rarely does the journey progress in the way or form that you anticipate. That is certainly my experience anyway. There are some key lessons and observations I made along the way, which we will dig deeper into and explore further in this chapter.

Although I have over 15 years' experience as I mentioned in the introduction, my journey as a business owner started in 2011. I was working as a Finance Director for a company in London at the time and simply got to the point where I had enough of the corporate merry go round and felt it was time for a change.

That's when I resigned by choice in the middle of the recession to embark on my entrepreneurial journey, because the fundamental issue I was seeing time and time again was that too many people go into business thinking that being good at what they do or good at the 'doing' part of their business is enough, and for a variety of reasons tend to overlook one of the most important aspects of running a business – the numbers!

So I set out with the intention of making numbers simpler for small businesses and helping them to better understand the underlying business growth drivers that are key for their ultimate success. But like many, to say I had rose-tinted glasses was a complete understatement, and I'm not too proud to openly admit that with the benefit of hindsight, as I believe that only by sharing our experiences can we truly help each other.

I can honestly say that I was completely unprepared for the emotional rollercoaster I would be embarking on when I chose to leave employment – which I have known all of my working life – to set up my own business. I was soon to find myself in at the deep end. My first 12 months in business can only be described as a complete and utter car crash. There's simply no other way to describe it.

There were a variety of reasons for this, such as not having a network to begin with and having to start from scratch; not really understanding the time it takes to nurture a potential client, especially when the offering was based on a high value retainer. I naively I thought that, as I had been doing my job for over 12 years at the time, transitioning from doing it for someone else to doing it for myself would be a breeze. Wrong!

I thought that by getting a website, online client enquiries would start flooding in miraculously and people would be queuing up to work with me. It is what I refer to now as the 'build it and they will come' mentality. Well, I built it and they didn't come! I started out with a particular business model in mind and figured that people would be happy to pay for the service at that level. What I didn't fully factor in is the huge role that relationships play in business.

People generally buy from people they know and trust. Well, they didn't know me so they didn't have the opportunity to know whether I was trustworthy or not. At the time I read that it takes approximately ten points of contact before a person will generally trust you enough to buy from you. Quite staggering really, so I had a huge challenge ahead.

What I also failed to factor in was that I never had to go out in any of my previous roles and generate business, so this concept was completely foreign to me. The idea of selling was also foreign to me. While employed, my pay would come in regularly and it really didn't matter whether it was a busy or quiet month at work because I would be paid irrespectively.

To defend myself a little bit, I did think I had a chunky bit of work lined up to kick off with, but because of circumstances beyond my control, that client work fell through at the very last minute.

So there I was, having resigned from my role as Finance Director and given up the six-figure salary that went with it to set up my new business, and the client I thought I had to kick things off with pulled the rug at the last minute. Oh no!

When the rug was pulled, as it were, I knew that my focus had to be to build a network of contacts, and fast, so I threw myself into the unfamiliar world of networking. Frequently, business owners in this situation are trying to do everything themselves because they think that is what they have to do. They are seeking perfection in all that they do. I have definitely been guilty of this way of thinking in the past.

Being congruent and authentic

About six months after setting up my business, I had a real crisis of confidence. I felt a bit dejected and deflated in that all my efforts over the previous six months hadn't really yielded what I considered to be any tangible results. Sure, I was busy 'doing stuff' and having meetings but I was failing miserably in my eyes to be where I thought I should be at this point.

It's also very easy to fall into the trap of judging your own success by the *perceived* success of others. I know I've been guilty of it. Social media hasn't helped at all in this regard. I say 'perceived' very deliberately as often there is a very big difference between the message most people share with the external world via social media about what's going on in their business versus the reality of the situation. In my experience, the reality is usually very different to what they are posting, with many people adopting the 'fake it until they make it' strategy. The simple answer is to block out this social media noise and not to succumb to the online envy that afflicts many. Recognise it for what it is and just focus on doing what you need to do to achieve the success you are seeking. That's how you ultimately become successful.

So I took some time out for a week from the day-to-day. It was really important for me to remove myself from the situation and reflect on the previous six months. What I realised in giving myself this space to think and reflect was that I was being too generalist in my approach. I was trying loads of different ideas, hoping that if I cast my net wide enough then I would catch something. The problem was that I was exhausting myself trying to get everything done and spreading myself very thinly. I was trying to be everything to everyone, but ended up being nobody to no one. Sound familiar?

This led me to start questioning whether I had just wasted six months entirely, whether I was cut out for creating my own business or if, in fact, I should just give up and go back to employment. Then I thought about the prospect of being an employee again and the thought filled me with dread. I am not the sort of person that likes to wallow and I really don't like feeling rudderless and lost, as I did at that point. I wanted to get back on track as soon as I could.

Start with your why

What I realised at that point was that I needed to relax more about things generally and stop putting so much pressure on myself to get results, and the biggest breakthrough came when I became clear on my 'why'. It is the answer to the question, 'Why am I doing this?' What is the core belief that I will build my business around and which will influence the products and services my business provides its valued clients?

The answer for me was and is: *'I believe that there is tremendous power at the intersection of innovation, entrepreneurialism and profit numeracy.'* Now this may not mean anything to some people and that is perfectly fine. It doesn't matter. What matters is what it means to me. You see, I believe that if a business can get the balance right between the idea or product offering, provide it in an innovative way and have an understanding of the numbers associated with it, then it creates a truly powerful combination from which to grow.

EXERCISE

What's your 'why'? Why do you do what you do?

_____

_____

_____

_____

Not doing what you think is expected of you – know what you want

When I became clear on my 'why' everything else started to fall into place. I became clearer on the area I wanted to focus on and the type of client I wanted to help. It enabled me to be bold enough to scrap most of what I had done with confidence, as I realised that some of my actions were down to what I thought I was *expected* to do, not necessarily what I *wanted* to do. Does that make sense? Sometimes we do things that we think others expect of us, but they don't necessarily make us happy.

It is this realisation that led me to write my first book *Profit Rocket* which became an Amazon Bestseller in the UK, and which rapidly enabled me to build the six-figure outsourced Finance Director offering that I set out to do. Now this all sounds good in theory and it was, until I realised a further six months later that I had focused my energies on building the wrong business! It hit me like a slap in the face. Doh!

I had a full client load at that point and no spare capacity, so very little time to focus on anything other than client delivery work. As my offering was based on my direct input, I was very time poor – there never seemed to be enough time. My clients were demanding and, as is often the case in the pursuit to ensure client satisfaction, scope creep began to happen where clients try to be cheeky and ask for more and more for no extra fees. I started to feel owned by my clients.

The possibility of an extended holiday for any more than a couple of days was a pipe dream, as taking the time out of the business for a well overdue holiday to recharge because of my client commitments every month simply wasn't possible. I began to feel resentful, and the reality was I wasn't enjoying or having fun with it at all. There had to be a better way.

I am Australian born so given the vastness of my birth country it won't be a surprise to hear that one of my highest values is freedom. I wanted to be geographically independent, not location bound for my business. I wanted to be able to run my business from anywhere in the world if I chose to. I realised then that achieving this simply wasn't possible with the 'time for money' business consulting model I was pursuing at the time. It is then that I realised that I needed to pivot the business and become more 'product' focused. I will explain what I mean by that shortly.

Connecting with your passion and defining your business vision

At the same time, I very much consider myself to be a heart-centred entrepreneur, which means I want to DO WHAT I LOVE, I want to MAKE A DIFFERENCE and I want to LEAVE A LEGACY. As I mentioned earlier, I knew what my 'Why' was but the missing ingredient was a Vision/Mission of how I would go about doing that, which would be the guiding light for all that I would do moving forward in growing the business.

Throughout 2013, I enlisted the help of a really great mentor who has co-founded and built not one, not two, but three billion pound businesses in the UK. Basically I was in a group of twelve mentees with him as the lead mentor. It was all about purpose beyond profit, lasting legacy and the dent you wanted to make. It was an amazing experience but most importantly what emerged for me from it is my BIG VISION or 'game' if you like, which is to help at least 100,000 small businesses globally to improve their profit by at least 20% over 12 months by the end of 2020, because I believe that profit is a 'means' not an 'end'. If I can enable the 'means' for small business owners then they can achieve more of their 'end', whatever it is for them, and that's how I make a difference.

EXERCISE

What's your business vision?

_____

_____

_____

_____

So I had defined my vision that really resonated with every part of my being, but the challenge I had was that I was working one-to-one with clients at the time. With the best will in the world, it just simply isn't possible to work with 100,000 of them on a one-to-one basis, so it caused me to pause and take a step back to really have a think about how I could leverage technology to deliver what I did at scale, and that's just what I've done.

Creating your own reality – taking focused action

At the time of writing this book, I've spent over 2 ½ years developing and creating a technology platform for helping business owners to profitably THRIVE, and it is called 'Profit Pod'. Profit Pod itself is an online profit improvement ecosystem that helps small businesses with typically between 0 and 9 employees – which means micro businesses and one person operations –

those that I believe are the heart and soul of our economy but don't necessarily get access to the affordable help they need to THRIVE. These businesses have been trading for at least 1 year so are through the start-up pains and tend to have annual revenue less than £2M ($3M).

It is designed to help these business where perhaps the owner isn't as confident on the 'numbers side' of their business as they should be, to make numbers simpler for them so they can grow their profits, or improve their losses, by at least 20% over 12 months through a proven five-stage process with GUARANTEED results. It's FREE to get started at: profitpod. com. Profit Pod is the implementation platform for the Profit Accelerator Formula from my first book *Profit Rocket*, which again features in this book because of the important role it will play as a key ingredient in building your business – but more on that in Chapter 5.

I've now taken Profit Pod from conceptual idea right through to full operational website. As I write this book, the site is currently in beta and performing well. There's still a long way to go but the early signs are very encouraging. Creating anything isn't without its challenges, as I was soon to find out once again.

Having never been involved in an IT build on this scale before I quickly realised two things: first, they cost more than anticipated, which I knew but never thought would happen to me. Well, it did happen. Second, they take longer than planned: I had wound down my consulting clients in anticipation of the build being ready when it was supposed to, but it wasn't, so as you can imagine it caused what I now affectionately refer to as a 'revenue gap'.

I didn't want to fill this gap by raising equity, as I had so far self-funded the build and didn't want the pressure of investors that would dilute my vision, or by resorting back to consulting to fill the gap. I tend to work with clients for 12 months or longer and I didn't want that extended lock in.

The challenges you face can often result in even better outcomes

There had to be a way to leverage what I had created with Profit Pod so far that would get me closer to my ultimate vision, but would also address a short to medium term need. That's when the light bulb went on for me, and I realised that I needed to influence the influencer more. My thinking was if I could provide access to the tools and infrastructure that business coaches, consultants, advisors and accountants need to better support the growth needs of their business clients, and at the same time help to address many of the common challenges faced by these businesses (see Chapter 1), so that they could build their own business or practice in a more profitable and time efficient way, then that would be a TRIPLE WIN in my view. The end client would benefit, the business support professional would benefit and it would move me closer to fulfilling my overall mission.

So that's precisely what I did. As part of creating Profit Pod I created four business growth diagnostics. It's these diagnostics that I spun out into a completely separate standalone offering under my Profit in Focus brand, and now enable business support professionals to harness their power. They are able to white-label them completely as their own, so the diagnostics become part of their business process for all of their clients, and they can fast track their own success.

These four business growth diagnostics tend to take the form of 35 questions, each with Yes, No or Maybe answers, and each of which typically take a client less than 5 minutes to complete, so no more than 20 minutes overall. Off the back of each set of diagnostic questions, a full tailored circa 20-page growth feedback report is produced based on how that client answered every single question for that diagnostic. The report for that diagnostic goes back out to their client branded entirely as their business, so the client only ever sees their brand as part of this process and not ours.

This helps our clients to ATTRACT, WIN and RETAIN more profitable business clients as it is perfect for 'qualified' lead generation and results in up to 140 client insights being gained in less than 20 minutes of a client's time. They can better support the 'growth' needs of their business clients, reduce levels of profit-draining non-chargeable time and easily identify new revenue opportunities to grow their business in a more profitable and time-efficient way.

You must first decide what you want from your business

I share my story for three reasons. The first is to demonstrate that things rarely go to plan or the way that you intended from the outset. The second is to demonstrate that by having an overall guiding mission, and an underlying product or service ecosystem (which we will cover in more detail in Chapter 4) to support it, when challenges arise you have choices in finding a solution that works best for you – after all, your business should be by 'design' not by 'default'. Your business should be the 'vehicle' you have consciously designed to get you to your intended business destination. This principle was critical in transforming the profitability and growth of the businesses I share with you in the case studies in the book, so it is fitting to share more about it for context in case it is relevant to you.

It's your game so set it up and play by your rules

The 10 main driving criteria for what I wanted from my business, which helped to shape the business I designed, are:

1. To be a 'productised' service offering that can scale easily.
2. Minimal time input in terms of delivery.
3. Online sales process.
4. Geographic independence – ability to operate anywhere in the world – even on the beach should I choose – and not

have it impact the business.

5. One that leverages technology as much as possible.

6. Recurring income streams.

7. Not to be a people heavy business nor have to manage a big team.

8. Be connected to my heart in doing what I love.

9. A business that makes a difference and leaves legacy.

10. Capable of achieving a successful exit at a material value.

EXERCISE

What are 10 key elements you want your business to make possible and enable for you?

Some questions to ask yourself here to stimulate ideas are:

i. How much revenue do you want to produce? Business Size?

ii. Profit it produces?

iii. Number of staff or team size?

iv. Location for it?

v. Lifestyle you want for you and your family?

vi. Type: service or product?

vii. Online or physical in nature?

viii. Group or individual?

ix. Time frame for it?

x. Build to sell or keep forever?

xi. Your time commitment?

1. _____

_____

2. _____

_____

3. _____

_____

4. _____

_____

5. _____

_____

6. _____

_____

7. _____

_____

8. _____

_____

9. _____

_____

10. _____

_____

In this chapter we looked at the second key ingredient and explored the importance of being congruent and authentic, not doing what you think is expected of you but instead knowing what you want. It's about starting with your 'why', which only comes from connecting with your passion when defining your business vision. Creating your own reality only comes from taking focussed action, and often the challenges you face can result in an even better outcome than you ever envisaged. Ultimately, your business is your 'game' so you must set it up and play by your own rules.

So, in the next chapter we will delve into the third key ingredient being '**Crystal clarity on your target market and positioning.**'

CHAPTER 3

Key ingredient #3
_____
Crystal clarity on your target
market and positioning

Times are changing. No longer is it enough for business support professionals to have a purely transactional relationship with their clients. They must be more proactive. The focus has shifted to a more 'value added' approach around supporting business growth, not just dealing with various elements in an isolated vacuum. The reality is that those business coaching, consulting, accounting and advisory businesses that embrace this and evolve for this new paradigm will THRIVE and those that don't will wither. It might sound harsh but it's true.

According to the recently published Small Business Survey 2014, 44% of SMEs across the UK had sought external information or advice in the 12 months preceding interview. Compared with 2012 when the same survey was last completed, there was a change in the type of advice required. The proportion seeking it for business growth increased by 10 percentage points, and there was an eight-percentage point increase in the proportion seeking it for efficiency and productivity. There was a corresponding decrease in the proportion seeking financial advice for the running of the business (down 10 percentage points). Thus advice was sought more for business growth, and less for business stability. Therein lies a massive opportunity for a new breed of dynamic, proactive and forward thinking business support professionals.

A NEW PARADIGM – 5 key SHIFT principles that you must embrace to THRIVE in changing times.

What are the 5 key shifts that many of you must make, if you haven't already, to embrace this changing client expectation landscape?

They are:

1. From *Reactive to Proactive*
2. From just *Transactional focussed* to more *Commercially focussed*

3. From *Inflexible* to *More Flexible*
4. From an *Employee Mentality* to a *Business Owner Mentality*
5. From *Introverted* to *More Dynamic*

Let's go through each principle one at a time.

 ## KEY SHIFT PRINCIPLE 1:
From being reactive to being more proactive

OLD PARADIGM

Being reactive means:

1. Waiting for clients to come to you – being passive.
2. Only dealing with issues when they become a problem.
3. Always reacting to a situation, not leading it.
4. Blaming others and shirking responsibility – "It's their fault, not mine".
5. Having an excuse for everything and often responding, when asked a question, by starting with the words "The problem is ..." – too focussed on the problem, not the solution.

NEW PARADIGM

Being proactive means:

1. Anticipating the needs of your clients by taking an active, not passive, interest in their business growth.
2. Regularly and consistently contacting your clients, at least once every three months, so you can keep your finger on the pulse of the issues they are facing.
3. Actively being the extended eyes and ears of your clients, looking out for potential opportunities and threats.

4. Always looking for ways to 'add more value'.
5. Becoming an advocate for your client.

Key Benefit

Clients will see you more as a trusted adviser and not just a service provider, which means you will likely become their first point of call for any big decisions.

KEY SHIFT PRINCIPLE 2:
From being mainly transactional focussed to being more commercially focussed

OLD PARADIGM

Being transactional focussed means:

1. Being more concerned with ensuring that boxes are ticked rather than with what it actually means for the client.
2. Being more worried about ensuring your 'back' is covered.
3. Being too overly risk focussed, often to the detriment of what's in the best interest of the client.
4. Seeing a client mainly as a number in your portfolio.
5. Often having tunnel vision and being single minded.

NEW PARADIGM

Being more commercially focussed means:

1. Embracing technology at every turn to do things better and more efficiently.
2. Introducing clients to each other where a potential collaboration or mutually beneficial partnership could result

between them.

3. Understanding and embracing the ethos that 'together we are stronger'.

4. Embracing 'value-based' pricing and avoiding competing on price for the services offered.

5. Understanding the profitability or not of each of their clients and feeling comfortable 'firing' the more problematic or less profitable ones. Having an awareness of client lifetime value.

Key Benefit

Clients will be clear on the 'value' that you add to their business, so decisions around how you engage are likely to be taken on more than just price alone, which ultimately will mean more profitable clients for your business.

KEY SHIFT PRINCIPLE 3: From being inflexible to more flexible

OLD PARADIGM

Being inflexible means:

1. Being a stickler for the rules.

2. Not thinking 'outside of the box' when proposing solutions.

3. Having a prescribed set of services with no flexibility.

4. Being very rigid about the way the work is done.

5. Not understanding or having empathy for any issues a client may be facing.

NEW PARADIGM

Being more flexible means:

1. Offering clients a menu of choice.

2. Enabling clients to scale up or scale down the support they get depending on the stage of their business.

3. Having a higher percentage of 'client growth' focussed support services than compliance type services.

4. Evolving and adapting to the changing needs of the marketplace.

5. Constantly looking for ways to 'add more value' for clients.

Key Benefit

Clients will know that you are best placed to support the changing needs of their business, which enables you to increase your fees generated from them as their business grows. They shouldn't need to look elsewhere, so client loyalty will likely be much stronger because of this.

KEY SHIFT PRINCIPLE 4:
From an employee mentality to a business owner mentality

OLD PARADIGM

Having an employee mentality means:

1. Working strictly from 9 to 5.

2. Not having empathy or understanding the way a business owner thinks.

3. Having team members who have no idea of what it's like to run a business.

4. Doing only what's needed on a job and clocking off at the end of the day.

5. Not necessarily understanding the bigger picture.

NEW PARADIGM

Having a business owner mentality means:

1. Constantly looking at things from the clients' perspective.

2. Having empathy, and knowing that it is not an easy journey as a business owner.

3. Ensuring that the team is educated around what it's like to run a business, as most of them will likely have never run a business before.

4. Ensuring that any team incentives are aligned to the success of both your business and that of your clients in a balanced way.

5. Being innovative across all areas of the business.

Key Benefit ⸾

Clients will have more confidence that you are looking at things from their perspective. This improves their confidence and trust that your advice is relevant, practical and based on the real world, not merely textbook theory driven.

⸾ **KEY SHIFT PRINCIPLE 5:**
From being introverted to being more dynamic

OLD PARADIGM

Being introverted means:

1. Being shy and quiet.

2. Being afraid to voice an opinion for fear of being judged.

3. Commonly being uncomfortable as the focus of attention.

4. Fitting the stereotypical mould.

5. Being scared to rock the boat in case the client gets upset. Not saying what needs to be said.

NEW PARADIGM

Being more dynamic means:

1. Having confidence in the value offered clients.

2. Being very personable, friendly and approachable.

3. Always having the best interests of clients at the heart of whatever advice is given. Not being afraid to say what's needed to be said.

4. Fostering the spirit of openness and collaboration – we are in it together.

5. Smashing any stereotypical moulds that may exist.

Key Benefit

Clients will know that they can come to you with anything and won't be judged. This means you can help shape solutions for them that also have the potential to benefit your bottom line in a positive way, provided it is in the best interests of the client of course!

Now that you hopefully have made the conscious decision to adopt the new paradigm way of working, the next stage is to really understand your competitive landscape and most importantly really get under the skin of the problems that your clients face so that you can tailor a very specific solution to solve those problems for them.

Niching for greater success.

A niche is a focused, targetable segment of the market. You are a specialist providing a product or service that focuses on the specific needs of an identified client group, which cannot or are not being addressed in such detail by the dominant providers in your industry. But it is important to understand that there is in fact a *difference between your identified niche and your target market:*

Your target market is a specific identifiable group of people you work with, e.g. women in the city, technology start-ups, creative agency owners, small and medium businesses in a particular revenue range.

Your niche is the service you specialise in offering to your target market.

Here are 7 reasons why it is important to have a niche:

1. **To avoid spreading yourself too thin**

    Instead of the risk of spreading yourself too thin in saying that 'everyone' is your potential client, niche marketing will help you to focus on a specific grouping of people, and particularly on what their needs and wants are. You will unlikely to be able to serve everybody, so it is important to focus on what you do best and aim it at a specific group of people who will likely buy what you offer.

    It is important to find out what is important to them, what blogs they read, their beliefs and attitudes, who the main influencers in that network are. Having these insights means that you can develop products or services specifically aimed at this group, based on your thorough knowledge and understanding of what they are interested in.

2. **It's easier to identify and target potential clients and partners to work with**

   As the pool of people that exists for a niche is smaller than its mainstream equivalent, it will be easier to identify potential clients and partners to work with, as you can be much more targeted and laser focused with your marketing efforts.

3. **It's easier to become an expert and wellknown in your niche**

   Niching means it will be much easier for others to understand 'what you do' and 'for whom', which will make it easier to position you as an expert in your field. As this group is more targeted and of a smaller size, you can rapidly become well known within this group of people. Your profile and overall visibility will increase within this group. It is a small world after all!

4. **More and better referrals**

   Since it will be easier for others to understand what you do and for whom, it in turn becomes much easier for them to refer more and better quality clients to you that fit the profile of your ideal client, as you have built up trust, credibility, visibility, and it is very clear as to what your specialism is.

5. **The more unique you are, the less competition you will have**

   There will be less competition, as you will provide the specific services or create the specific products for the specific people you are seeking to help in a specific way that meets their needs. The BIG advantage of becoming more unique is that usually it can't be easily replicated by your competition!

6.   Marketing becomes much easier

Effective niche marketing should really help with your marketing, positioning and branding as you will attract the 'right people' much more easily and quickly. People with similar interests tend to behave and are attracted to similar things. This means that many of your clients will do all the hard work for you as they will refer you more and more, because your profile, credibility and influence is readily apparent within your tribe.

7.   More repeat business

As you are able to provide an increasingly better service or product, based on your specific client's needs, it is likely that you will get more repeat business – people will come back for more, and as an added benefit will often start spending more with you as your relationship grows with them.

EXERCISE

Create Your Ideal Client Persona

If you understand the complete profile of your target ideal client, including what they like, their buying habits and the problem for which you have a solution, then if you centre your marketing activities around reaching that profile of client, they will be more inclined to buy from you because your product or service meets their identified need.

Instead of being a cold prospect when they are contacted, they are more likely to be a warm prospect and therefore more inclined to be interested in what you are offering. Better quality prospects will translate into better results.

Here are ten basic questions that can help you define your target ideal client:

1. **Who buys** your product or service?
2. Are they predominately **female or male?**
3. From what **age group?**
4. What are their **spending habits?**
5. What's their **geographic location?** Local? Regional?
6. What **media** do they use most often? Magazines? Newspapers? TV shows?
7. What's the general **attitude** toward the product or service?
8. **Education** level?
9. What are their **product or service needs?**
10. **Lifestyle** indicators (home, car ownership, where they like to eat, what they like to do on weekends, do they like travelling, for example)?

1. _____

_____

2. _____

_____

3. _____

_____

4. _____

_____

5. _____

_____

6. _____

_____

7. _____

   _____

8. _____

   _____

9. _____

10. _____

   _____

Case Study 1

Mark's ideal client is an owner-managed business that has revenue in the range of £75,000 to £500,000, they are based within the M25 loop of London and are in the age group of 35 to 50. They have typically been in business for three years or more and are active users of LinkedIn. They have reached a plateau and don't know how to grow the business to the next level. He works with businesses in the creative industry that have a B2B (business to business) offering and they tend to have less than nine employees on their team. Mark's ideal clients are predominantly male.

Case Study 2

The ideal client for Susan and Brian's mastermind groups are small businesses in the local area wherever a mastermind group is established. They tend to be service oriented businesses that have been trading for at least a year, which means they are through the start-up pains and have a track record or foundation from which to improve. Their ideal clients are frequent users of Facebook and tend to be members of local chambers of commerce or federations of small business. Their ages range from 28 to 45. They are owner-managed businesses and tend to have a small team of no more than five staff. Their yearly revenue tends to be in the sub £300,000 region. They are highly ambitious and go getting in nature though. Susan and Brian's ideal clients can be either male or female –they like to keep the master-minds balanced in that regard.

Case study 3

The ideal client profile for Joanne, Elizabeth and Craig's new growth advisory offering is female with annual revenue in the £100,000 to £2M range. They tend to be service-based businesses operating within the new media, technology, PR and marketing type space. They are currently profitable but are now at the stage where they want to take the business up to the next level. They are highly ambitious, have a thirst for knowledge so frequently attend business seminars and events to quench this thirst. They are members of at least one 'women in business' network. Geographical location within the UK isn't important but they will be UK based. This is because they embrace video technology and other cloud or online tools readily to ensure that their client delivery is borderless in that regard. As a result, their ideal clients are technology savvy so appreciate this point of difference.

EXERCISE Defining your uniqueness

Creating a point of difference in the eyes of potential clients is imperative for your success, as is communicating that point of difference widely, which could involve including the vision you have for your business in material that goes out to clients. It's about letting the world know what your business stands for and why they should do business with you.

This is known as your USP (Unique Selling Proposition) and it is one of the fundamental components of any successful marketing campaign. Put simply, it's a summary of what makes your business unique or different and valuable in the eyes of your target market. It answers the question: 'How does your business offering benefit your clients more than anyone's else can?'

Implementation Steps

1. Describe your target market audience – before you can even start marketing your products and services, you must know whom you are targeting. In this step, you should be very specific.

2. Explain the problem you solve – from your potential clients' perspective, what is the specific need, challenge they face or outcome they are seeking that your business can solve or provide for them?

3. List your biggest distinctive benefits – in this step, list three to five of the biggest benefits a client receives from choosing to work with you that they could not get from another business (i.e., what sets you apart from your competition). You must always put yourself in your clients' shoes when doing this. The benefits you list should articulate why your products and services are important to them and deliver compelling reasons why they would choose you over another provider.

4. Define your promise – an important component of a successful USP is around making a solid pledge or undertaking to your clients. Write down the promise you do or will make to your clients in this step.

5. Combine and rework – once you've completed steps one to four, gather all of the information you wrote down and combine it into one powerful paragraph. There will likely be some recurring ideas, thoughts and messages from the initial steps, so merge common themes where they exist so that overall the paragraph flows and makes sense.

6. Refine it further – in this step, take your paragraph from step five and condense it even more into just a sentence. You want your final USP sentence to be as specific and simple as possible, yet deliver a powerful punch.

In this chapter we have delved into the 5 key SHIFT principles that you must embrace to THRIVE in changing times. The new paradigm means shifting from being *reactive* to more *proactive*, from just being *transactional focussed* to more *commercially focussed*, from being *inflexible* to more *flexible*, from having an *employee mentality* to having a *business owner mentality* and from being *introverted* to being more *dynamic*. We also looked at the key benefits for your business for making these shifts.

We also looked at the importance of niching for greater success and explored the seven main reasons for doing this. Having crystal clarity on your target market and positioning is achieved by spending the time to create the persona of your ideal client, which then allows you to help shape and define your uniqueness.

It is only by creating a point of difference in the eyes of potential clients that fit your ideal client persona that you are able to tailor your marketing and communications for maximum impact – because you will be speaking directly to their wants and needs. It's about letting your target market know what your business stands for and why they should do business with you.

In the next chapter we will explore the fourth key ingredient being **'Your client service ramp and offering ecosystem.'**

CHAPTER 4

Key ingredient #4

Your client service ramp
and offering ecosystem

So far on our journey together, we have looked at the common challenges faced by business support professionals and covered the first three key ingredients, so you should be very clear on what you want your business to be and also have crystal clarity on your target market and positioning.

The next step comes in detailing the buying decision-making journey your ideal client goes on, so that you can create an ecosystem of various products and services to help them on that journey that ultimately leads to them buying your core or signature product or service.

The rules have changed

Unless you have been living in a cave somewhere there's no avoiding the changing landscape that the advent of social media has had. Love it or hate it social media is here to stay, as we have now moved to a more online way of operating. The old ways of approaching things have changed and you need to adapt and evolve the way you attract and interact with either potential clients or existing clients to stay relevant to the changing ways they consume, absorb and digest the volumes of information that are now at their fingertips. It's very easy for you to be lost or overlooked by them in all the online 'noise' that is now common.

In 2011, Google released an eBook called ZMOT in which they coined the phrase 'Zero Moment of Truth.' The term refers to the moment in a typical buying process – from first becoming aware of a product or service to where a decision to actually purchase is made. So... what is the zero moment of truth (ZMOT)? It refers to the point in the buying cycle when a consumer researches a product or service, often even before the seller knows that they exist. The number of people researching products or services online prior to buying has been on the rise in recent years. This is directly related to advances in the internet and mobile technology, making it much quicker and more convenient to do so. In 2011, the average buyer used 10.4

sources of information before making a buying decision. It was half as many sources just a year earlier in 2010 (ZMOT, 2011). This trend is expected to continue.

While much of the research behind the ZMOT has been focused on B2C (Business to Consumer) companies, B2B (Business to Business) companies have something to learn from the principles here as well. Let's look at a simple illustrative scenario of someone encountering the ZMOT during a typical B2B buying cycle:

1. Danielle, a director at a creative agency, sees a PPC (Pay-per-click) ad on Google for business growth coaching for creative agencies. The ad functions as the stimulus in this example because it captures her interest and curiosity enough for her to want to discover and learn more.

2. Danielle decides to do some research. This is the zero moment of truth, it is when she looks at your client testimonials and the services on offer to find the right type of support for her business needs.

3. Next comes the first moment of truth, where Danielle decides to buy the initial business review on offer as the next step.

4. The second moment of truth, which is the experience that Danielle has after buying the initial business review in determining whether to take the next step and so on.

You are who Google says you are.

The most important lesson from Google's research about the zero moment of truth is that business owners need to be aware of the ZMOT, and prepared for it – irrespective of whether they have a B2C or a B2B focused offering.

'If you're available at the Zero Moment of Truth, your customers will find you at the very moment they're thinking about buying, and also when they're thinking about thinking about buying.' (ZMOT, 2011)

Here are three ways that will help ensure your product or service passes the ZMOT test:

1. Make sure information about your product or service is readily available. Clients will be looking for the following types of content to help them make a decision: **client testimonials, service brochures** and **case studies**. They want reviews that are coming from your clients, not directly from you.

2. Focus on **optimisation** of both the desktop and mobile versions of your website. Most research starts with an online search through Google via a computer or phone as a starting point, so it's important that your site is optimised for SEO to improve your 'discoverability'. Try doing searches on Google for your brand, reviews of your brand, and opinions on the best product or service in your industry or marketplace. This will show you exactly what a potential client sees (or does not see) when they're researching your product or service.

3. Think video always. It's not a coincidence that YouTube is now the second most popular search engine on the internet (ZMOT, 2011). Consumers are looking for more visually presented information to help them decide. For a B2B this means incorporating your **business pitch, video testimonials, product tours or demonstrations** and **webinars** into your digital strategy.

Putting an effective strategy in place to cater for the zero moment of truth principles should not drastically change your current marketing strategy. After all, the fact that people research what they are going to buy before they buy it isn't earth-shattering news. But it is really important that you are aware of what information potential clients need during this process, and how easy it is for them to access as part of the overall product or service ecosystem for your business. It's all about the client journey. This is about really knowing the wants, desires and needs of your ideal client and ensuring that the buying process you take them through is completely aligned with it.

For a coaching, consulting or advisory business a potential client will typically go through the following buying process:

1. Experiencing a business 'pain' or have a business 'need' and are seeking a solution for it.
2. Research either online and/or asking their network for suggestions.
3. Initial contact is made; a conversation takes place.
4. A next step suggestion is made for trialing your business in a way that has low perceived risk for them but delivers value and leaves a positive lasting impression to build their confidence in proceeding to the next stage and spending more with you.
5. If expectations are met or exceeded then move to the next stage for a larger engagement, hopefully resulting in them engaging with your core business offering.

Rarely will a potential client that is not known to you spend a lot of money without first going through a discovery and familiarisation process to build confidence, trust and rapport. The higher the price of your core offering, the longer this process will take.

Defining your client service ramp

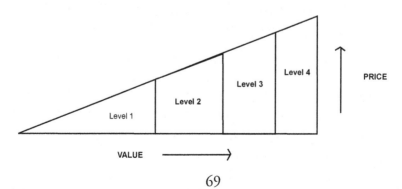

Level 1 – Explore

These tend to be FREE and gift like in nature and designed to show value for the potential client without any obligation from their side, and demonstrate your expertise and specialism. Like with any gift, you should ask nothing in return – not even an email address for them to access the information at this level. It's about embracing all forms of digital media in a smart way – digital assets that can be created once and have enduring benefit. The key characteristic here is that whatever you create should take none or very little of your time to maintain once initially created, so it is imperative that you leverage technology here to achieve this in the most cost effective and time effective way.

Here are 5 suggestions for 'explore' level products:

1. Blog regularly
2. Publish articles online
3. Create short e-books
4. Actively participate in online communities
5. Run webinars

Level 2 – Taster

At this level the client has seen value in your 'explore' assets and now wants to 'taste' more. Here a value exchange is needed though. To give them more value, you want potential clients to reciprocate by either providing their contact information so you can engage further or to actually buy a taster type product.

Here are 5 suggestions for 'taster' level products:

1. Offer a business diagnostic – either for FREE or PAID FOR
2. Produce longer e-books and guides
3. Offer taster workshops or business boot camps

4. Publish a book
5. Run an event

Level 3 – Discovery

This is the level where you go through the process of uncovering and discovering the specific need of a client so that you can ultimately suggest a solution that matches or is aligned with their needs. For example, it may involve conducting an initial *Business Blitz Review* to quickly get underneath where the growth pain points are in a business so that you can have a meaningful one-to-one conversation with that potential client to dig deeper and create an action plan.

Level 4 – Core

Your core service or offering is the signature product, programme or solution that you offer. It should ideally have the following characteristics:

1. Be a full and remarkable solution for your ideal client
2. Be about implementation, not ideas
3. Be priced at a level is profitable for your business

The important thing is that as a potential client progresses through each of the client service ramp levels, they are spending more money with you to get the benefit or outcome you are providing in order to build trust and confidence, which will likely lessen the resistance to buying your core proposition. Having these levels is a great 'qualifying' process in its own right as people will 'disqualify' themselves by not progressing to the next level, which saves you time, energy and resourcing by enabling you to only focus on those clients that want what you offer.

It's important that whatever you define for each of the levels here is completely aligned and relevant to what you want your business to be. Remember, your business should be 'by design' not by default. Let's revisit the case studies, which may stimulate some ideas for you on how you can structure your client service ramp based on what these business owners did with their businesses.

Case Study 1

You will recall that Mark was an independent business growth coach and consultant. His ambitions are to double his revenue but to spend less time on what he saw as wasted hours, having coffee meetings which routinely didn't result in any new business. His issue was really around not sufficiently pre-qualifying the conversations he was having and in expecting people to agree to his core offering too soon in their buying process.

Let's see what Mark's Client Service Ramp now looks like.

Level 1 *– Blog and article writing predominantly but has strong social media presence so he is 'discoverable' by potential clients when they go to research him on Google.*

Level 2 *– Offers a Business Health Check for £99 – the client gets a full 20-page feedback report on how they answered each of the 35 questions. At this point he then determines whether the client is suited to go to the next level or not based on their responses and scores from the diagnostic.*

Level 3 *– Business Blitz for £395 where he bundles in all four business growth diagnostics from our service together with one hour of his time with the client to run through the results and dig deeper into the issues uncovered. This allows him to then tailor the next steps of the core offering for that potential client, as he is getting 140 insights on their business as part of the blitz. This only takes 20 minutes of the client's time to complete, but most importantly none of his time at this stage. This is in sharp contrast to the way he used to work, wasting sometimes up to three* *hours on a potential client who didn't end up buying.*

Level 4 – He has rejigged his core offering and settled on an offering that he'll charge at £500 per month.

ⓘ Case Study 2

Susan and Brian's challenge was in better leveraging their networking community to feed new participants into their mastermind programme. Let's see what their Client Service Ramp now looks like.

Level 1 – Free networking community. They regularly produce tip videos and articles as part of their vibrant community.

Level 2 – They now use FREE webinars with great effect. At the end of each webinar they give a call to action to attend one of their ½ day taster workshops, which they have a set schedule for. The cost for this workshop is £99. As an incentive for the webinar participants to take immediate action, what Susan and Brian now say is that if clients book within 48 hours then not only can they attend the taster workshop, but also as a bonus they will have a Business Health Check done on their business (valued at £99 also). This sees the client getting a full 20-page feedback report on how they answered each of the 35 questions. So instead of doing an early bird discount to encourage action, Susan and Brian are smartly doing an early bird 'value add'. Giving more value for rewarding those who take prompt action.

Level 3 – From the attendees to the taster workshop they then offer them the ability to upgrade to the Business Blitz Review – like Mark from Case Study 1 – for £395, and they derive the same end benefits that Mark does.

Level 4 – There has been no change to their core master-

mind offering which costs participants £300 per month. There has been an important change in the dynamic here however, through the adjustments made at Levels 2 and 3. This is because now, for everyone in their mastermind group, Susan and Brian have 140 growth insights into the clients' businesses, which enables them to tailor their support based on the needs of the group. It also becomes an important benchmark from which to track improvements. They plan to conduct Business Blitz Reviews with their participants every 12 months. This enables them to compare it to the previous results for each client, which is a seamless way of quantifying the improvements being made for participants as the 'value' they've received by being supported in the mastermind. This is hugely beneficial to Susan and Brian for attracting new delegates, as they can promote the value impact they have for the businesses that work with them.

Case Study 3

As the lead on the project, Joanne took control of shaping this. Let's see what their client service ramp now looks like for their advisory business.

Level 1 – *Actively participate in online communities, providing tips and producing guides.*

Level 2 – *Offers a FREE Business Health Check (Valued at £99), which enables them to pre-qualify who possible candidates are for progression to the next level based on their responses and scores from the diagnostic.*

Level 3 – *They then offer a Business Blitz Review (like Case Studies 1 and 2) for £297. If the potential client agrees to it then one of their business growth associates will conduct the blitz. For them this ensures that they have a structured and systematic way of generating qualified leads so that* *their business growth associates are only engaging in pro-ductive and chargeable work. Win/Win.*

Level 4 – They have developed a tariff of services which is modular in nature, making it easy for them to tailor a bespoke solution to suit the growth needs of their clients. Whilst a typical 'accounting only' client generates on average £1,500 of revenue per year, they expect that a typical business advisory client will generate on average £3,000 of revenue per year.

EXERCISE: Defining your client service ramp and offering ecosystem

Now it's your turn: think about your ideal client based on the profile you defined in this last chapter, and consider all that you want from your business. Write down some ideas for each of the client service ramp levels that work for attracting your ideal clients but still enable you to build the business you want. Brainstorm at least three ideas for each.

Level 1 – Explore

1. _____

2. _____

3. _____

Level 2 – Taster

1. _____

2. _____

3. _____

Level 3 – Discovery

1. _____

2. _____

3. _____

Level 4 – Core

1. _____

2. _____

3. _____

In this chapter, we've looked at the principles of ZMOT and how to make sure your products or services pass the ZMOT test. We explored why you are what Google says you are, and how it is important to put your best foot forward here. I explained how a potential client that is not known to you would rarely spend a lot of money without first going through a discovery and familiarisation process to build confidence, trust and rapport. The higher the price of your core offering, the longer this process will take.

As such, you must be able to describe in detail the buying decision making journey your ideal client goes on so that you can create an ecosystem of various products and services to help them on that journey, which ultimately leads them to purchasing your core or signature product or service. Your client service ramp and offering ecosystem should have four levels to match the ideal client journey, being:

Level 1 – Explore
Level 2 – Taster
Level 3 – Discovery
Level 4 – Core

It is very important that whatever you define for each of the levels that it is completely aligned and relevant to what you want your business to be. Remember, your business should be 'by design' not by default.

So, in the next chapter we will grapple with the fifth key ingredient being **'Pricing profitably.'**

CHAPTER 5

Key ingredient #5

Pricing profitably

Profit is the aim of the game for most businesses, otherwise what is the point of having to deal with all the curve balls that are frequently thrown at us as business owners? If you are in business purely for the love of what you do, then you can do what you do for enjoyment, perhaps as a hobby – you don't necessarily need the additional burden of dealing with everything that comes with running a business if that is the case.

The purpose of a business is to make money. It is how that profit is used that differentiates a private business from a charitable organisation. The profits in a private company are generally used by the owner for whatever they choose, whereas the profits of a charity are used to re-invest and fund new services. Don't be mistaken however; they must both make money to survive.

Intelligent or profitable pricing is **one of the most important** elements for any successful business. Yet many business owners fail to understand and educate themselves enough about the various pricing components and strategies before launching their new business.

Smart business owners will weigh many factors before setting prices for their products and services. Business owners must understand their market, distribution costs and competition. Remember, the marketplace will generally respond rapidly to technological advances and competitive forces. As such, businesses must proactively monitor the factors that affect pricing and be ready to adjust quickly as needed.

Revenue versus profitable revenue

There is a very important distinction between revenue and profitable revenue. Revenue is basically derived from whatever you charge your clients for your products and services. You can charge anything you want, but market forces usually cap it. By market forces I mean supply and demand in the segment of the market your business operates within.

Just because you have decided to charge a certain amount for your products and services is not a guarantee that you are making a profit on that revenue. Many business owners make the mistake of confusing revenue and profitable revenue, and it is usually because they have adopted an ostrich mentality, or the fact that they are scared of anything numbers related, that they haven't taken the time to actually sit down and work out whether they are making a profit on what they are charging or not.

Many of these same business owners are adopting the philosophy that as long as money is flowing into their bank account, all is well. The reality is that just because money is flowing into your bank account from revenue doesn't necessarily mean you are making a profit on it.

I remember as a young boy of twelve I became interested in vegetable gardening in a big way. I grew up in a rural area near Adelaide in Australia, surrounded by dairy farms and market gardens. I started with a small plot of about two metres by one metre, but soon ran out of space. I wanted to grow more and a bigger area to grow them in, so I had an idea to grow seedlings and sell them to my Mum's nurse friends. I also knew some of them were into vegetable gardening, so it was an opportunity for generating some cash to buy more stuff I needed for my garden. I charged $1.50 per punnet, when the nurseries (garden centres) usually sold them for $2.50. I sold 37 of them in that lot for a sum total of $55.50. I was completely chuffed to bits.

But did I actually make any profit on that? I needed containers, soil, seeds, water and fertiliser to produce those punnets of vegetable seedlings. I largely used materials I already had to do this, but I did have to buy more supplies.

My point is that I was so focused on the amount of revenue I generated that to this day I don't know whether I actually made money or lost money in my little enterprise. I didn't sit down and work out what the costs would be to produce each punnet and whether what I was charging Mum's nursing colleagues was more than it cost me to grow the seedlings, leaving enough of a difference, i.e. the profit. I have never forgotten this lesson, and thankfully I learned it very early on, so was able to take this into my next enterprise at sixteen, which saw me generate enough profit to buy my first car.

There are two approaches that can be used to work out if your revenue is profitable to you or not. They are called the top-down approach and the bottom-up approach. I have tried to break it down in a simple way to illustrate and explain the process of completing the exercise. It is like anything: it can be made as complex or kept as simple as one chooses. I don't know about you but I like to try and keep everything as simple as possible.

Top-down approach

As the name suggests, it entails starting at the top with the selling price and working down from there. It is more commonly used than the bottom-up approach. The three steps to do this are:

1. Ascertain market price – i.e. what price the market will accept.
2. Allocate desired profit margin.
3. Determine if the product or service can be provided within the net amount.

For the purposes of this exercise let's assume that you are looking to introduce a new product with the following information:

Scenario 1 – Top Down

- Price (P) = £500 per unit
- Desired Profit Margin = 40%
- Variable Cost (VC) = £350 per unit

STEP 1:   The market price is £500
STEP 2:   The desired profit margin is 40%, which means £200 per unit profit is desired (£500 x 40%)

STEP 3:   The implied maximum cost allowance = £300
           (£500 – £200)

Upon further investigation, the cost to produce this new product was shown to be possible at a variable cost of £350 per unit. As £350 is greater than the implied maximum cost of £300, you are faced with one of three options:

1. Review all the information again to ensure that no further favourable changes can be made to either the market price or cost to produce.

2. Reduce profit margin expectations, which means that more products will need to be sold to achieve break-even when overheads are factored in.

3. Or you can pass on the new product idea completely because the numbers don't stack up to meet your expectations. The important thing here is that by doing this exercise a rational and informed decision can be made on how best to take the business forward in a profitable way. It is not reliant on chance or luck.

Bottom-up approach

The bottom-up approach, as the name suggests, is the reverse of the top-down approach. The three steps to do this are:

1. Determine the cost.
2. Add the desired profit margin or mark up.
3. This determines the selling price.

Scenario 2 – Bottom Up

- Variable cost = £400 per unit
- Desired mark up = 25%

STEP 1:   The cost in this case has been determined at £400 per unit

STEP 2:   Work out the desired mark up amount, which is £100 in this case (£400 x 25%)

STEP 3:   Add the results from Step 1 and 2 together for an indicative selling price for that product of £500 (£400 + £100)

You can then test whether this £500 selling price is acceptable to the market in which you operate. If it is, then great, you could be on to something that meets your profit expectations and the market wants. Everybody is happy.

Your pricing strategy

There are many different pricing strategies from which you can choose, but you need to select carefully. An ill-considered decision in this area can place a heavy burden on your business and have severe consequences. Let's run through six such strategies now:

Pricing strategy 1 – Price bundling

This involves giving your clients the option of buying several elements of what you offer for one price. A consultant may offer a ticket to an event and a one-to-one discovery session with them at a price that is somewhat lower than what the two would cost if bought individually. Bundling has many advantages for service-oriented businesses. This is because the cost structure of most service businesses is such that the cost of providing an extra service is less than providing the second service alone. A second benefit of bundling that appeals to busy clients is that, by buying related services from one service provider, they can save time and money by interacting with only one rather than multiple providers. Third, bundling effectively increases the number of connections or touch points you have with your clients.

Pricing strategy 2 – Multiple pricing

Similar to price bundling, multiple pricing involves selling multiple units of an item for a single price. Buy-one-get–one–free offers are an example. A grocery store or supermarket that offers two boxes of biscuits at a single price, for instance, is using a multiple pricing strategy. Whereas price bundling is more commonly employed for big-ticket items, multiple pricing is usually used to sell lower cost 'explore' or 'taster' type products, of the sort we covered in the previous chapter.

Pricing strategy 3 – Competitive pricing

You may choose to base your own prices on the prices of your principal competitors. If you choose to do this, however, you should make sure that you are looking at competing businesses of similar size and strength to yours. Bigger businesses can often buy in larger volume and therefore their cost per unit will be less. Instead, where possible price products or services on the basis of like for like comparisons and then highlight other competitive factors where you are superior to the competition, such as personalised client service and superior results.

Pricing strategy 4 – Value based pricing

In a market where price is not a client's most important consideration, you can do very well employing this strategy. The key to making 'value based pricing' work is to determine the value that your product or service has to your ideal client and aligning your price to that. What is it worth to them? It moves away from the more time or cost based pricing approach. Clients come to you for results, so they want to pay for you helping them to get that result. They don't want to pay for your time, which is transactional and an old paradigm way of thinking, but instead they want to buy the result they are seeking. It's really about your pricing reflecting your 'secret sauce' or 'value add' that you bring to the table for them that your competition can't or don't to the same extent as you.

Pricing strategy 5 – Pricing below competition

Pricing below competition involves setting your prices below those of your competitors. It is commonly used as way to win new business or entice clients away from your competition, but be very careful here. It is a very dangerous game if not managed carefully as it could undermine your market entirely by lowering the perceived value of what you offer in the eyes of the very clients you are seeking to attract, ultimately damaging your profit margins. There is usually little margin for error so you must be 'very hot' on your numbers and monitor the impact very closely to pursue this strategy successfully.

Pricing strategy 6 – Odd pricing

Odd pricing is used almost universally now in some way or other. It is the practice of pricing goods and services at prices such as $99 (rather than $100) or £997 (rather than £1,000) because of the belief that clients will often round the price down rather than up when deciding whether to buy or not. This element of pricing psychology has become so universally employed that some question its value, but despite this the practice remains widespread.

16 key questions for determining your pricing strategy

There are numerous things that you must consider when determining the pricing strategy for your business. These range from the needs and desires of your target clients to general economic conditions. The following questions are among the most important to consider when arriving at a pricing strategy.

1. Is the price of your product or service of significant importance to target clients?
2. How popular is your product or service being offered?
3. What pricing and marketing strategies are compatible

with your business's other characteristics (service mix, location, service reputation, etc.)?

4. Are there opportunities for special market promotions?
5. What are your competitors charging for similar products or services?
6. Should your competitors' temporary price reductions be matched?
7. Will prices generate a satisfactory profit margin after calculating operating expenses and reductions?
8. When reducing prices on products or services, what are your competitors' likely reactions?
9. Are there legal factors to consider when establishing price?
10. Should 'odd pricing' or 'multiple pricing' practices be introduced?
11. Should your marketing efforts highlight sales of selected high-profile products or services to attract clients?
12. If discount incentives are offered, how will they impact your net profits?
13. To what extent will the delivery or execution of the product or service sold meaningfully add to your operating costs?
14. Are economic conditions in the area of your operations good or bad?
15. How does your pricing fit in with your client service ramp focused on the buying process for your client?
16. What is the perceived value of your product or service to your ideal client and how aligned is your pricing to that?

Profit under the spotlight

In business the emphasis is on getting more CLIENTS, growing REVENUE and generating PROFIT. But there is a subtle distinction that needs to be made here. These areas are actually 'results'. They are the end product of certain activities. Strictly speaking they can't be altered directly. You can't simply get more clients, revenue or profit.

There are in fact other factors that determine these results, so to change these results we need to understand and focus on the underlying factors that influence these results. Let's look at the steps involved in closer detail.

STEP 1:   You identify through marketing or other means people that are **potentially** interested in your products and services.

STEP 2:   You have sales conversations with them and they decide whether to buy your products and services or not. If they do proceed, it is at this point that they become a CLIENT.

STEP 3:   Once they have decided to buy from you a trans action will take place at the time they buy, whatever they are seeking.

STEP 4:   They will pay a certain price for that transaction to occur and receive the products and services they are seeking. It is the transaction and price paid for that transaction that results in the REVENUE generated from that client.

STEP 5:   Provided you have priced your revenue properly, you will generate a margin on that revenue. It is the margin on revenue you generated that ultimately determines the PROFIT you make from that particular client.

So it follows that if the desired result is to increase overall PROFIT then you need to be talking to more people who are potentially in the market for the products or services you provide, and you need to get more of them to say 'yes'. Of those that say 'yes' and become a client, you need to encourage them to buy more frequently from you, spend more with you each time they do buy from you and ensure that you are making a decent margin on the revenue you are generating. That's how you grow profits. I hope you can see the underlying logic of this.

If we take it a step further, from the five steps we covered, there are five key areas to focus on to grow profits, which can be summarised as follows:

KEY FOCUS 1:   Number of leads (L)
KEY FOCUS 2:   Conversion rate (C)
KEY FOCUS 3:   Number of transactions (T)
KEY FOCUS 4:   Average sale value (S)
KEY FOCUS 5:   Margin (M)

Profit Accelerator Formula

Each of the five key focus areas interacts in what I call the Profit Accelerator Formula as follows:

Number of leads (L)
X
Conversion rate (C)
=
NUMBER OF CLIENTS
X
Number of transactions (T)
X
Average sale value (S)
=
REVENUE
X
Margin (M)
=
PROFIT

KEY FOCUS AREA 1 – Number of leads (L)

This is the total number of potential clients that you have contacted or that contacted you last year – also known as prospects or potentials.

KEY FOCUS AREA 2 – Conversion rate (C)

Your conversion rate is the percentage of people who did actually buy from you as opposed to those who could have bought. For example, if you had ten conversations with potential clients and only three of them became clients, you'd have a conversion rate of three out of ten, or 30%.

KEY FOCUS AREA 3 – Number of transactions (T)

Some of your clients will buy from you weekly, others monthly, others on the odd occasion and others just once in a lifetime. What you want to know now is the average – not your best and not your worst, but the average number of times one of your clients buys from you in a year.

KEY FOCUS AREA 4 – Average sale value (S)

Every product or service a business provides has a price associated with it, which represents what must be paid by a client to buy it. The price represents the sale value for that item. However, if a client buys more than one thing at the same time, then the sale value for that particular transaction is the sum total of all items purchased at that point.

KEY FOCUS AREA 5 – Margins (M)

Margins represent the percentage of each and every sale that is profit. In other words, if you sold something for £100 and £25 was profit, then you've got a 25% margin. Remember, this is after **all costs** are taken out.

I wrote extensively about this key formula and its power in my first book *Profit Rocket*. It makes an appearance again in this book because it should be a foundational element and cornerstone for every business. The elements of this formula form five of the ten key ingredients we cover in this book and we will explore them further over the coming chapters.

It provides you with the five key magic numbers that you need to focus on to grow your business. Making these focus areas a priority in your business and directing your energies towards improving them consistently over time is ultimately the key to growing a highly profitable business coaching, consulting or advisory business. It is very easy to become overwhelmed by business numbers if that isn't your strongest skill, but knowing and applying this formula eliminates all that and focuses your attention on what matters. I promise that if you focus your energies on this, your profit will take care of itself. But first we need to have a starting point from which to begin and measure improvements from.

Determining your business baseline for improvement

For each of the five key focus areas take your last set of annual accounts and break the results down into the five key focus areas of the Profit Accelerator Formula we have covered:

1. Number of leads
2. Conversion rate
3. Number of transactions
4. Average sale value
5. Margins

Number of leads

You may already have a customer relationship management (CRM) system that you use to track this information, from which you can simply extract the figures. It's okay if you don't, a simple solution for getting this is to start tracking all of your leads over a 90-day period. Keep track of it on a simple spreadsheet or even in a notebook. With this exercise you are seeking a starting point, a benchmark from which to improve.

If you don't already have a system in place to capture this

information on an on-going basis, there are many inexpensive off-the-shelf software packages that you can buy to help in this area. These software and Cloud solutions can have either simple or more advanced functionality. Find the right package that works for you and is aligned to the needs of your clients and business.

Asking a simple question such as 'How did you hear about us?' can provide valuable insight into where best to target your activities moving forward.

Conversion rate

If you aren't currently tracking your conversion rate, you need to start doing so. It's okay if you haven't yet, we can work backwards to identify what it is by looking at your **number of active clients** in a 12-month period and dividing that by your total number of leads identified in the above step. Remember, the goal here is to have a starting point.

Number of transactions

This figure will usually come from your accounting system. It is basically the number of invoices you have produced and sent out in that year. Once you have this number you simply need to divide it by the total number of unique clients you have invoiced at least once in the same period.

Average sale value

This figure can be worked out by working backwards; by taking the total revenue figure for the year and dividing it by the average number of transactions (i.e., the number of invoices produced and sent out for your products or services for each unique client).

Margins

You calculate this by taking the net profit figure from the last set of annual accounts we are using for this exercise, which is usually at the bottom of the Profit and Loss statement, and dividing it by the total revenue as shown on the same report.

🕐 Case Study 1

For Mark, the information he gathered for this exercise was as follows:

Number of leads: 208 (4 per week) (A)
Number of clients: 12 (B)
Total revenue = £45,000 (C)
Number of invoices generated: 129 (D)
Net profit = £10,500 (after paying himself a £24,000 wage) (E)

Reverse engineering the base line Profit Accelerator Formula from which he would start to improve involved the following:

Profit Accelerator Formula applied

Number of leads (L)	*208 (A)*
Conversion rate (C) =	*5.8% (B / C)*
NUMBER OF CLIENTS X	***12 (B)***
Number of transactions (T) X	*10.75 (C / S / B)*
Average sale value (S) =	*£348.84 S (C / D)*
REVENUE X	***£45,000 (C)***
Margin (M) =	*23.3% (E / C)*
PROFIT	***£10,500 (E)***

Case Study 2

For Susan and Brian, the information they gathered for this exercise was as follows:

Number of leads: 240 (A)
Number of clients: 36 (B)
Total revenue = £115,000 (C)
Number of invoices generated: 383 (D)
Net profit = £29,500 (E)

Profit Accelerator Formula applied

Number of leads (L)	240 (A)
X	
Conversion rate (C)	15.0% (B /C)
=	
NUMBER OF CLIENTS	**36 (B)**
X	
Number of transactions (T)	10.65 (C / S / B)
X	
Average sale value (S)	£300.00 S (C / D)
=	
REVENUE	**£115,000 (C)**
X	
Margin (M)	25.65% (E / C)
=	
PROFIT	**£29,500 (E)**

Case Study 3

For Joanne, Elizabeth and Craig, the information they gathered for this exercise was as follows:

Number of leads: 4,750 (A)
Number of clients: 485 (B)
Total revenue = £745,000 (C)

Number of invoices generated: 550 (D)
Net profit = £157,940 (E)

Profit Accelerator Formula applied

Number of leads (L)	*4,750 (A)*
X	
Conversion rate (C)	*10.2% (B /C)*
=	
NUMBER OF CLIENTS	**485 (B)**
X	
Number of transactions (T)	*1.13 (C / S / B)*
X	
Average sale value (S)	*£1,354.55 S (C / D)*
=	
REVENUE	**£745,000 (C)**
X	
Margin (M)	*21.2% (E / C)*
=	
PROFIT	**£157,940 (E)**

EXERCISE

What's your baseline?

Number of leads =   _____ (A)

Number of clients =   _____ (B)

Total revenue =   _____ (C)

Number of invoices generated =   _____ (D)

Net profit =   _____ (E)

Profit Accelerator Formula applied

Number of leads (L)	_____	(From A)
X		
Conversion rate (C)	_____	(B / C)
=		
NUMBER OF CLIENTS	_____	**(From B)**
X		
Number of transactions (T)	_____	(C / S / B)
X		
Average sale value (S)	_____	S (C / D)
=		
REVENUE	_____	**(From C)**
X		
Margin (M)	_____	(E / C)
=		
PROFIT	_____	**(From E)**

In this chapter we have covered the important distinction between revenue and profitable revenue and explored both the top-down and bottom-up approaches for profitable pricing. We ran through six pricing strategies that you may choose to adopt, and then shared with you 16 key questions for determining your pricing strategy and put profit under the spotlight. Here you were introduced to the incredibly powerful Profit Accelerator Formula and its components were explained. You were provided with detailed instructions for determining your business baseline for improvement, so hopefully you have now done that, following the three case study examples.

So, in the next chapter we will continue to the sixth key ingredient being '**Consistent lead generation**.'

CHAPTER 6

Key ingredient #6

Consistent lead generation

Number of leads (L)
X
Conversion rate (C)
=
NUMBER OF CLIENTS

Consistent high quality lead generation is a challenge for many business support professionals but ultimately one of the ten key ingredients that result in your success. So what is meant by 'number of leads'?

This is the total number of potential clients that you contacted or who contacted you in the last year, also known as prospects or potentials.

Most business owners confuse responses, or the number of potential buyers, with results. Just because the phone is ringing or enquiries are arriving in your inbox doesn't necessarily mean revenue will result.

And what is even more amazing is that very few businesses even know how many leads they get a week, let alone from each and every marketing campaign.

A lead generation strategy is basically your marketing strategy. The purpose of a marketing strategy is to reach new pools of potential clients that may be willing buyers of your products and services.

Before we get into some ideas for things you can do to improve your number of leads there are seven things to keep in mind when putting a lead generation strategy into action.

Be targeted in your marketing activities

A scattergun approach rarely works and can be a complete waste of your marketing money. You should be very clear on the profile and demographics of your ideal client. You need to determine their habits and behaviours, which includes understanding the things they like to do, read and watch. The goal here is to focus your marketing activities on those areas that your ideal client will be exposed to as they go about their daily lives.

Set your objectives before undertaking the strategy

This involves understanding the reasons why a particular marketing activity is being undertaken. If you don't set objectives then how can you know whether the strategy has been successful or not?

Be clear on the outcome you're seeking

You must have a desired outcome in mind. It involves being clear on what you are expecting your target audience to do as a result of the communication. Is the desired outcome that they sign up for a mailing list? Purchase a particular product or service? Register for a newsletter? Be clear on the action and response you expect.

Test and measure

This is a big one and an area in which most businesses fail miserably. It is important to determine whether a particular strategy will deliver the intended results or not before embarking on a massive campaign. This involves starting with smaller test campaigns to see how effective the main campaign might be.

Often subtle changes to the wording or colours used can have a dramatic impact on the results. Try them on a small scale to begin with and measure the results. For example, say you are looking at producing a flyer. You may have six versions of an advert with different wording, look and feel. It is best to do restricted tests within a distribution area, which represents a subset of the overall market you are targeting. The responses to each small test campaign with different wording and content will vary and from that you can identify what works best, refine it further and keep testing until you crack the winning formula that delivers a more positive result.

You should do this testing, measuring and refining cycle before committing to a huge campaign. When you think about it, what is the point of committing to the time and expense of jumping in the deep end with, say, a 100,000 run of flyers when the outcome is completely unknown? By doing smaller runs of, say, 500 flyers and measuring the results as you go, you are not throwing away unnecessary money. Marketing is a science as well as an art. It needs to be approached in a methodical and deliberate manner.

The key message to remember here is that the only way you can know whether a marketing campaign is working or not is to test the options and then measure and track the results. How will you know otherwise? It's just like when we try a new food. We usually have a taste first and, if we like it, we feel confident enough to eat more. The same principle applies here.

Ensure you are getting a decent return on your marketing investment

Many business owners make the mistake of viewing marketing as a cost. It is a common reaction in leaner times that marketing is one of the first areas that a business will cut back on. For me, this represents flawed thinking and is a silly decision to be making. Targeted and effective marketing represents the future sales pipeline of your business. Why then would you choose to cut back on the very area that will generate future revenue?

In my opinion, marketing must be viewed as an investment and, as with anything, there can be good and bad investments. Take a share portfolio for instance. Contained within that are a number of different types of shares, and some of those will be performing better than others in the portfolio.

It is only by tracking and measuring the performance of those shares that you can determine which are performing better and which aren't. Part of portfolio management involves letting the good investments run and getting rid of the bad. When investing, we are told that a balanced portfolio is the desired objective. This means holding a number of different shares in the portfolio to diversify our risk so that if one investment performs badly, then the good performance of other shares helps offset the bad performance. It's all about not having all of our eggs in the same basket. If we were to drop that basket and all the eggs break, we'd have nothing left.

The exact same investing principles apply to marketing. We need to have a balanced portfolio of marketing activities within our business so that we are not reliant on just one. We need to know what is working and what isn't by constantly monitoring the results; by testing and measuring.

We want to stop the marketing activities that aren't working and increase the marketing activities that are working. It should also be noted that people and markets evolve. Something that is working one month may not necessarily work as effectively in another month. Having a portfolio of activities helps balance out this impact.

The distinction here is between conscious marketing and unconscious marketing. Conscious marketing is deliberate targeted marketing and should be seen as an investment. Like any investment, we expect to receive a return that is acceptable to us. Unconscious marketing that is not deliberate, not targeted and not measured is actually a cost and should be stopped immediately.

If a strategy isn't delivering the intended results after giving it a good go – cut your losses and move on. Time is money

Don't be scared to stop doing something that isn't working. As with investing, if a share price is performing poorly and you are losing money on it, there reaches a point where you need to cut your losses and redirect your resources into finding an alternative. Frequently, business owners fall into the trap of doing what they have always done. They continue to do something because they have become used to doing it.

Don't get emotionally attached to a particular marketing activity. It must be viewed as an investment decision. If something is not delivering the intended results then stop doing it immediately, cut your losses and move on to identifying the next initiative to try. It would be lunacy to continue doing the same thing simply because it is what you have always done and to expect a different result.

Be creative and have fun with it!

You should see marketing as a game. It is a game that you can have fun with, thinking up new ideas that are innovative and creative for reaching your ideal clients. Be bold and go for it. The real value to a business lies in the execution of the ideas. Remember, conscious marketing!

5 key LEAD IMPROVEMENT strategies

STRATEGY L1 – Strategic alliances and joint ventures

To grow and prosper as a business, more often than not, you will reach a point where it becomes necessary to find creative new ways to expand and develop into new markets or grow within your existing market. Strategic business alliance relationships are a really effective way of doing this.

This involves working together with other businesses that are not your competitors and have common objectives. This is where you help each other get more business for both of your businesses. It's all about creating mutual benefit and a win/win situation for all involved.

The pooling of knowledge, resources and talent with a collective and truly aligned outcome in mind is at the heart of all successful strategic alliances. When strategic alliances have the 'right fit' and are executed properly then they can be incredibly effective, but if the fit isn't right from the outset then it can often turn into a complete mess.

Implementation steps

1. Identify potential partners both within and outside of your existing network, where you see that there could be a strategic advantage for working together.

2. Approach and have a conversation with them to explore the possibilities – it mustn't be one-sided though. It needs to be balanced and ensure the objectives of all parties are met.

3. Define and document how any collaboration will work before officially starting anything.

Success tip

There are five key areas that are important when entering into a new strategic alliance to maximise the chances of a successful outcome. They are: Compatibility, Motive, Readiness, Planning/Outcome and Agreement. If you want to gauge how well your intended strategic alliance ranks for these areas then you should check out the *Strategic Alliance Fit Indicator* on our profitinfocus.com website.

STRATEGY L2 – Ask your existing clients for referrals and set up a referral scheme

Client loyalty is a key aspect of any business's strategy, not least because client retention is far more cost-effective than securing new clients. If your clients are loyal to you then they will likely bring you new business. You just need to ask them for it. A happy client is one of the biggest assets there is – they will advertise you to people they know, for free!

According to the latest Global Trust in Advertising Report by Nielsen, 92% of consumers trust recommendations from family, friends and close colleagues above any form of advertising. If a friend tells you about a bad experience with a certain brand or product, it will likely deter you from buying something from that brand. Similarly if someone you know raves about a certain restaurant or coffee shop, for instance, you might check it out for yourself.

The concept of loyalty should never be overlooked by a business, especially as it can also be used to win new business and to reward existing clients through referral schemes. Although loyal clients may have willingly recommended the business anyway, an incentive can give an extra helping hand and works in two ways: it acts as another boost to existing clients' good opinion of a business and can also drum up new business through word of mouth. Some reports suggest that incentivised referrals can increase client acquisition success by up to 20%. It can also see up to a 25% higher average order value. Interesting, isn't it!

Implementation steps

1. Be clear on what type of referral you want to receive from your clients to make sure they fit your ideal client profile.

2. Make sure that whatever your referral reward is, it has relevant value for your intended recipient.

3. Ask existing happy clients if they would consider referring people they know to you. Set up a system within your business for asking for referrals.

Success tip

One of the key components for success here is always putting yourself in the shoes of your clients and asking yourself, 'Is this motivating me to refer my friends or network to this business?' Make it a quick and easy process for your clients to refer people to you. Consider a double incentive that involves rewarding both the referrer and the new client also with a special incentive. Above all, make sure an incentive makes sense for your business and results in a profitable outcome.

STRATEGY L3 – Raise Your Profile

This is all about raising the profile your business. It is about becoming 'known' in your marketplace as the 'go to' resource for what you do. Here you could perhaps write a book, utilise press releases, generate more PR or become a speaker, which are a just few suggestions from many possibilities.

Press releases, both online and traditional offline, generally cost very little to produce and distribute but can make a massive difference for the exposure of a business if done in the right way.

When it works well, money can almost never buy advertising with an equivalent impact and reach as a properly executed press release. Editors want relevant, informative and interesting content for their readers. Produce that for them and they will love you, and your business will benefit from the exposure as a result.

Implementation steps

1. Be clear on your business message or what you would want to speak about. Taking a very close look at where your expertise and your passions lie may reveal that what you think you want to talk about and what you're actually qual-

ified to talk about with credibility are two different things. Often, it's better to give a talk on a very specific area of specialism than to present a more general talk about a topic on which you have less to offer.

2. Do your homework. Pay attention to the kinds of people who write for publications or speak at events and get to know the event organisers. If you can establish yourself here, you'll be more likely to be invited to write or speak, or at least more likely to be accepted if you ask. People are far more likely to pay attention to someone that they've met before.

3. Execute. Nothing beats action. Do and build things that show you're good at what you do in a real, tangible way, and people will take notice.

Success tip

The key objective here is to present yourself as an expert. If you're highly knowledgeable on a topic, nobody will likely know about it unless you demonstrate that knowledge. By writing a book, releasing articles and/or speaking publicly on your specialism area within your area of expertise, you can position yourself as an authority within your industry.

STRATEGY L4 – Put on a seminar or event by yourself or collaboratively with others.

Become a speaker on your subject matter – once again this could provide instant credibility if you know what you're talking about and, of course, deliver a solid presentation. The success of a seminar, workshop or event depends on how well it is organised. To help you with it, here are some straightforward tips that will ensure it runs smoothly and achieves its goals.

1. **Define your objectives** – From the outset, stop and ask yourself the following question: 'Why am I putting on this

event?' Clearly defining your expectations for the project will help you make the right decisions throughout the planning process. You should also take your target audience into account. After all, seminars for partners are very different events to those arranged for colleagues or clients.

2. **Choose the theme and format** – The first thing to do is come up with a catchy title that will create a framework for your project and a buzz among participants. You then need to decide if you're planning on a weekend, one-day event or a dinner, all while keeping the objective of your seminar very much in mind.

3. **Create a budget**

4. **Choose the date and location** – To attract the maximum number of guests to your event you need to get the date right, which means avoiding bank holiday weekends, school holidays and major national sporting occasions.

5. **The event schedule** – The day itself needs to be carefully planned. And there's no better way of doing that than drawing up a detailed schedule, which you can then send out to your guests. As well as arranging time slots for conferences and activities, make space for a break in the middle of the day and give yourself a little time to play with in the event of something unexpected happening.

6. **Get the word out before the event** – Regardless of whether your participants are employees or external to the business, it's important to get a communication plan in place. You also need to plan sufficiently ahead, making sure people attend by getting invitations out to them and asking them to reply. That way you'll know how many people will actually be coming.

7. **The day before, confirm everything** – On the eve of the event, try to visit your venue and go through a few dry runs if need be, especially if you've booked a performer. Contact the people who will be appearing at your event and let them know the schedule.

8. **Follow up after the event with all delegates** – There's still work to be done even when your event is over. One week afterwards send your participants a little word of thanks and an account of the day, which you can perhaps even brighten up with photos taken during the seminar. The day after the event you should also create a satisfaction questionnaire to get some feedback from participants.

STRATEGY L5 – Optimise your website and embrace digital marketing

This a big subject area in its own right and includes such things as Search Engine Optimization **(SEO)** for your website, Pay-Per-Click **(PPC)** advertising and much more.

SEO is about maximising the visibility of your website in the 'natural' or 'organic' (un-paid) search engine results.

PPC stands for **pay-per-click**, and is a form of online marketing whereby advertisers pay a 'fee' to the search engine company each time one of their ads is clicked. It's effectively a way of buying visits to your website. Search engine advertising is one of the most popular forms of **PPC.**

One of the best ways to get your business noticed online is to make sure it appears near the top of search engine results for a given search term (remember ZMOT from Chapter 4). There are lots of tips and tricks you can use to do this, including:

1. **Keyword research** – It's essential to research and embed the right keywords in your website. Search engines will use these to rank your site. Using a free tool like the Google Keyword Research Tool allows you to see how often the keywords you want to use are searched for each month. The tool will be able to tell you which search terms are most popular for people searching for what your business does.

2. **Use tags properly** – Search engines also use 'tags' to rank

websites. Some of the most important tags are the title tag and the Meta tags. These tags are applied behind the website and you use keywords in them to help search engines rank your website. The page title and Meta description tags are commonly used by Google to rank search results.

3. **Look actively for opportunities to embed relevant keywords** – It's important to embed your keywords in your website in such areas as headings, content and pictures because the more often your keywords are used in the content of your website, the easier it is for search engines to find you. However, you must avoid 'keyword stuffing' at all costs, as Google could penalise your site for this, harming where you appear. Keyword stuffing is when you tag everything on your website with your chosen keywords in an attempt to have the pages of your website ranked as high as possible on search lists.

4. **Use commonly available webmaster tools to improve your search ranking** – Google has a number of different webmaster tools they make readily available which you can use to improve the ranking of your website.

5. **Measure your website's performance with analytics** – There are a range of different analytical tools you can use to get information about your website. These include insights on where people have been before and where they go after browsing away from your website. Google Analytics is a good and free example of this. A good analytics tool will also help you produce regular reports you can use to optimise your site further.

6. **Create useful content** – You should include full descriptions of your products and services and their specific features and benefits on your website. Not only is this information vital to your clients, it also helps search engines find you. Including useful information such as research papers, blogs and information guides means people will likely see your website as a valuable source of information, not just a place to buy products or services. Including downloada-

ble tip sheets is another very effective way to increase your business's position in the search engine rankings.

7. **Link to other websites** – Another way to improve your search engine rankings is to link to other useful websites related to your own.

EXERCISE: Your LEAD IMPROVEMENT Plan

Rank the lead improvement strategies below from 1 to 5 in the order of priority that you will give them in your business, 1 having the highest priority and 5 the least.

Strategy	Lead improvement strategy description	Priority rank
L1	Strategic alliances and joint ventures.	
L2	Ask your existing clients for referrals and set up a referral scheme.	
L3	Raise your profile.	
L4	Put on a seminar or event by yourself or collaboratively with others.	
L5	Optimise your website and embrace digital marketing.	

In this chapter we have covered the first of the five key focus areas from the Profit Accelerator Formula, and discussed some strategies for lead generation. It's great to get a lot of leads, but you've also got to take into account the conversion rate, which ties in nicely with the seventh key ingredient covered in the next chapter: **'Conversion success.'**

CHAPTER 7

Key ingredient #7

Conversion success

Number of leads (L)

X

Conversion rate (C)

=

NUMBER OF CLIENTS

There's no point having discussion after discussion with potential clients unless a good proportion of them actually become clients! This then is all about how successful you are in getting potential clients to become actual clients: whether they choose to work with your business.

Your conversion rate is the percentage of people who did actually buy from you as opposed to those who could have bought. Imagine you're on a basketball court and you have a limit of ten attempts to get the ball through the hoop. If only four of those balls go through the hoop then the conversion rate between the actual number of times it went through the hoop and the total number of tries you had is 40% (four successful attempts out of a total of ten possible attempts).

In business, the conversion rate is worked out by dividing the number of people who say 'yes' to buying what you offer against the total number of people you have communicated with about potentially buying from you. Do you see the distinction?

This is a literal gold mine of opportunity for any business. A business simply needs to get more of their leads or potential clients to say 'yes' and become actual clients by buying something from them! You may not realise this yet, but by doubling your conversion rate, you double the number of clients that are buying from you! Let me show you how this works. Say you have 100 leads and a conversion rate of 10%, and you manage to double that conversion rate to 20%, your actual number of buying clients increases from 10 to 20. That's double!

Frequently many businesses over estimate their conversion rate as they fail to measure it. The actual reality is usually much lower when measured.

Do you measure your conversion rate? Do you know what it is with a high degree of accuracy? If you don't then you need to take ACTION to do something about it.

5 Key CONVERSION SUCCESS Strategies

STRATEGY C1 – Follow up and follow up again

Have a communication strategy around dealing with new enquiries and leads. Make people feel special and loved. Be diligent in following up but don't be annoying!

Also, always do what you say you are going to do. If you promise somebody something by a particular time and date, then make sure you stick to that promise. It builds trust, rapport and confidence in the prospect so they know that you won't let them down.

A good follow up system will have three sets of emails:

1. An email to send immediately after wrapping up your original work with the client
2. An email for 1–3 months after your first project is done
3. A 'keep in touch' email for touching base after that

Success tip

The key to success here lies in having a robust CRM (Customer Relationship Management) system, so much of the follow up can be scheduled and automated ensuring nothing falls through the cracks.

STRATEGY C2 – Add perceived value and sell on emotion and dreams

Connect with the emotional needs of the client and the reasons for a purchase. Clients generally need an emotional connection to buy. Sell the idea of the destination or outcome they will achieve if they purchase what you are offering them.

Buying decisions are nearly always the result of a change in the client's emotional state. While information may help change that emotional state, it's the emotion that's important, not the information.

All buying decisions stem from the interplay of the following six emotions:

1. Greed – 'If I make a decision now, I will be rewarded.'
2. Fear – 'If I don't make a decision now, I'm toast.'
3. Altruism – 'If I make a decision now, I will help others.'
4. Envy – 'If I don't make a decision now, my competition will win.'
5. Pride – 'If I make a decision now, I will look smart.'
6. Shame – 'If I don't make a decision now, I will look stupid.'

Every successful sales conversation encompasses one or more of these emotional states. When enough of these emotions are present, a buying decision becomes inevitable.

The second component is about adding perceived value to your clients by bundling in things that are important to them but cost you very little to deliver. For example, creating content, e-books, podcasts and guides on your area of expertise obviously takes time to initially create, but once created will cost you very little to deliver beyond that, and will have high perceived value from your target client.

Success tip

If you are creating content in any form to add perceived value for your client, always assign a value to it that reflects what knowing and having that information is worth to your client. Don't undervalue your knowledge and expertise.

118

STRATEGY C3– Target better prospects

As we covered in Chapter 3, if you understand the complete profile of your target ideal client, including what they like, their buying habits and the problem for which you have a solution, then if you centre your marketing activities around reaching that profile of client, they will be more inclined to buy from you because your product or service meets their identified need.

Instead of being a cold prospect when they are contacted, they are more likely to be a warm prospect and therefore more inclined to be interested in what you are offering. Better quality prospects will translate into better results.

Here's a reminder of the ten basic questions that can help you define your target ideal client:

1. **Who buys** your product?
2. Are they predominately **female or male**?
3. From what **age group**?
4. What are their s**pending habits**?
5. What's their **geographic location**? Local? Regional?
6. What **media** do they use most often? Magazines? Newspapers? TV shows?
7. What's the general **attitude** toward the product or service?
8. **Education** level?
9. What are their **product or service needs**?
10. **Lifestyle** indicators (home, car ownership, where they like to eat, what they like to do on weekends, do they like travelling, for example)?

Once your target audience has been defined, you can then focus on exploring and finding the best ways to reach them based on their identified characteristics.

> ## Success tip
>
> Be as specific as you can be here – the more specific the better. Be creative and think outside the box in terms of ways you can better target your ideal clients based on this information. Get this right and your conversion rate should improve dramatically.

STRATEGY C4 – Leverage all of your visual media

This is all about leveraging the media assets you have created in your business and creating new ones. There is no point having them saved on your computer somewhere where no one can see them. We have now moved to a much more visual focused era thanks to the advent of social media. Your focus here should be heavily on video.

Video has been an incredibly powerful marketing tool for quite some time now. However, in more recent times it has become a medium that's much more accessible for businesses of all sizes with clear benefits to both clients and businesses. If you haven't discovered video for yourself, there's no better time than now to get started.

Videos play varying roles throughout each phase of the client lifecycle, and the best approach is to create specific videos that support each phase:

Stage 1: Attraction

At the initial enquiry phase, first impression counts for everything – and video can be a personal and engaging way to attract clients. The visual and storytelling nature of video lends itself to engaging potential clients in a way that evokes emotion, reminds them of a need or desire, or teaches them something new.

Stage 2: Inspiring action

After hooking the interest of a potential client, you need to inspire action. Videos can play a pivotal role in converting this interest into sales as they provide a quick, easy to digest and simple way to bring your service offering to life and turn a prospect into a paying client. Videos at this stage should include product or service features and show case studies in greater detail, as the potential client is looking to understand your offering better in this phase.

Brief client testimonials explaining the problem your product or service solves can also give the potential client the connection they need to feel compelled to buy what you offer. For coaching, consulting and advisory businesses consider giving people a behind-the-scenes look at your business so potential clients build a personal connection to you and your team.

Stage 3: Building loyalty

It's imperative to keep the communications regular after a client has bought what you offer so they become loyal clients who will be more likely to spread the word about your business to others within their network. In this way, video can be a great touch point. When done well, videos can make your clients feel a strong connection with your business. With increased loyalty comes the powerful strength of word of mouth. Online videos are now so easy to share and they are a great way to raise the visibility of your brand and bridge the client lifecycle so it becomes full-circle.

Success tip

Ensure your videos are as short, punchy and concise as you can make them. Instructional videos may be longer, for instance, but just be aware that people's attention spans are generally short so you may want to consider breaking longer videos into shorter sub 3-minute chunks

---

> to make it more digestible for the viewer. Every time you create a new media asset, share it far and wide.

STRATEGY C5 – Get sales training

While some people may possess a more extroverted 'sales personality' that makes them a natural at selling, effective selling is still a skill that must be learned and developed. Sales training is very important: it can help you develop and practice the skills you need to succeed and increase your confidence levels while selling. When a sales person stands in front of a potential client, being prepared, having high levels of product or service knowledge, pricing and a presentation can make all the difference to their results. After all, the better trained an individual is, the better they should be able to perform.

The seven steps of a sale tend to be as follows:

1. Planning and preparation
2. Introduction and opening
3. Questioning
4. Presentation
5. Overcoming objections
6. Closing – negotiation
7. Follow-up and administration

Getting training around each of these areas should help improve your conversion rate as your skills develop. To begin, ask your colleagues and friends if they would recommend anybody they have done sales training with as your first point of call, as there are many businesses dedicated to the provision of sales training.

Sales success tips

1. It's not what you say; it's what your client believes.

2. Never go into a sales call not knowing how you're going to close the sale.

3. Set dedicated time aside regularly to focus on business development or 'prospecting'.

4. Believe in yourself and that what you're doing is helping your clients.

5. Show up and show up on time. Your credibility will be damaged instantly if you don't.

EXERCISE: Your CONVERSION SUCCESS plan

Rank the conversion success strategies below from 1 to 5 in the order of priority that you will give them in your business, 1 having the highest priority and 5 the least.

Strategy	Conversion success strategy description	Priority rank
C1	Follow up and follow up again.	
C2	Add perceived value and sell on emotion and dreams.	
C3	Target better prospects.	
C4	Leverage all of your visual media.	
C5	Get sales training.	

In this chapter we have covered the second of the five key focus areas from the Profit Accelerator Formula. It is the combination of the two factors we covered in this chapter and the preceding chapter that determines the number of clients who have purchased from your business over a set period of time.

Remember that the number of clients is a RESULT, which can't be influenced directly. This result can only be changed by focusing on what influences the result, which in this case is the number of leads being generated and the conversion rate being achieved.

Number of leads (L)
X
Conversion rate (C)
=
NUMBER OF CLIENTS

The result represents the number of different clients you deal with. You work it out by multiplying the total number of leads by the conversion rate.

Remember, it's not about getting more clients. You can't change that number directly. It's about getting more leads and then improving your conversion rate. These are the areas to focus on to improve your results.

In the next chapter we will look at the eighth key ingredient, which is '**Repeat business.**'

CHAPTER 8

Key ingredient #8

Repeat business

NUMBER OF CLIENTS
X
Number of transactions (T)
X
Average sale value (S)
=
REVENUE

Some of your clients will buy from you weekly, others monthly, others on the odd occasion and others just once in a lifetime. What you want to know now is the average – not your best and not your worst but the average number of times one of your clients buys from you in a year.

To demonstrate this simply, take the act of grocery shopping. We all usually do it daily, weekly or monthly. Each time we go to do our grocery shop this is considered a 'transaction' and it is usually triggered by us running out of certain foods. We need to replenish our fridges and cupboards with the food and things that satisfy our needs for the next period until we run out again. But we generally have a choice of which of the big supermarket chains we go to shop at. If we aren't happy with something or have a bad experience, then we go elsewhere. The same principle applies with your business.

The number of transactions represents the frequency with which a client buys from you in a given period. If their experience with you is bad, then they will look elsewhere and are therefore less inclined to buy from you again. Oh, and be warned, they will usually share that they have had a bad experience with many other people, which could be fatal for your business, especially now in the age of social media.

People are generally less inclined to share details of a positive experience and definitely more likely to share details of a bad experience. It's human nature. Where's the drama and story in a good experience? Bad experiences tick all the boxes in that respect. The old rule of thumb was that if someone had a good buying experience they might tell one other person, but if they had a bad experience then they would tell ten. Now think about Facebook and Twitter alone, where most people usually have hundreds or thousands of friends or followers. In this case, all that person needs to do is post a message about their bad experience and those hundreds and thousands of friends or followers will know instantly. What happens if those contacts then share that with their networks?

My point here is that we are in a new age of interacting online and one bad experience could quite easily be spread by posting a simple message online which can be seen by hundreds, thousands or even millions of people. Think about the impact something like that could have on your business. It is really quite sobering, which is why you must be on top of your game.

The aim of the game here is to encourage your clients to keep coming back and spending their money with you regularly by providing them with a compelling reason to do so.

5 key REPEAT BUSINESS strategies

STRATEGY T1 – Make realistic promises and over deliver

Surprise your clients by consistently exceeding their expectations. They will not only keep coming back for more, time and time again, but they will become raving fans and advocates for your business.

If you set expectations high and fail to live up to them you will lose trust and credibility. Make realistic promises and over deliver, always.

Over-delivering and WOW-ing your clients makes for good business. It increases client satisfaction, encourages repeat business, and sparks word of mouth which all mean more referrals.

You can never add too much value for your clients!

Success tip

Make OVER-DELIVERY a habit. What specifically can you do TODAY to over-deliver on something you've committed to or promised? Why not DO that – whatever it is – today?

STRATEGY T2 – Timetable of communication

Just like you would stay in touch with your friends or family, you need to stay in touch with your clients. Find out what they are doing and share with them what you are doing. Have a structured strategy for regularly and consistently communicating with your clients, which may involve sending out a regular newsletter, tips or phoning them every now and then.

Keep your clients updated on the areas they have expressed an interest in on your products and services in the past, and remind them why they need to come back and buy from you again by providing them with a compelling reason to do so. This counters the 'out of sight, out of mind' phenomenon.

Implementation steps

1. On a 12-month calendar mark out the key dates and times that you will share with your clients.

2. Brainstorm creative ways of staying in touch that add value for your clients – but that are not always sales focused. You want to aim for a balance, so provide valuable content etc. that doesn't have an overt sales call to action four times in every five. People quickly tire and switch off if they think they are being sold to all of the time.

3. Leverage technology as much as you can to make this process as streamlined and simple as possible. Being consistent is key here.

Success tip

You want to aim to be in contact with your clients at least on a quarterly basis as a minimum. Aim to 'add value' always in the content you provide and don't forget about birthdays and anniversaries – they are great reason to get in touch with your client and offer them something special that may encourage them to buy from you again.

STRATEGY T3 – Increase your range of products, services or packages

Don't be a one-trick pony. By understanding the needs of your clients, you can expand your range to include other products and services that you know your clients will be more inclined to want and need, thereby encouraging them to undertake more transactions with you.

Implementation steps

1. Review your entire product or service offering and ensure it matches the identified needs of your clients. Are there any missing gaps in providing a full and remarkable solution? Can you complement what you offer further?

2. Get client feedback – survey them and offer them an incentive for doing so and providing valuable insights.

3. Create them, price them and launch them.

Success tip

Be creative here. Look at what your competitors and others are doing for some inspiration if you need it. Above all, always be focused on creating products, services or packages that are focused on solving the identified problems of your target clients. Remember, they don't necessarily have to be your products, you could leverage other people's products and generate revenue from them.

STRATEGY T4 – Create related subscription type plans

Most businesses tend to focus on one-off payment type products and services and are in the perpetual cycle of:

1. Identify new clients
2. Secure new clients
3. Deliver product or service

Rinse and repeat. What many don't realise is that it usually costs significantly more to acquire new clients (marketing costs etc.) than it does to sell more to existing clients, so obviously it is wise to have a balanced approach that also sells more to existing clients.

Subscription type plans entail clients paying an amount of money each period (for instance monthly is common) towards accessing some sort of benefit from your business over time. Not only can they increase the lifetime value of a client but can have the added benefit of a smoother and more consistent cashflow predictability compared with the usually more lumpy 'one-off' type business model.

Implementation steps

1. Review your entire product or service offering and identify any opportunities for creating subscription-type products or services that complement your service in matching the identified needs of your clients.

2. Get client feedback – survey them and offer them an incentive for doing so and providing valuable insights.

3. Create them, price them and launch them.

Success tip

Make sure you are pricing for a profitable outcome. To do this you need to be very clear on what the costs of delivering the new product or service are, ensuring that what you charge is more than both the variable and incremental fixed costs for delivering it. Refer back to Chapter 5 if you need to revisit the profitable pricing principles for this.

STRATEGY T5 – Create and implement a post-purchase sales process

Many businesses are leaving money on the table by neglecting existing clients in favour of increasing new business. Post-purchase products and services are aimed at offering clients ways to stay engaged with your business beyond simply making a purchase.

The buying cycle shouldn't end after a purchase is complete. To develop advocates for your business, you must first nurture their positive buying experience, combat any buyer's remorse, and provide ways for clients to share their experiences with others. Often this is achieved by implementing various strategies that aim to maximise the value of the purchase.

Here are some ways to get started:

1. Confirmation messages.

Send these messages immediately, and customise the content in order to provide optimal value to the client.

2. Value-add messages.

Ensure that your clients get the most out of their purchase. Send them tips and tricks that maximise the value of their purchase, which helps nurture an overall positive experience with your business. In doing so, you'll not only deepen your relationship with the client, but also drive repeat business.

3. Reviews and testimonials.

Make it convenient for them to provide feedback, but be very deliberate with your timing. This type of message should always be sent after the client has had the opportunity to experience your product or service, but before they no longer feel an association to the purchase.

4. Sharing

Make it easy for clients to tell their friends and colleagues about their recent purchase with you through social sharing. In doing so clients become advocates for your business, and social media networks help amplify that reach in spreading the good message.

5. Cross-sell and up-sell.

Inform your clients of additional products or services or add-ons that complement what they bought. Be informative and make it easy for the client to research and buy what you're recommending. When sending messages along these lines, remember that relevancy and selectivity are very important. You must never bombard them with irrelevant promotional type material.

When considering ways to improve your post purchase process, keep in mind the impact that a positive experience can have on a client. If a communication isn't adding to the experience, it may be time to re-evaluate whether you should be sending that particular message at all.

Success tip

Document your post purchase sales process and ensure it is consistently applied. Leverage technology as much as

---

possible to streamline and automate the delivery of this whilst being conscious of not losing too much of the personal touch.

EXERCISE: Your REPEAT BUSINESS Plan

Rank the repeat business strategies below from 1 to 5 in the order of priority that you will give them in your business, 1 having the highest priority and 5 the least.

Strategy	Repeat business strategy description	Priority rank
C1	Make realistic promises and over deliver.	
C2	Timetable of communication.	
C3	Increase your range of products, services or packages.	
C4	Create related subscription type plans.	
C5	Create and implement a post-purchase sales process.	

So, in this chapter we have now covered the third of the five key focus areas from the Profit Accelerator Formula. As with all the five key focus areas we are covering, this area by itself is a goldmine of opportunity. Most businesses never collect a database of their past clients, let alone write to them or call them and ask them to come back, so there is often a lot of room for improvement to be made in repeat business.

In the next chapter we will explore the ninth key ingredient, being '**Increasing client spend.**'

CHAPTER 9

Key ingredient #9

Increasing client spend

NUMBER OF CLIENTS
X
Number of transactions (T)
X
Average sale value (S)
=
REVENUE

This is one area that at least some business owners do measure. Every product or service a business provides has a price associated with it, which represents what must be paid by a client to buy a product or benefit from a specific service.

The price represents the sale value for that item. However, if a client buys more than one thing at the same time, then the sale value for that particular transaction is the sum total of all items purchased at that point. To demonstrate this simply, think again about your weekly grocery shopping. You typically buy more than one item when you do your food shopping but only one transaction happens for that visit to the supermarket.

The value of the transaction is worked out by adding up the individual prices of the groceries that are in your shopping trolley. It is the total amount you must pay to go home with everything you put in your trolley on that occasion.

As a simple demonstration, assume a customer buys the following items on a visit to a shop:

Item 1 sells for – £2.00 (A)
Item 2 sells for – £5.00 (B)
Item 3 sells for – £10.00 (C)
Item 4 sells for – £15.00 (D)
Item 5 sells for – £8.00 (E)

Total Sale Value = £40.00 (A + B + C +D + E)

In this case, the total value of the transaction occurring is £40.00. This is the total amount that the customer must pay for this visit to the shop.

Once again, some clients will spend more with you and some will spend less each time they buy. What you want to know is the average that a client spends with you in total each time.

Simply add up your total revenue and then divide it by the number of transactions (or invoices generated) incurred in generating that revenue.

For example, assume the following five transactions have happened in a set period:

Transaction 1 – £40.00
Transaction 2 – £60.00
Transaction 3 – £100.00
Transaction 4 – £125.00
Transaction 5 – £75.00

Step 1: Add up the total value of the transactions, which in this case comes to £400.00.

Step 2: Identify the number of underlying transactions involved to generate this revenue, which in this case are five transactions.

Step 3: Divide the revenue figure by the number of transactions involved in generating that revenue. In this case there were five transactions involved, so £400 revenue divided by five transactions results in an average sale value of £80 per transaction in that set period.

That's how you arrive at the average sale value amount.

5 key INCREASING CLIENT SPEND strategies

STRATEGY S1 – Increase your prices

This might seem a bit obvious so is easily overlooked, but by raising your prices you increase your average sale value. Clients will be paying you more for the same products or services every time they buy from you.

This isn't always an easy thing to do in reality and there may be market forces at play that prevent you from doing it, especially if you operate within a competitive marketplace. This is beneficial to your business provided the rise in prices, and resultant revenue, isn't offset by a reduction in transactions occurring.

The following are some tips for successfully executing a price increase:

1. **Give your client lead-time.** Provide your client with enough notice to allow them to make adjustments if needed on their side and to buy at least once more at the existing price.

2. **Avoid showing any client favourites.** Ensuring pricing integrity is always essential, but especially during a period of price change. It is important to not treat particular clients more favorably than others in pricing during an increase. Different levels of pricing are fine provided there is a sound reasoning for it which is defendable, so that a client who is not receiving the price break can understand and accept the price change.

3. **Give advance warning.** Do not allow your clients to find out about a price increase from your invoice. Any pricing changes must be communicated properly. Information regarding a price change should only appear on a client invoice after every person involved has been personally notified.

4. **Team awareness.** Make sure each client service representative and anyone else who comes in contact with the client

is fully aware of when the price increase is going to be communicated. One of the most significant areas for possible confusion is when the client hears conflicting information from different departments. Everyone within the client service area of your business should be fully aware of the price increase, the reasoning for it, and the process of implementing it. Here a FAQ guide is often useful for the team to ensure that when clients do ask them about elements of the pricing increase, they are able to give them an accurate and consistent answer.

5. **Believe in the price increase.** After all, to ensure you are paid what you are worth, you must charge what you are worth.

6. **Have an open-phone/open-door policy.** Any time a price increase happens, it is important for you to be willing to answer a phone call from a client or to make phone calls to key clients.

7. **Monitor the impact on your business.** Before and after the price increase it's important to monitor the sales patterns at an individual client level. This is to ensure that you can quickly catch any adverse changes that occur as a result of the price increase.

Success tip

You will know whether or not there is scope to raise prices in your business, but it is a really good place to start for improving your average sale value amount.

STRATEGY S2 – Up-sell

This involves suggesting a product or service that is one bracket higher than the one under consideration. Selling a higher-priced product or service means a higher sale value.

To up-sell effectively, you need to do three things:

1. It's important to approach the up-sell not from the perspective of boosting revenues, but rather from the perspective of finding more ways to better serve your client. In other words, your job is to find additional ways to add real value for your client. In this way your clients and prospects will feel your sincerity.

2. You must thoughtfully probe the clients for more opportunities to serve. Put their needs first.

3. Script your delivery so that:

   i.   The benefits of your offer are clear.
   ii. There is relevance to the client or prospect.
   iii. There is a reason to act now.

Success tip

The ultimate goal here is to get your client to spend more than they ordinarily would have by offering them upgraded or more premium 'value adding' products and services that come at higher price, but result in the best overall solution for your client.

STRATEGY S3 – Stop discounting

Think about what discounting is. It is where you reduce or lower the amount you are charging for a product or service. If you stop doing it, then the amount of revenue you receive is higher.

However, discounting can not only damage your profitability, but it can also lower your value in the eyes of the client. It is time to stop any discounting that you may do and embrace the world of premium value.

Here are three easy tips to help you sell at a premium price.

1. Be proud of your premium position

Most business owners are afraid to propose more expensive solutions compared to those of their competitors, but the best performers use their premium pricing as an indicator of where they sit in their marketplace. It's about explaining to your potential clients that, 'If you are only looking for the lowest cost solution then we're not it. We are about delivering results.' By confidently leading with the message that your offering is more expensive than those of your competitors for the reasons you share, you will immediately appeal to any potential clients that want the value of the results you deliver for themselves.

2. Understand the challenges

Potential clients are never really buying your product or service. In fact, they often don't care about what you have. Their priority is usually on finding the best solutions for the challenges they are facing. So be sure to speak their language by spending the first part of a meeting discovering and learning about these issues. You can then better understand what they need and therefore then propose the best solution for which they are more likely to buy.

3. Pile on the goodies

Many businesses focus on selling a bare bones solution in order to keep the cost low. The problem with this strategy is that half-baked solutions almost never achieve the results your clients want. Rather than scaling back your offering, pile on every added value goodie (that costs you very little to deliver) that you can to raise the price. Clients are seeking solutions for their challenges, and the right clients are willing to pay for them. Make your offering irresistible to your ideal client by piling on massive value.

STRATEGY S4 – Propose three options

You are not a mind reader, so stop assuming you know exactly what your potential client wants from you. After you have fully discussed and understood a potential client's challenges, present solutions at three different levels. For example, this could mean offering Silver, Gold and Platinum options.

It turns out that the human brain is wired for good/better/best pricing options. We like choices and options. However, 'option overload' paralyses us so keep the options down to a manageable number.

Here's a common objection to this strategy: Research shows about 2/3 of buyers choose the mid-price product or service. So why not just offer that most popular option? Simple – a premium product or service establishes your credibility. It shows you recognise and offer top quality, creating a halo effect for your other products and services.

The lowest-end solution should be the cheapest solution that will still solve the clients' challenges. The highest-end solution should be well above your clients' stated budget but include every possible bell and whistle to solve the client's challenges and more. The middle option solution should be somewhere in between.

By offering three options, you are not only creating choice, which potential clients love, but you are also giving them the opportunity to select the highest-level option. If you have focused on value throughout the sales conversation, at least 20 percent of your prospects will likely choose the top option.

Implementation steps

1. **Show, tell and sell.** Once you have created your packages that build on each other, walk your clients through their options.

2. **Create an at-a-glance way of showing each tier's features.** A simple chart will suffice here as a visual reminder.

3. **Offer an upgrade path.** Consider whether the mid-range or premium offering be combined into a value package that saves money over the standalone individual prices.

Success tip

Clients need comparisons to establish value. Offering this enables them to 'comparison shop' without leaving your office. If you don't offer choices, they'll find them elsewhere. Don't underestimate what clients want. Clients will upgrade in areas that matter to them – but again, only if you give them the option.

STRATEGY S5 – Educate on value, not price

Avoid having to compete on price wherever possible, as this usually ends up with you having to discount or lower prices, which is the complete opposite to the objective of this exercise – to increase average sale value.

Help your clients to understand why they should buy from you and not another business, and explain the reasons why the price of an item represents real value to them.

Once you have learned your prospect's most critical needs, problems and challenges, it is time to understand the value of solving those challenges. Remember, the value of what you offer is directly proportional to the value of the challenges you are solving for your client.

For example, if your potential client's challenges are costing them £500,000 in lost revenue, and your solution can help solve those challenges, then the 'value' of your solution to them is £500,000. Therefore, when your solution costs £50,000, it is a no-brainer for them and represents a significant return on their investment in you.

Here are some tips that can help:

1. Emphasise the VALUE they are getting.
2. Explain why your product or service is more expensive.
3. Make it easy to compare what you offer with others in terms of better quality, results and service, and price usually becomes almost irrelevant.

Success tip

Remember that people generally want a good deal, not the cheapest. They will be much happier spending the money to get a solution that does exactly what they want, rather than spending less on a solution that only does half the job. Explain the reasons why some people charge less and emphasise what they will miss out on if they do it 'on the cheap'.

EXERCISE: Your INCREASING CLIENT SPEND Plan

Rank the increasing client spend strategies below from 1 to 5 in the order of priority that you will give them in your business, 1 having the highest priority and 5 the least.

Strategy	Increasing client spend strategy description	Priority rank
S1	Increase your prices.	
S2	Up-sell.	
S3	Stop discounting.	
S4	Propose three options.	
S5	Educate on value, not price.	

In this chapter we have covered the fourth of the five key focus areas from the Profit Accelerator Formula. It is the combination of the two factors we covered in this chapter and the preceding chapter together with the 'number of clients' result that determines the revenue amount for a set period of time.

Remember that the revenue amount is a RESULT, which can't be influenced directly. This result can only be changed by focusing on what influences the result, which are the four key focus areas we have covered so far.

You need to multiply the total number of clients you dealt with by the number of times they came back on average, and then by the average amount they spent with you each time. That's your revenue.

Put simply:

NUMBER OF CLIENTS
X
Number of transactions (T)
X
Average sale value (S)
=
REVENUE

Their revenue is another figure most business owners will know, but they most probably have no real idea how they got to it. Of course, you want more, but you can't simply *get more revenue!* However...

What you can encourage is more transactions and a higher average sale value from the total number of clients you deal with.

Making sense? It's a subtle but important difference.

In the next chapter we will explore the final of the ten key ingredients, '**Margin improvement.**'

CHAPTER 10

Key ingredient #10

Margin improvement

REVENUE
X
Margin (M)
=
PROFIT

Margins represent the percentage of each and every sale that's profit. In other words, if you sold something for £100 and £25 was profit, then you've got a 25% margin.

Your *gross profit* is the difference between the selling price of a product or service and the direct costs taken to produce that product or service. These direct costs are also known as *variable costs*.

Variable costs are those costs that tend to fluctuate with the level of sales. They include, but are not limited to, such things as direct labour, raw materials, sales commissions and delivery expenses. They represent the *costs of the goods sold*.

At a whole business level, gross profit is the difference between the sum of all the revenue generated from all sources and the sum of all the variable costs or direct costs of the goods sold to generate that revenue. It doesn't take into account the overheads or fixed costs of the business at this level. The *gross profit margin* expresses the gross profit amount as a percentage of revenue.

Net profit is the next layer down. It is the residual profit after all the fixed costs or overheads of the business have been deducted.

Fixed costs are those costs that do not tend to fluctuate with the level of sales. They include, but are not limited to, such things as rent, equipment leases, insurance, interest on borrowed funds, business taxes and administrative salaries.

The *net profit margin* expresses the net profit amount as a percentage of revenue. It is the net profit margin that we are most interested in here; at the end of the day that is what's left for you after all costs including taxes have been deducted.

150

At a very basic level, margins can be influenced in two ways:

1. Increasing prices at a greater rate than the associated costs of providing that product or service.
2. Reducing your costs at a rate that is greater than any reduction in prices.

Let's apply three scenarios to the following example to demonstrate this influence in action:

Total revenue	£350,000	(A)
Total costs of goods sold	£200,000	(B)
Gross profit	£150,000	C (A–B)
Total fixed costs/overheads	£80,000	(D)
Net profit	£70,000	E (C–D)
Net profit margin	20.0%	F (E/A x 100)

Scenario 1

You decide to put all your prices up by 10% only. There is no impact on your costs by taking this action. This means that your revenue figure would increase by £35,000 (A x 10%) to £385,000. Let's see the impact this would have on your net profit amount and margin.

Total revenue	£385,000
Total costs of goods sold	£200,000
Gross profit	£185,000
Total fixed costs/overheads	£80,000
Net profit	£105,000
Net profit margin	27.3%

This results in an increase in your net profit margin from 20.0% to 27.3%, which is 7.3 whole percentage points and represents a 36.4% (7.3% divided by 20%) improvement. A result you should be happy with.

Scenario 2

You manage to negotiate better terms with your suppliers, which results in a 20% reduction in your costs of goods sold. There is no impact on your revenue by achieving these savings. This means that your costs of goods sold figure would decrease by £40,000 (B x 20%) to £160,000. Let's see the impact this would have on your net profit amount and margin.

Total revenue	£350,000
Total costs of goods sold	£160,000
Gross profit	£190,000
Total fixed costs/overheads	£80,000
Net profit	£110,000
Net profit margin	31.4%

This results in an increase in your net profit margin from 20.0% to 31.4%, which is 11.4 percentage points and represents a 57.1% (11.4% divided by 20%) improvement. Surely this result would make you even happier!

Scenario 3

Now let's look at the impact if you manage to increase your prices by 10% as in Scenario 1, and you also manage to negotiate better terms with your suppliers resulting in a 20% reduction in your costs of goods sold, as illustrated in Scenario 2.

This means that your revenue figure would increase by £35,000 (A x 10%) to £385,000 and that your costs of goods sold figure would decrease by £40,000 (B x 20%) to £160,000. Let's see the impact this would have on your net profit amount and margin.

Total revenue	£385,000
Total costs of goods sold	£160,000
Gross profit	£225,000
Total fixed costs/overheads	£80,000
Net profit	£145,000
Net profit margin	37.7%

This results in an increase in your net profit margin from 20.0% to 37.7%, which is 17.7 percentage points and represents an 88.3% (17.7% divided by 20%) improvement. Let's look at what this means for the money going into your pocket.

Previously, you were left with £20 in every £100 of revenue being generated by your business. After completing the two initiatives to improve your margins, you are now walking away with £37.70 in every £100 of revenue being generated. That's **an extra £17.70** in every £100 of revenue without doing anything extra in providing the underlying product or service. That extra profit is better off in your pocket than someone else's pocket, is it not?

Now that should be something to get excited about. If it doesn't give you enough incentive to start looking at your margins more closely, I don't know what will!

Before we go into the 5 key MARGIN IMPROVEMENT strategies I just wanted to take a slight diversion and share the 7 biggest profit draining mistakes that many business support professionals make, and give you some proven strategies for avoiding them.

Profit draining mistake 1 – Failing to factor in fixed costs when pricing

When deciding on pricing, many business owners make the mistake of only focusing on the gross profit margin and tend to forget about allocating something for their overheads or fixed costs. They then wonder why they don't make any profit.

Remember there are generally two types of costs: variable costs and fixed costs. We covered these earlier in this chapter.

Tip

To ensure that your business is pricing profitably, it is key that the break-even point, whether it be expressed in revenue or volume, is known and understood. Break-even analysis takes into consideration all three elements of price, variable costs and fixed costs.

Profit draining mistake 2 – Thinking as long as money is flowing into the business bank account they are making money

Just because money is flowing into the business bank account doesn't necessarily mean that a profit is being made on it.

Many businesses fail to look at all the factors when agreeing to do work at a given price level. Usually the price is set by market forces but many business owners fail to even do a basic analysis to work out whether they can deliver the service at a cost, less than the revenue received, whilst generating a sufficient profit margin above all costs – both fixed and variable.

Tip

Use a top-down approach when assessing a product or service offering. This starts by taking what the market will pay for the product or service and then deducting from it the desired profit margin. You then need to determine if you can deliver the product or service at a total cost equal to or less than this result. If you can, then it is a green light to proceed. If you can't then you may need to look at other ways to deliver it more efficiently, reduce the desired profit margin or simply not proceed with the opportunity at all. We covered this in further detail in Chapter 5.

Profit draining mistake 3 – Thinking it is job done once a client has been invoiced

It is not the end of the story when an invoice is sent to a client for payment. A business must ensure that the payment is collected in accordance with its payment terms.

There is no point in invoicing a client if payment is not collected for it. Remember, a profit isn't actually earned until the amount for the invoice is physically received as cash in the business bank account.

A business must be proactive in the collection of its invoices.

Tip

A common metric for assessing how good your business is at collecting payments for its invoices is called Debtor Days Outstanding. This metric is used to show the average number of days that have elapsed between the invoice date when it was produced and the present date if payment still remains outstanding. The closer the Debtor Days Outstanding calculation is to your payment terms (e.g., 30 days is common) the better your business is at collecting payment for its invoices.

Profit draining mistake 4 – Not paying close enough attention to cash flow

In business, cash is king! In some ways, managing cash flow is the most important aspect of running a business. If at any time a business fails to pay an obligation when it is due because of the lack of cash, the business is technically insolvent.

Insolvency is the primary reason businesses go bankrupt. Obviously, the prospect of such a dire consequence should compel businesses to manage their cash with care. Moreover, efficient cash management means more than just preventing bankruptcy. It improves profitability and reduces the risk to which the business is exposed.

Businesses suffering from cash flow problems have no margin of safety in case of unanticipated expenses. They may also experience trouble in finding the funds for innovation or expansion. Finally, poor cash flow makes it difficult to hire and retain good employees.

Tip

The key to successful cash management lies in:

- Making realistic projections of cash-needs, including timings
- Monitoring collections and disbursements
- Establishing effective billing and collection measures
- Sticking to a budget – not overspending

Profit draining mistake 5 – Not producing and reviewing financial reports regularly

Many business owners I talk to have an ostrich mentality when it comes to the numbers side of their business. They just hope that everything will be fine.

When I ask these same business owners when they last spoke to their accountant, or where the last set of financial statements are, many reply that they only speak to their accountant at year-end time or respond with, 'What set of financial statements?' or 'In the bin.'

I find these situations really alarming because apart from the legal obligations when it comes to record keeping, the business could be at serious risk with this avoidance mentality. Numbers aren't to be feared. If working with and understanding numbers isn't a strong skill set of a business owner then they should seek help immediately.

> **Tip**
>
> You really need to have a handle on the numbers, whether you do it yourself or enlist the help of an expert, because knowing and understanding the numbers is key to long-term business success. Management accounts should be prepared and reviewed each quarter as an absolute minimum, with a preference for monthly accounts.

Profit draining mistake 6 – Not having a budget

A budget is a comprehensive plan that estimates the likely expenditure and income for a business over a specific period, typically on twelve-month cycles. It is a financial road map for a business that is derived from an underlying business plan.

Budgeting describes the overall process of preparing and using a budget. A budget is a hugely valuable tool for planning and controlling finances. The process itself helps a business to determine the most efficient and effective strategies for making money and expanding its asset base.

To be successful, budgets should be SMART, which means specific, measurable, achievable, realistic and timed.

> **Tip**
>
> A budget is a key business tool and should be prepared annually at a minimum and then chunked down into monthly and quarterly targets. Performance against budget should be actively monitored throughout the budget period, also on a monthly or quarterly basis. Unless a business tracks its actual results against the budgeted results it has no way of knowing whether it is on track to achieve its annual targets or not.

Profit draining mistake 7 – Wasting money unnecessarily

I guarantee that almost every business is wasting money on something unnecessarily, whether it is through paying more than they should be, buying the wrong type of input or buying things that the business doesn't actually need.

Take business utilities for instance. Most businesses simply accept what the utility companies offer them in terms of tariff. They don't know if it is even the right tariff and usually get locked into a higher tariff than is needed. I would advocate using one of the free impartial utility management services to manage this side of a business. I would advocate the same approach when it comes to office stationery and procurement.

Producing regular financial statements and having a handle on the numbers is key to effective cost management. Unless a business knows exactly what it is spending its money on, it is difficult to know where and how improvements can be made.

> **Tip**
>
> It is imperative to review your supplier terms regularly to ensure that what is being paid is competitive. Be sure to shop around for the best deal, be clear on why something is being purchased and take advantage of early payment discounts, if available, and cash flow permits.

5 key MARGIN IMPROVEMENT strategies

STRATEGY M1 – Streamline your operations – outsource the non-essential and review all of your costs.

Reduce unnecessary management. It's about systemising the routine and humanising the exceptions. It's all about becoming more efficient. This means reducing duplication where possible and automating as much as possible. Make sure you are taking advantage of any advances in technology that are relevant for your business that can help you deliver your products and services more efficiently to your clients. This will generally translate into lower costs.

Don't outsource something just because you don't want to do it. Sometimes there are things you don't want to do but they are fundamentally important to your 'core' business.

Before selecting which functions you will outsource, thoroughly review your business and be clear on your strengths and values. You must identify your core competencies and capabilities and focus on developing them so you stand out in your market. Outsourcing any aspect of these 'core' functions will likely be a big mistake because it could mean that you cease to offer anything to your clients that they couldn't get elsewhere. It would mean losing an element of your uniqueness.

The types of tasks that are suitable for outsourcing fall into three general categories:

1. Highly skilled, or executive, expertise. For example, you may not need to pay a full salary for a Finance Director or CFO; you could instead get access to these vital skills and expertise through someone that only comes in a couple of times each month. They can provide the financial analysis support needed and can ensure that the book-keeper is doing everything properly.

2. Highly repetitive tasks. Some examples of this are accounts payable, data entry and shipping products.

3. Specialised knowledge. An example here might be the IT support for your website. Once again, you may not be able to afford or actually need a full-time IT person, so in this case it would be easier and more cost effective to change to an outsourced provider that can adapt to your changing IT needs as your business grows.

Once you decide what to outsource, look for the right service providers:

1. Ask for referrals from your own professional network. Talk to other business owners, colleagues and professionals about how and where they have outsourced successfully. Share best practice.

2. Check professional networking sites such as LinkedIn for contractors in the field in which you are trying to outsource work. You might also use other social media platforms such as Facebook, Google + or Twitter to advertise what you are looking for.

3. Connect with contractors and freelancers on websites that you can easily find by doing a search on Google. These websites allow business owners to create a listing that describes their requirement, and contractors respond with their qualifications and rates.

4. Review qualifications and experience. Check references carefully and conduct interviews by phone or online. Ask to see samples of their work.

Success tip

When you find a person or business you are ready to outsource your tasks to, have a contract prepared that describes the work, deadlines, expectations and rate of pay. Communicate clearly and often with the providers to whom you have outsourced work. Evaluate the outsourcing process regularly.

STRATEGY M2 – Stop any advertising or marketing that doesn't work

As we covered earlier, marketing must be viewed as an investment, and as with anything there can be good and bad investments. Any marketing that isn't working is a cost.

To reiterate, the distinction here is between conscious marketing and unconscious marketing. Conscious marketing is deliberate targeted marketing and should be seen as an investment. Like any investment, we expect to receive a return on our investment that is acceptable to us. Unconscious marketing that is not deliberate, not targeted and not measured is actually a cost and should be stopped immediately.

If a strategy isn't delivering the intended results after giving it a good go – cut your losses and move on. Time is money.

Don't be scared to stop doing something that isn't working. As with investing, if a share price is performing poorly and you are losing money on it, there reaches a point where you need to cut your losses and redirect your resources into finding an alternative. Frequently, business owners fall into the trap of doing what they have always done. They continue to do something because they have become used to doing it.

Implementation steps

1. List out all of the current marketing or advertising campaigns you are currently undertaking.
2. Determine the effectiveness of each in the context of return on the money invested.
3. Stop those campaigns that aren't delivering the desired results.

Success tip

Don't get emotionally attached to a particular marketing activity. It must be viewed as an investment decision. If something is not delivering the intended results then stop doing it immediately, cut your losses and move on to identifying the next initiative to try.

STRATEGY M3 – Sell more higher-margin products or services

If you sell more products or services with a higher profit margin than those products or services with a lower profit margin, then your overall profit margin will improve. Introduce more high-margin goods and services wherever you can and sell more of them.

Implementation steps

1. List out all of your products or services.
2. Determine the profitability of each so you are very clear what margin you are making on each of them.
3. Sell more products/services where the margin you achieve is higher than your average overall business profit margin. So, if the net profit margin of your business is 30%, then by selling more products that have a 40% margin, your overall business margin will improve.

Success tip

Revisit Chapter 5 for the principles of pricing profitably and do check out the Profit Learning Portal listed at the end of book as an extra learning resource. If you don't have any higher margin products or services to sell then create some!

STRATEGY M4 – Get rid of difficult or less profitable clients.

Sack your less profitable clients. You will find that your less profitable clients will usually provide you with the most headaches.

It is the 80/20 principle, which in this case means that 20% of your clients will constitute 80% of your headaches and will be less profitable. Not only will sacking them make your life easier, it will free up the time and resources required to focus on seeking out more profitable clients.

Implementation steps

1. List out all of your current clients.
2. Determine the profitability of each so you are very clear who your most profitable clients are.
3. Devise a plan to target more clients that are more profitable to you and get rid of the less profitable ones.

> ### Success tip
>
> If you don't feel confident enough or equipped to tackle this then seek out the help of a more proactive, forward thinking accountant to help you complete this exercise. The payback will definitely be worth it.

STRATEGY M5 – Do it right the first time

The costs of rework in business can be huge and are often ignored and overlooked. Think about a business that charges based on their time. Say they can charge a client for one hour, but because of mistakes the work needs to be redone and this takes, say, a further two hours to do. Three hours have been spent to get one hour of chargeable revenue, which will obviously impact the bottom line profit.

By doing work once and doing it properly, you avoid the costs associated with re-work, which can be disproportionate to the revenue received.

Some rework is inevitable if there are new processes, equipment, materials, or staff involved; a learning curve is needed. But over time once done properly, quality should improve and the number of errors should fall.

The reality is that it is unlikely that work will ever be completely error free. Mistakes happen. The key is to focus on reducing and minimising any double handling or rework by automating your processes and having quality checks to reduce errors.

Success tip

Explore cases where rework was needed and seek out the source. Perhaps it was a client change order, the specification was unclear, there was a communication breakdown, or there was material variance. Each has a different solution. There is a direct relationship – when rework goes down, profits go up.

EXERCISE: Your MARGIN IMPROVEMENT plan

Rank the margin improvement strategies below from 1 to 5 in the order of priority that you will give them in your business, 1 having the highest priority and 5 the least.

Strategy	Increasing client spend strategy description	Priority rank
M1	Streamline your operations – outsource the non-essential and review all of your costs.	
M2	Stop any advertising or marketing that doesn't work.	
M3	Sell more higher-margin products or services.	
M4	Get rid of difficult or less profitable clients.	
M5	Do it right the first time.	

We have now covered the fifth and final of the five key focus areas of the Profit Accelerator Formula, and the final element of the ten key ingredients for building a highly profitable business coaching, consulting or advisory business. We have explored how margins determine the profit being made after all costs have been deducted. Remember that the profit amount is a RESULT, which can't be influenced directly. This result can only be changed by focusing on what influences the result, which are the five key focus areas we have covered so far.

Every business owner wants more profit, not realising that they can't simply just get more profit, but they can earn greater margins on the revenue they are generating to improve the amount of profit that results.

Put simply:

REVENUE
X
Margin (M)
=
PROFIT

And that's it. This five component Profit Accelerator Formula that we have covered over the past five chapters can be applied to any coaching, consulting or advisory business. In fact, it can be applied to any business on earth!

By simply breaking down your business and marketing efforts into these five key focus areas of the formula and understanding how each affects the other, you're halfway there – and way ahead of 90% of businesses out there.

The true power of combining each of the five key focus areas from the Profit Accelerator Formula together with the other five key ingredients we covered earlier in the book is yet to come, and all will be revealed in the next chapter when we revisit our case studies. Prepare to get really excited!

CHAPTER 11

Powerful results

We've now covered the ten key ingredients that are vital for building a highly profitable business coaching, consulting or advisory business. So, let's check back in with our three case studies to see what they did, and most importantly what impact it has had on their business profitability and results. Be sure to make note of any ideas or insights that stand out for you which you think could be interesting for your business. Inspiration may come from just one or perhaps elements from all three of the case studies, so keep an open mind.

Case Study 1

You will recall that Mark is an independent business growth coach and consultant who constantly battled with the 'feast' versus 'famine' nature of his business. His ambitions were to double his revenue and to spend less time on what he saw as wasted hours having coffee meetings that routinely didn't result in any new business. His issue was really around not sufficiently pre-qualifying the conversations he was having and in expecting people to agree to his core offering too soon in their buying process.

Mark's Client Service Ramp now looks like the following:

Level 1 – Explore: Blog and article writing predominantly, but has a strong social media presence so he is 'discoverable' by potential clients when they go to research him on Google.

Level 2 – Taster: Offers a Business Health Check for £99 – the client gets a full 20-page feedback report on how they answered each of the 35 questions. At this point he then determines whether the client is suited to go to the next level or not based on their responses and scores from the diagnostic.

Level 3 – Discovery: Provides a Business Blitz for £395 where he bundles in all four growth diagnostics from our service together with one hour of his time with a client to run through the results and dig deeper into the issues uncovered, and to demonstrate his expertise to the potential client in a way that adds value for them. This only takes 20 minutes of the potential client's time, and allows him to then tailor the next steps of the core offering for that client, as he gets 140 insights on their business as part of the diagnostic exercise. Most importantly, however, none of his time is wasted at this stage, which is in sharp contrast to the way he used to work, sometimes spending up to three hours on a potential client that went nowhere.

Level 4 – *He has rejigged his core offering and settled on an offering that will be charged at £500 per month.*

His baseline Profit Accelerator Formula was:

Number of leads (L)	208
X	
Conversion rate (C)	5.8%
=	
NUMBER OF CLIENTS	**12**
X	
Number of transactions (T)	10.75
X	
Average sale value (S)	£348.84
=	
REVENUE	**£45,000**
X	
Margin (M)	23.3%
=	
PROFIT	**£10,500**

Mark really focussed his energies on targeting better prospects. He set up the business growth diagnostics as a key part of the Level 2 and 3 elements of his Client Service Ramp, he got some sales training and he rejigged his core offering to always present three options.

Here is what happened:

Number of leads per year	208 (Unchanged)
Conversion rate	5.8% to 14.9%
Average number of transactions	10.75 to 6.3
Average sale value	£349.84 to £493.19
Margin	23.3% to 28.5%

Profit Accelerator Formula applied

Number of leads (1)	208
X	
Conversion rate (2)	14.9%
=	
NUMBER OF CLIENTS	**31**
X	
Number of transactions (3)	6.3
X	
Average sale value (4)	£493.19
=	
REVENUE	**£96,320**
X	
Margin (5)	28.5%
=	
PROFIT	**£27,451**

Now let's look at what this meant for Mark. As a result of the changes he made to his client service ramp and strategies he used, the following happened:

- His number of clients increased by 19, from 12 to 31, which is a 158.3% improvement
- Total revenue increasing by £51,320, from £45,000 to £96,320, which is a massive 114.0% improvement
- Net profit increased by £16,951 from £10,500 to £27,451, which is a staggering 161.4% improvement

Key outcome points

- Even though the same time and energy was spent on lead generation as before, by targeting better prospects, better 'qualifying' them using the business growth diagnostics, and by getting more potential clients to say 'Yes', Mark's number of clients improved.

- Enhancing his core offering to suit his ideal client saw the overall amount he was charging each client increase, which had a positive effect on the 'Average sale value' per transaction. Previously he served 12 clients on his core offering, now he serves 31 for less effort, but look at the impact that this has had on his revenue – more than doubling it!

- Despite more than doubling the amount he paid himself from £24,000 to £60,000, the residual profit even after this extra salary means that he now has a reserve of revenue to help smooth out the previous feast and famine challenges that he used to suffer from.

Case Study 2

Susan and Brian run a business networking community and separately a mastermind group. You will recall that they wanted to better leverage the way that the networking community feeds into the mastermind group in terms of paying participants. They had some ideas but have been struggling to make ground with it, primarily because their current approach is just too time intensive and they don't think it can be easily replicable in its current form. Their challenge was in consistent lead generation and determining what their client journey should be to ensure that they scale in a profitable way.

Susan and Brian's Client Service Ramp now looks like the following:

Level 1 *– Free networking community. They regularly produce tip videos and articles as part of their vibrant community.*

Level 2 – *They now use FREE webinars with great effect. At the end of each webinar they give a call to action to attend one of their ½ day taster workshops, which they have set a schedule for. The cost for this workshop is £99. As an incentive for the webinar participants to take immediate action, a deal is offered where if they book within 48 hours then not only can they attend the taster workshop, but also as a bonus they will have a Business Health Check done on their business (valued at £99 also). This sees the client getting a full 20-page feedback report on how they answered each of the 35 questions. So instead of doing an early bird discount to encourage action, Susan and Brian are smartly offering an early bird 'value add', giving more value and rewarding those who take prompt action.*

Level 3 – *From the attendees of the taster workshop they then offer the ability to upgrade to the Business Blitz Review, in the same way as Mark from Case Study 1, for £395, and they see the same end benefits as him.*

Level 4 – *There has been no change to their core mastermind offering which costs participants £300 per month. There has been an important change to the dynamic here now, however, due to the changes made at Levels 2 and 3, because now for everyone in their mastermind group they have 140 growth insights into their businesses, which enables Susan and Brian to tailor the support they give based on the needs of the group. It also becomes an important benchmark from which to track improvements. They plan to conduct Business Blitz Reviews with their participants every 12 months. This enables them to see each participant's progress against their previous results, which is a really seamless way of quantifying the improvements being made as the 'value' clients have received for being supported in that mastermind group. This is hugely beneficial to Susan and Brian for attracting new delegates, as they can promote clearly the value impact they have on the businesses that work with them.*

Their baseline Profit Accelerator Formula was:

Number of leads (L)	*240*
X	
Conversion rate (C)	*15.0%*
=	
NUMBER OF CLIENTS	**36**
X	
Number of transactions (T)	*10.65*
X	
Average sale value (S)	*£300.00*
=	
REVENUE	**£115,000**
X	
Margin (M)	*25.65%*
=	
PROFIT	**£29,500**

Susan And Brian really focussed their energies on promoting their FREE 1 hour webinars on a variety of growth related topics that would appeal to their target market, and filling the taster workshops off the back of that through leveraging their networking community. Forming strategic alliances and partnerships played a huge role here, too. They also set up the business growth diagnostics as a key part of Levels 2 and 3 of their Client Service Ramp.

Here is what happened:

Number of leads per year	240 to 325
Conversion rate	15.0% to 25.8%
Average number of transactions	10.65 to 10.57
Average sale value	£300.00 (Unchanged)
Margin	25.65% to 30.5%

Profit Accelerator Formula applied

Number of leads (1)	325
X	
Conversion rate (2)	25.8%
=	
NUMBER OF CLIENTS	**84**
X	
Number of transactions (3)	10.57
X	
Average sale value (4)	£300.00
=	
REVENUE	**£266,394**
X	
Margin (5)	30.5%
=	
PROFIT	**£81,241**

Now let's look at what this meant for Susan and Brian. As a result of the changes they made to their client service ramp and the strategies they used, the following happened:

- Their number of clients increased by 48, from 36 to 84, which is a 133.3% improvement
- Total revenue increasing by £151,364, from £115,000 to £266,394, which is a massive 131.6% improvement
- Net profit increasing by £51,741, from £29,500 to £81,241, which is an amazing 175.4% improvement

Key outcome points

- By being smarter about their approach and better 'qualifying' using the business growth diagnostics, combined with getting more potential clients to say 'Yes', their number of active mastermind group participants improved from three active groups to seven active groups. Each of their mastermind groups has twelve clients, so they now have 84 clients with a strong pipeline for this trend to continue.

- Their core pricing remained the same but the increased volume and better conversion rate saw their revenue more than double.

- The improvement in their margin as more client volume was introduced demonstrates that the way they have now structured their business is working: they are capable of scaling in a profitable way. If their margin was declining as they introduced more volume this would have been a warning sign that something was wrong. However, thanks to their close attention to the profitable pricing principles we discussed earlier in the book they have avoided it happening, and it has paid off for them handsomely.

Case Study 3

You will remember that Joanne, Elizabeth and Craig run a three partner accountancy business and have 15 staff. They are proactive and forward thinking in nature and see the changing trend in the market place, which is that the more traditional transactional focussed work is diminishing due to advances in automation and technology.

They wanted to remain ahead of the game, so they were keen to build the growth advisory side of their business further. They see that each business area is a feed for the others in providing a more rounded, growth focussed support solution for their clients, which is especially important as more of their team capacity is going underutilised with the changing market conditions.

Joanne took the lead on this expansion project. Their challenge however was in articulating what their growth service offering should be for it, and also ensuring that it could be delivered in both a consistent and scalable way so that the client experience with their brand continued to be positive, and so that the utilisation levels of their teams was high. They planned to do this by bringing on board associates to deliver under their brand, but were not clear of the best way to approach this. They invested in getting the support they needed to make it happen.

Here's a reminder of how their Client Service Ramp now looks for their advisory business.

Level 1 *– Actively participate in online communities, providing tips and producing guides.*

Level 2 *– Offer a FREE Business Health Check (valued at £99), which enables them to pre-qualify who might be possible candidates for progression to the next level based on responses and scores from the diagnostic.*

Level 3 *– They then offer a Business Blitz Review for £297. If the potential client agrees to it then one of their business growth associates will conduct the blitz. For them this ensures that they have a structured and systematic way of generating qualified leads so that their business growth associates are only engaging in productive and chargeable work. Win/Win.*

Level 4 *– They have developed a tariff of services that is modular in nature, which makes it easy for them to tailor a bespoke solution to suit the growth needs of their clients. Whilst a typical 'accounting only' client generates on average £1,500 of revenue per year, they expect that a typical business advisory client will generate on average £3,000 of revenue per year.*

Profit Accelerator Formula applied

Number of leads (L)	*4,750*
X	
Conversion rate (C)	*10.2%*
=	

NUMBER OF CLIENTS	*485*
X	
Number of transactions (T)	*1.13*
X	
Average sale value (S)	*£1,354.55*
=	
REVENUE	*£745,000*
X	
Margin (M)	*21.2%*
=	
PROFIT	*£157,940*

Here is what happened:

Number of leads per year	4,750 to 2,445
Conversion rate	10.2% to 15.9%
Average number of transactions	1.13 to 1.56
Average sale value	£1,354.55 to £1,860.25
Margin	21.2% to 28.3%

Profit Accelerator Formula applied

Number of leads (1)	2,445
X	
Conversion rate (2)	15.9%
=	
NUMBER OF CLIENTS	**389**
X	
Number of transactions (3)	1.56
X	
Average sale value (4)	£1,860.25
=	
REVENUE	**£1,128,163**
X	
Margin (5)	28.3%
=	
PROFIT	**£319,270**

Now let's look at what this meant for Joanne and team. As a result of the changes they made to their Client Service Ramp and the strategies they used, the following happened:

- Their number of clients decreased by 96, from 485 to 389, which is a fall of 19.8%, but this was a positive thing (as will be explained in the key outcome points below).

- Total revenue increased by £383,163, from £745,000 to £1,128,163, which is a very meaningful 51.4% improvement.

- Net profit increased by £161,330, from £157,940 to £319,270, which is an incredibly impressive 102.1% improvement – a more than doubling!

Key outcome points

- You will have noticed that the leads for their business fell from 4,750 to 2,445, which is a 48.5% reduction. On the face of it, you may think this is a bad thing. It's not. What they did after reviewing their client portfolio was to sack a whole swathe of their least profitable clients who caused them headaches, and focussed their energies on attracting more clients who fit their ideal client profile.

- They also put in place a referral system to leverage their existing established client portfolio, which rewarded quality referrals using a FREE Business Health Diagnostic as the incentive for the client being referred, and other growth focussed incentives for the client making the referral. It worked really well. What they recognised is that it was a return on investment decision. By covering the cost of having each diagnostic produced themselves (worth £99 for the recipient, but the cost to them was approximately £50 per processed report), they knew that for approximately every seven well targeted prospects they let complete the business health check diagnostic at no charge, one would become a client, which would be worth approximately £3,000 per year in revenue. The total investment they made in producing the 20-page diagnostic reports for each of the seven businesses was approximately £350, but this resulted in £3,000 of revenue. This meant spending £350 to get an added benefit to the business of £2,650 above this investment and achieving a 757% return on investment for them – it was a complete no-brainer and it worked.

- The combined impact of what they did in implementing the various strategies and the business growth diagnostic solution meant they were dealing with fewer clients, yet their overall revenue improved by 51.4% along with a more than doubling of their profit. They didn't need to take on any additional employed staff as they now had an associate model way of working that allowed their overhead structure to remain very lean. They had a business formula and client service offering ecosystem that worked, so for the next year the focus would shift to improving each of the five focus areas from the Profit Accelerator Formula.

So there you have it, these are the impressive results that the three case study businesses achieved by implementing in a structured and methodical way the 10 key ingredients we have covered together in this book. You will have noticed a common theme amongst all of them. Each embedded the business growth diagnostic solution into their business by making it the key component of the Level 2 – Taster and Level 3 – Discovery elements of their client service ramp and offering ecosystem.

Here's a summary of the top seven benefits that each of the case study businesses received in doing this, which enabled them to achieve their impressive results and overcome the very challenges that I shared in the first chapter.

1. For every client they put through the Business Health Check Diagnostic they have 35 immediate insights into the strengths and weaknesses of that business. This extends to 140 insights for every client they put through all of the four growth diagnostics.

2. These 140 insights were all gleaned in less than 20 min-

utes of the client's or potential client's time. This avoids the pitfall of the normal approach to client discovery, which usually takes far too long and puts off potential clients. This is also in stark contrast to the curse of the typical unqualified FREE consultations, which sees up to three hours of time wasted on a consultation that was never likely to go anywhere in the first place. The difference for the case study businesses is that all of the insights are gleaned and reports produced without it taking ANY or their or their team's or their associates' time. This enables them to focus on revenue producing activities more often.

3. The business growth diagnostics now enable all of the case study businesses to receive a consistent flow of 'qualified' new leads because they are now able to position themselves as providing a 'growth' focussed support solution for their defined target market. They are now only having conversations with pre-vetted highly targeted potential clients who fit their ideal client profile, which improves the chances that these conversations will result in a 'Yes'.

4. Every client is given a WOW experience and has a positive first impression. They have taken a process and systems led approach, which means that they can now replicate and scale what they do in a more profitable, consistent and time efficient way. This is vital for building a highly profitable business coaching, consulting or advisory business. The level of rework, the hidden cost that damages many business as we covered in Chapter 1, required in the initial diagnosis phase is now non-existent.

5. What they are all finding is that having up to 140 insights on each of their clients means that even if the client isn't ready to 'buy' immediately to progress to the next stage of support, they can now instead drip feed tailored solutions, by email and social media, to the 'problems' that they know these potential clients are experiencing. These are the problems which were identified as part of the diagnosis to build trust and rapport, and to position themselves as a trusted growth advisor, so that when the clients are ready

to 'buy' they come back to them first – the client now sees them as a proactive problem solver.

6. Also 140 insights on each client means potentially 140 up-sell or cross-sell opportunities for each client in providing solutions for the identified problems in a tailored way. If they don't have expertise in website optimisation for instance, they can now build up a panel of trusted providers that can do the bits they can't, they have agreed a revenue share with each of them, and many deliver the service under their umbrella or brand so that they maintain the primary relationship with the client always. The message for you from this is that whenever your clients have a problem, you want them to come to you first, before going to anyone else, always. The insights from the diagnostics are perfect for doing this.

7. Data is now power for each of these case study businesses. They can create aggregated insights from the diagnostic results to spot trends or opportunities that are common to more than one business in their client portfolio. This enables them to better tailor material and content to their target audience, whether it be through one to one delivery or in workshop or group based settings, and to look at refining or enhancing their client service ramp and offering ecosystem even further to deliver a first class service for their clients – one that encourages high retention but also a high referral rate because of the high client satisfaction levels.

Obviously, the three case study businesses I have shared with you have fully committed and invested in doing what it takes to really put their business on the right foundations to reap the benefits as they have. There is no magic wand solution to making this happen though so working through each of the ten key ingredients we have covered is the key to ultimate success.

For whatever reason, you may not feel ready or able to make all of the changes needed for implementing the key ingredients into your business. If that's the case then it's completely okay.

Ultimately, you need to do things in your own time and at your own pace. At the very least I encourage you to embrace the Profit Accelerator Formula and really embed the five key focus areas into how you grow your business. It will be transformational if you do, I promise you that.

Your mission should you choose to accept: the 25% improvement challenge

If you have at the very least done the exercises at the end of each chapter you will be well on the way to having your business baseline for improvement for the Profit Accelerator Formula. This is the starting point you need. The mission I challenge you with is to focus on improving each of those areas by 25%.

So what does a 25% improvement for each actually mean?

- If you are talking to 100 potential clients across the course of a year, put strategies in place to improve this by 25 to 125 – it's achievable.

- If you usually get 20 people in every 100 to say 'Yes' to becoming a client, work on getting an extra 5 in every 100 to say 'Yes' – it's achievable.

- If on average a client buys from you 4 times a year – work on getting them to buy from you 1 extra time each year – it's achievable.

- If on average a client spends £1,000 each time they buy from you, work on adding more value so that they now spend £1,250 each time – it's achievable.

- An finally, if £20 in every £100 is now profit – look to improve this by £5, so that £25 in every £100 is profit – it's achievable

That's what a 25% improvement would look like. All very achievable if you look at each of the five focus areas one by one and approach them as individual projects.

Let's look at the impact that these achievable improvements would have overall on your business.

BEFORE – The business baseline:

Number of leads per year	100
Conversion rate	20%
Number of transactions	4
Average sale value	£1,000
Margin	20%

Profit Accelerator Formula applied

Number of leads (1)	100
X	
Conversion rate (2)	20%
=	
NUMBER OF CLIENTS	**20**
X	
Number of transactions (3)	4
X	
Average sale value (4)	£1,000
=	
REVENUE	**£80,000**
X	
Margin (5)	20%
=	
PROFIT	**£16,000**

AFTER – The RESULT

Number of leads per year	100 to 125
Conversion rate	20% to 25%
Number of transactions	4.0 to 5.0
Average sale value	£1,000 to £1,250
Margin	20% to 25%

Profit Accelerator Formula applied

Number of leads (1)	125
X	
Conversion rate (2)	25%
=	
NUMBER OF CUSTOMERS	**31**
X	
Number of transactions (3)	5.0
X	
Average sale value (4)	£1,250
=	
REVENUE	**£193,750**
X	
Margin (5)	25%
=	
PROFIT	**£48,438**

An achievable 25% improvement in each of the key focus areas would mean the following:

- Your number of clients would increase by 11, from 20 to 31, which is a 55.0% improvement

- Total revenue would increase by £113,750, from £80,000 to £193,750, which is a massive 142.2% improvement. **That's a more than a doubling of your revenue!**

- Your net profit would increase by £32,438 from £16,000 to £48,438, which is a mind-blowing 202.7% improvement. **That's more than a tripling of your profit!**

Now if that isn't enough to get you excited and interested in the power of what we have covered so far, I don't know what will! If you love and take care of the right profit drivers in your business, your business will love and take care of you back!

So will you be bold, decisive and commit to the 25% improvement challenge?

We've even create a bonus download that contains 135 strategy suggestions for improvements you can make to achieve this, which you can access in the book resources at www.profitinfocus.com/profitable-professional.

Make your commitment declaration for accepting this challenge and doing this here NOW:

Your commitment declaration

I hereby declare that I am 100% committed to doing what it takes to improve my business profit and do accept the 25% challenge. I understand that there is no magic wand here, and it won't necessarily always be an easy journey, but I am definitely committed to dedicating the time, energy and resources needed to make this happen. I will be sure to keep the reason for doing this firmly in mind as inspiration for me to keep going no matter what. I recognise and truly understand that I am ultimately responsible for my success. I am excited to get started!

Signature: _____

Print name: _____

Date: _____

EXERCISE

Write down 5 key ideas, insights from the case studies or actions that this chapter has triggered for you to implement in your business.

1. _____

_____

_____

2. _____
   _____
   _____

3. _____
   _____
   _____

4. _____
   _____
   _____

5. _____
   _____
   _____

You have now seen the power of the 10 key ingredients in action, so are hopefully feeling excited about how you can do the same for your business as our case studies did for theirs.

In the final chapter, I'll outline a 12-step blueprint so that you can make this a reality, and also recap what we have covered on our journey together throughout the book.

CHAPTER 12

12-step blueprint and closing remarks

Well done on sticking with me this far. Hopefully you are feeling really excited about the potential opportunity that lies ahead in applying the 10 key ingredients in your business. However, you are probably wondering how to go about doing it, so I have put together a 12-step blueprint to help you get started.

STEP 1

Being clear on why you are doing what you do and defining exactly what you want from your business.

In life we generally do everything for a reason. It is so important that you understand the reason you are in business. It will be a source of motivation, energy and commitment when you encounter more difficult patches. You must be clear on what you are trying to achieve with your business and why it is important to you.

We have to make choices every single day in business; being clear on your 'why' will help focus your attention on making the right choices that support where you are heading and will help you to filter out the opportunities or distractions along the way.

I shared with you my 'why', which is '*I truly believe that there is tremendous power at the intersection of innovation, entrepreneurialism and profit numeracy*'. I use this as my guiding principle in whatever I do in my business. I only look at introducing products and services that are aligned with my 'why' and to what I am consciously 'designing' my business to be. Focus brings clarity and clarity brings better results.

STEP 2

Get your financial records in order

If your accounts are in a state of disarray, you must take action and get them cleaned up and sorted out. If you can't do it yourself then get a resource to help you. Remember, there are rules, laws and obligations around keeping up-to-date and accurate financial records for your business, which include filing all relevant returns accurately and on time. Failure to do so risks huge fines and maybe even imprisonment. It is a very serious issue that must be dealt with so you must stop being an ostrich and remove your head from the sand immediately.

Aside from this very important reason, we need to identify a firm base from which to move forward, and be able to work with this information to identify the relevant figures to use in the Profit Accelerator Formula. Any skeletons in the cupboard must be uncovered and dealt with. How can you improve something if you don't know what it is to start with?

STEP 3

Information gathering

Take your last set of annual accounts and break the results down into the five key focus areas of the Profit Accelerator Formula we have covered:

1. Number of leads
2. Conversion rate
3. Number of transactions
4. Average sale value
5. Margins

Here's a reminder of how to get this information:

Number of leads

You may already have a customer relationship management (CRM) system that you use to track this information, from which you can simply extract the information. It's okay if you don't, a simple solution for getting this is to start tracking all of your leads over a 90-day period. Keep track of it on a simple spreadsheet or even in a notebook. You are seeking an indication from this exercise, which will form a benchmark from which to improve. You need a starting point, though.

If you don't already have a system in place to capture this information on an on-going basis, there are many inexpensive off-the-shelf software packages that you can buy to help in this area. These software and Cloud solutions can have either simple or more advanced functionality. Find the right package that works for you and is aligned to the needs of your clients and business.

Asking a simple question like 'How did you hear about us?' can provide valuable insight into where best to target your activities moving forward.

Conversion rate

If you aren't currently tracking your conversion rate, you need to start doing so. It's okay if you're not, we can work backwards to identify what it is by looking at your **number of active clients in** a 12-month period and dividing that by your total number of leads identified in the first part of this step. Remember, the goal here is to have a starting point.

Number of transactions

This figure will usually come from your accounting system. It is basically the number of invoices you have produced and sent out in that year. Once you have this number you simply need to divide it by the total number of unique clients you have invoiced at least once in the same period.

Average sale value

This figure can be worked out by working backwards; also by taking the total revenue figure for the year and dividing it by the average number of transactions (i.e. number of invoices produced and sent out for your products or services for each unique client).

Margins

You calculate this by taking the net profit figure from the last set of annual accounts we are using for this exercise, which is usually at the bottom of the Profit and Loss statement, and dividing it by the total revenue as shown on the same report.

You will recall from what we have covered in Chapter 10 that margins represent the percentage of each and every sale that is profit. In other words, if you sold something for £100 and £25 was profit, then you have a 25% margin. Remember, though, that this is after **all costs** are taken out.

STEP 4

Your ideal client profile and defining your uniqueness

If you understand the complete profile of your target ideal client, including what they like, their buying habits and the problem for which you have a solution, then if you centre your marketing activities around reaching that profile of client they will be more inclined to buy from you, because your product or service will meet their identified need.

For this step, you need to at the very least answer the ten basic questions we covered for the exercise in Chapter 3 together with the 'Defining your uniqueness' exercise in the same chapter. The more detailed the better for this because every little extra bit of insight you can gain should improve your marketing results and success.

STEP 5

Your client service ramp and offering ecosystem

Rarely will a potential client that is not known to you spend a lot of money without first going through a discovery and familiarisation process to build confidence, trust and rapport. It's all about the client journey. This is about really knowing the wants, desires and needs of your ideal client and ensuring that the buying process (client service ramp) you take them through is completely aligned with it.

Complete the exercise in Chapter 4 for defining what the elements are for each of your four client service ramp levels, being:

Level 1 – Explore
Level 2 – Taster
Level 3 – Discovery
Level 4 – Core

STEP 6

Product or service profit check

You need to list all of the products and services you provide (and the ones that you are looking to introduce) and work through each one to determine whether you are making money or not on that product or service. The reason for doing this is that it makes absolutely no sense to promote and seek greater volumes of sales for something that makes you very little or no money.

Without addressing the profitability issue for that product or service first, you are only making the problem worse when greater volumes are involved.

To do this exercise, you need to compare the price you are charging against all of the costs associated with providing that product or service at a gross profit level. Go back and re-read

Chapter 5 if you need to remind yourself about the principles involved here.

The aim is to make sure that the revenue received for each of your products or services is more than the costs incurred, with sufficient left over at a gross profit level to contribute towards your fixed costs or overheads.

Remember, your gross profit represents the speed and ability of your business to cover its fixed costs and overheads and have something left over after all costs have been covered. The more gross profit you can generate, the quicker you can pay for your overheads and have something left over for you to enjoy, i.e. the net profit.

Get your house in order first on the profitability of your products and services before doing anything else. Make changes accordingly, based on the findings of this review. Again, if you don't feel confident enough to do this exercise yourself then get the help you need to complete the task.

STEP 7

Set the improvement targets

By this point you should have enough information to be able to identify all of the figures required for the Profit Accelerator Formula, which will form the foundation and business baseline from which you will benchmark all your future improvements.

You will also have looked carefully at the profitability of each of your products and services and made changes where appropriate and necessary to ensure you are maximising the gross profit amount for each of them.

You now need to set the targets for improvement for each key focus area. I have put together some tools to help you do this so it should make your life much easier. Simply go to my website

www.profitinfocus.com/profit-tools and follow the instructions to download them for free.

The simple tools I have created enable you to enter different percentage improvement amounts for each key focus area and it will show you what the target will be to achieve that result for each specific focus area. It also shows you the combined impact of all of the improvements. The other version you can access here allows you to enter absolute amounts for each of the focus areas if you know what you want your targets to be, and the tool will automatically calculate your revenue and profit target based on what you have entered. It will make more sense when you see them, so I suggest you download as soon as you can.

STEP 8

Selecting the strategies that will form your plan

In Chapters 6 through 10 you were asked to rank the five strategies presented for each area in order of priority from 1 to 5 based on the attention you would give them in your business. Take the top three strategies (the ones you ranked 1, 2, or 3) from each and look into each of those strategies more closely, research more if you need to, identify the resources you will need to carry out the strategies, and write down what you will need to do to make them happen.

The aim is to emerge with **at least one key initiative for** each area. Of course, you can have more than one initiative on the go at the same time – as some of the case studies did – if you feel you can handle it, but I would suggest for starters that you stick with one initiative per focus area for which you can carefully track and monitor the results.

The difficulty with having multiple strategies running at the same time, especially when it comes to marketing, is that unless a system is in place to carefully track the results, it can be difficult to know what has worked and what hasn't. It's like deciding to do a detox or change your diet; you normally make more than one change at the same time. When you get the results you are seeking, you are in the dark as to what worked and what didn't from the changes you made, as you only see the cumulative effect.

You need to be able to identify exactly what works and what doesn't so that you can do more of what is working and stop doing what isn't working.

STEP 9

Pulling your plan together

You should now have at least one strategy identified for each key focus area of the Profit Accelerator Formula and have a plan for how you will approach each of them. You will also have clear and tangible targets set in Step 7 to aim for and have a clear timeframe for when the targets should be achieved.

STEP 10

Margin focus first – cost review

Your business house needs to be in order before targeting any additional volume, because if you aren't coping and making sufficient levels of profit now, then the situation may not necessarily improve with additional volume – particularly if you are not delivering your products and services as efficiently and cost-effectively as possible.

Look at all of your overheads and identify ways to save money or restructure any processes so that they are done more efficiently. Negotiate better terms and prices with every supplier that you possibly can. Don't leave any stone unturned. Revisit the ideas in Chapter 10 for some added inspiration.

If you need to alter your prices, now is the time to make those changes before undertaking any of your new marketing activities or initiatives that encourage extra sales volume. The time spent will be worth it.

STEP 11
Take action

You may decide to focus on one area only, or some combination of the Profit Accelerator Formula areas. That is down to you, your confidence and how quickly you want to accelerate the growth of profits in your business.

STEP 12
Track everything and monitor it all closely

This really is key. How do you know if something is improving unless you are tracking it? How do you know you have arrived at the destination unless you know where you intended to go and monitored your progress along the way?

Set up whatever systems are required to track and monitor your results. Get the help you need if you can't do it yourself.

So there you have it, a 12-step blueprint that you can use to help get you started, as promised. Now, as we are reaching the closing stages of the book, I think it is worth reflecting on some of the key messages we have covered.

Being in business isn't always easy. Don't be hard on yourself

Remember, if being in business were an easy thing to do then everyone would be doing it. We can sometimes be our own harshest critic. Go easy on yourself and take some time out regularly to reflect on the successes you have had and will have in the future. Celebrate those successes as you would celebrate the successes of others. Be kind to yourself.

It's okay to make mistakes

We are all human and we sometimes make mistakes. That's okay. In fact, some would argue that we are not truly living unless we make a lot of mistakes. At least it means that we are trying new and unfamiliar things that push us out of our comfort zones. It's what we do after we make a mistake that sets us apart. If we learn from our mistakes and make improvements off the back of them, then that's called moving forward, but if we ignore the lessons and keep repeating the same mistakes, then that's called stupidity.

Balance between 'working in' and 'working on' the business

An important but necessary distinction needs to be made between 'working in' and 'working on' a business. 'Working in' roughly translates to the 'doing part' of the business, i.e. the delivery of the products or services to your clients. 'Working on' the business means focusing on activities and strategies to help move the business forward to its ultimate intended destination. You want to be 'working on' the business as much as possible and employing other resources where possible to do the 'working in' bit.

Overcoming the common challenges many face

Avoid FREE or unpaid consultations at all costs if you can unless there is a very deliberate reason to do them. Remember, with FREE consultations there is no financial commitment being made by the prospect. They have no vested interest or skin in the game. People tend to value things more when they pay something, however small, towards it. It shows that they are committed and can be used as a pre-qualifier in its own right.

It's important to consider that if somebody is not prepared to pay a relatively small amount to get access to your expertise, then you need to ask whether they would commit to the fees associated with a bigger piece of work from you. Applying the 'reverse risk' strategy as the case study businesses did is very effective.

Just as you expect your precious time to be respected, you must extend the same courtesy to your clients and prospects by keeping the initial client discovery to less than 20 minutes for best success. Ensuring you take a systematic and process driven approach to lead generation that adds value to all parties will ensure you have a constant stream of 'qualified' leads at your disposal. Embracing social selling and digital asset techniques plays a big part in this, especially in moving clients through each level of the client service ramp you should have now defined for your business.

A business by design, not by default

'Keep it real' and be congruent and authentic to your beliefs and ambitions always. It's not about doing what you think is expected of you, but building a business around doing what connects you with your passion and what you love every single day. It's about creating your own reality and taking decisive action to make it happen. Look at challenges as opportunities to grow and tackle those challenges we face with gusto. In the context of finding ways to always move you closer to your business vision, they can often result in even better outcomes. At the end of the day, your business is 'your game'. You're in charge, so you can set it up and play it by your own rules.

Niching for greater success

Don't be a generalist. Niching successfully means avoiding spreading yourself too thin. It's becomes easier to identify and target potential clients and partners to work with and also to become an expert and well known in your niche. It encourages more and better referrals, and since you are more unique there will likely be less competition – marketing therefore becomes even easier.

Knowing your ideal client is vital

Understanding the complete profile of your target ideal client, including what they like, their buying habits and the problem for which you have a solution, is key. If you centre your marketing activities around reaching that profile of client, they will be more inclined to buy from you because your product or service meets their identified need.

Instead of being a cold prospect when they are contacted, they are more likely to be a warm prospect and therefore more inclined to be interested in what you are offering. Being more precise on who you market to and more deliberate with your messaging to them will translate into better business results.

The rules have changed

Unless you have been living in a cave somewhere there's no avoiding the fact that the advent of social media has changed the landscape of business. Love it or hate it social media is here to stay, as we have moved permanently to a more online way of operating. The old ways of approaching things have changed and you need to adapt and evolve how you attract and interact with either potential clients or existing clients in order to stay relevant to the changing the ways they consume, absorb and digest the volumes of information that are now at their fingertips. It's very easy for you to be lost or overlooked by them in all the online 'noise'. Remember, you are who Google says you are, so make sure that message is the right one.

It's all about the client journey

Rarely will a potential client that is not known to you spend a lot of money without first going through a discovery and familiarisation process to build confidence, trust and rapport. It's all about the client journey and the touch points they have with you along the way. Remember, the Google ZMOT study suggests that it now takes on average 10.4 touch points or interactions before a person will buy from you. This is about really knowing the wants, desires and needs of your ideal client and ensuring that the buying process (your client service ramp and offering ecosystem) and the touch points you create as part of that are completely aligned with the way they interact and buy.

Numbers aren't scary

Numbers aren't scary mythical monsters; they can't hurt you physically but they do have tremendous power when harnessed properly. The numbers side of a business is one of the most important elements, and if ignored it is important to recognise and acknowledge that there will be consequences.

Just like when you learned to ride a bike, at first you didn't know how to do it. At the time, it was a foreign concept to you, but you knew that you wanted to learn and were determined to do so. There is no difference between that experience and learning more about the numbers side of your business. It all comes down to your desire and willingness to take action.

You need to start somewhere and get help from the many resources available to take you from the training wheel stage through to the confident bike rider stage. We can't expect to be good at something instantly, and as with any new skill we need to practice and practice some more. That is the only way we will get better at something.

Getting the price right

Profit is the aim of the game for most businesses. Profitable pricing is one of the most important elements in ensuring this. There is a very important distinction between revenue and profitable revenue, which you must understand.

The reality is that simply charging a certain price for your product or service is not a guarantee that you are making a profit on it. You must factor in all of the costs both fixed and variable to ensure that the revenue you are generating more than covers all of your costs and leaves you with a healthy profit. It's about being very clear on your pricing strategy and your positioning within your market.

Your attitude and mindset are very important

We all have a belief system that is individual and unique to us. It shapes our take on the world and all the various components that make it up. Our belief system is a collection of thoughts and ideas that we believe to be true which have been accumulated through what we have been told, taught and experienced during our lives.

Some of our beliefs work for us and some work against us. The game is to encourage the development of behaviours and attitudes in terms of mindset and beliefs that work for us, and to tackle and overcome those that work against us. It is being aware of what currently limits us and forms potential obstacles to our desired success that's important.

It's all about having a greater awareness of yourself. That means being aware of your tendencies to react in a certain way when it comes to money. Faced with this awareness, how you react then becomes a conscious choice. You are in a position to make an informed response on how to balance your behaviour and likely reaction accordingly. Remember, evolution not revolution. Rome wasn't built in a day – take things one step at a time.

Marketing is an investment

The exact same principles that apply to investing apply to marketing. You need to have a balanced portfolio of marketing activities within your business so that you are not reliant on just one marketing activity. You need to know what is working and what isn't by constantly monitoring the results through testing and measuring.

You want to stop the marketing activities that aren't working and do more of the marketing activities that are working. People and markets evolve, so something that is working one month may not necessarily work as effectively in another month. Having a portfolio of things that you do helps balance out this impact.

The distinction between conscious marketing and unconscious marketing is important. Conscious marketing is deliberate targeted marketing and should be seen as an investment. Like any investment, we expect to receive a return that is acceptable to us. Unconscious marketing that is not deliberate, not targeted and not measured is actually a cost to your business and should be stopped immediately.

Small improvements can lead to dramatic results

You were introduced to the Profit Accelerator Formula and I shared with you many ideas to help stimulate your thinking about each of the five key focus areas which are: number of leads, conversion rate, number of transactions, average sale value and margins.

I covered this area in a lot of detail so won't go over it again. The important message for you is that small improvements in the specific key focus areas I shared with you can absolutely transform the level of profits your business is generating. The power is in focusing on the right detail.

I have demonstrated to you how many small achievable changes to these areas can double or triple your profits, and have provided you with a 12-step blueprint to help you move forward and put what you have learned into action.

Here's a reminder of the 10 key ingredients that we have covered that are vital for building a highly profitable business coaching, consulting and advisory business.

1. Doing things differently
2. A business by design not by default
3. Crystal clarity on your target market and positioning
4. Your client service ramp and offering ecosystem
5. Pricing profitably
6. Consistent lead generation
7. Conversion success
8. Repeat business
9. Increasing client spend
10. Margin improvement

Speed versus velocity

I want you to think about the distinction between speed and velocity for a moment. What do both words mean? What is the difference between their meanings?

Sure, both are related to how fast something is moving, but there is a subtle yet important difference that is very relevant to the choices you make every day in your business.

Speed is defined as 'how fast something is moving', whilst velocity is defined as 'how fast something is moving in a given direction'. Let's think about what this really means. Take a hamster wheel for example – what does a hamster typically do? A hamster runs like crazy as the wheel turns, gets worn out in the process and stops because it can't go on any longer.

Having exerted all that energy, has that hamster actually gone anywhere? The answer is 'no'; it is still at exactly the same point as it was when it started, despite running like crazy as the wheel turned. The hamster has moved with speed, no doubt, but gone nowhere despite all its efforts. It could try to run faster or could slow down if it chose. In either case it wouldn't matter, because it would still remain in that same spot irrespective of the speed at which it moved.

It's time to be completely honest with yourself. Are you, or do you sometimes feel like, that proverbial hamster stuck in a hamster wheel? Is it symbolic of your business, running like crazy, busy doing things but not actually moving forward? Are you effectively treading water and growing wearier and more tired the longer you do it? Could not the same energy you are using in the wheel be transferred to activities that represent 'velocity'? This means refocusing your energies and attention to activities that move your business forward. That is what velocity is all about. It's about speed in a specific direction.

Every day you choose to focus your energies either on things that support your business moving forward or those that cause it to stand still. The more you focus your energies on activities that move your business forward, the greater velocity you will achieve.

So we reach the end of our journey together. You now find yourself at an important intersection. You have to make a choice. There are only two possible paths to take. On the signpost in front of you, the one pointing left reads 'doing what you've always done' and the other, 'evolution not revolution'. The first path represents your choice to continue making excuses for not doing what is needed to build the business you want. It represents doing what you have always done and getting the same results, good or bad.

The difference between these two paths is represented by the distinction between speed and velocity. The path to the left is the choice to be that proverbial hamster running like crazy in its little wheel while actually getting nowhere, because what you don't realise at this point as you look up at the signpost is that this path is circular and will eventually lead you back to this very intersection.

The other path, to your right, involves seeking help, having the real desire to improve your understanding and investing in what is needed for building a highly profitable business coaching, consulting or advisory business that is perfectly aligned with you and what you want for your life, family and loved ones. It's about recognising your current limitations and doing something positive about it. It represents uncharted territory for you, but you know that you have done new and unfamiliar things in the past. You soon learned how to do them, and became better and better as you practised and your confidence grew. This is the path that represents velocity. It is using the same energy but focusing it on activities that keep your business moving forward.

Now, let's not dress it up to be anything it's not: you are faced with having to choose between continuing to do what you have always done, or stepping up to the next part of your business journey head-on and moving forward from there. I can't make that decision for you – nobody but you can. You must make a decision, though. So what will it be – speed or velocity? Doing what you have always done and getting the same results – good or bad – or taking proactive and decisive action towards making more profit today?

NEXT STEP: You've read the book now take the scorecard!

Introducing the **The Profitable Professional Scorecard**. You've read about the 10 key ingredients for building a highly profitable business, coaching, consulting or advisory business, now it's time to see how your business scores for each of the ingredients covered in it.

The Profitable Professional Scorecard will help you gauge how effective you are in implementing the underlying elements that support each of the ten key ingredients for building a congruent, aligned and financially successful business. It is quick and easy to complete, taking less than five minutes to answer the questions and provides you with an 18-page detailed tailored recommendations, feedback and key actions report.

See more about the scorecard here:
www.profitinfocus.com/scorecard

About the Author

Kelly is a qualified accountant and profit specialist with over fifteen years' working experience, predominantly in the UK, with the latter years at Finance Director level. His experience extends across industries and businesses of varying sizes – ranging from smaller enterprises through to the £3 billion fund of a listed company.

Kelly is really passionate about helping business owners and entrepreneurs to be as profitable as they can be, so he has taken all of the experiences and the things he has learned throughout his career to date to write this book. He resigned from his well-paid job as a Finance Director – in the middle of arguably the worst recession in living memory – to set up his company Profit in Focus so that he could provide the vital tools, insights and resources that business owners need to transform the profits of their business.

Profit in Focus is a real, honest and genuine reflection of his open attitude to business, which is one of professional integrity, commitment and always seeking out collaborative opportunities for mutual benefit. Kelly is known for helping businesses to profitably THRIVE and his mission is to leave his clients and the people he helps feeling INVINCIBLE.

Connect with Kelly:

Email: kelly.clifford@profitinfocus.co.uk
Linked In: http://www.linkedin.com/in/kellymclifford
Twitter: @kellyclifford_
Website: http://www.profitinfocus.com

Help spread the word

If you have enjoyed reading the book, then please tell others about it. More details on how you can spread the word are on the website so be sure to check out how you can help. You will also find on the website instructions for accessing the bonus tools mentioned in the book

Visit **www.profitinfocus.com/profitable-professional**

Profit Secrets Webinar

'Learn how to better "qualify" potential clients to boost your profits.'

You're invited to join our FREE Profit Secrets Webinar and discover how to send your profit soaring!

On this webinar, you'll

- Discover the 6 common challenges that many businesses face when it comes to attracting new business clients.
- Learn why 'qualified' lead generation is where your focus should be and how it is key to your profitability.
- See ways to help eliminate time wasted on 'free' consultations or on unpaid consulting/advising.
- Get 7 profit boosting strategies that you can implement immediately.
- Discover a way to quickly and cost effectively 'qualify' your prospects to generate more revenue and boost your bottom-line profit.

This webinar is for ANY business coach, consultant, advisor or accountant who wants to grow their profits. It's for busy business support professionals who battle with the curse of FREE or unpaid consulting/advising and want to discover a better way of working to ensure a constant flow of 'qualified' leads for their business. It's for hard-working business people that want to take action and make bigger, faster and easier profits than ever before.

Join us, and dozens of other business support professionals, for this FREE 45 minute webinar that will unlock the secrets behind boosting your profits.

See more here and RESERVE your spot – before all the places are gone: www.profitinfocus.com/growth-webinar

Business Growth Diagnostic Service

At Profit in Focus we help business support professionals (including business coaches, consultants, accountants and advisors) to ATTRACT, WIN and RETAIN more profitable business clients through our unique white-labelled business 'growth' diagnostic and reporting service that helps their business to differentiate itself from their competition, provides a unique 'growth' focused point of difference that is perfect for 'qualified' lead generation.

It results in up to 140 clients insights being gained in less than 20 minutes of a client's time so they can better support the 'growth' needs of their business clients. This enables them to reduce levels of profit-draining non-chargeable time and to easily identify new revenue opportunities to grow their business in a more profitable and time-efficient way.

See more at **www.profitinfocus.com/growth**

The four included business growth diagnostics that all of the case studies in this book used with success are:

Business Health Check Indicator

Is your business profit friendly? This *Business Health Check Indicator* will help you gauge how effectively you are managing seven key areas that are important in building an enduring, profitable and successful business with 20-page detailed tailored recommendations, feedback and key actions report.

The *Business Health Check Indicator* looks at:

- Business strategy and direction
- Sales and marketing
- Team effectiveness
- Financial management
- Systems and processes

- Business protection
- Personal satisfaction

Total questions: 35

Brand and Marketing Impact Indicator

Are you making the biggest splash possible in your market? The *Brand and Marketing Impact Indicator* will help you gauge how effective you are across seven key areas that are important for making an impact in your marketplace. Includes a 20-page detailed tailored recommendations, feedback and key actions report.

The *Brand and Marketing Impact Indicator* looks at:

- Brand
- Market
- Messages
- Results
- Collateral
- Data
- Website

Total questions: 35

Financial Awareness Indicator

How well do you know your business numbers? The *Financial Awareness Indicator* will help you gauge how aware you are of and how good you are at managing the seven key areas that are important in building a financially successful business with a 20-page detailed tailored recommendations, feedback and key actions report.

The *Financial Awareness Indicator* looks at:

- Money in
- Money out

- Funding
- Pricing
- Trends
- Planning
- Performance

Total questions: 45

Sales Efficiency Indicator

How efficiently are you managing your sales? The *Sales Efficiency Indicator* will help you gauge how efficiently you are managing five key areas that are important in maximizing the sales opportunities in your business. Includes a 16-page detailed tailored recommendations, feedback and key actions report.

The *Sales Efficiency Indicator* looks at:

- Sales strategy
- Customer needs
- Measurement
- Sales process
- Opportunity management

Total questions: 25

Your Next Step

To see more about the service itself, download the interactive service brochure directly here that explains how it works, gives samples of the reports, describes the key benefits and outlines your options for getting started: **www.profitinfocus.com/growth-brochure/**

Profit Learning Portal

Introducing the Profit Learning Portal

This is an online portal where you will learn the foundational elements of finance and profit that are vital for building a profitable, successful and enduring business.

This comprehensive portal contains seven key modules:

1. Understanding your balance sheet
2. Understanding your profit and loss statement
3. Managing your cash flow
4. Profitable pricing
5. Break-even analysis
6. Business planning
7. Budgets

Each module is supported by video content (37 videos across all 7 modules) and includes a comprehensive workbook with case studies, worked solutions and a CD of various tools and templates to help you in your business. This toolkit enables you to learn and understand these areas from the comfort of your own home or office. It's a resource you can refer to time and time again.

In total the portal contains over 35 downloadable tools and templates to help you in your business. The combined value of the provided tools is more than £1,000 (approx. $1,400) alone!

SPECIAL READER OFFER

To say thank you for reading my book and to provide you with a real incentive to keep up the forward moving momentum, I am offering you as a valued reader the limited opportunity to get GOLD LEVEL access for SILVER LEVEL price – saving you US$140 or approximately £100.

See the below website for details:
www.profitlearningportal.com/profitable-professional/

Profit Pod

Profit Pod is an online profit improvement ecosystem that helps small businesses that typically have between 0 and 9 employees, annual revenue less than £2M ($3M) and have been trading for at least 1 year but perhaps who aren't as confident on the 'numbers side' of their business as they should be, to make numbers simpler for them so they can grow their profits, or improve their losses, by at least 20% over 12 months through a proven 5-stage process with GUARANTEED results. It's FREE to get started at: **www.profitpod.com**

See the short 3-min animated **About** intro video here: https://profitpod.com/aboutus

See the short 3-min animated **How It Works** video here: https://profitpod.com/howitworks

The Profit Pod advantage:

- **It'll save time** – takes up just ONE HOUR of your time per month

- **It'll boost your bank balance** – GUARANTEED to increase your profitability by AT LEAST 20% over twelve months

- **It's easy to implement** – guides you through a step-by-step process, HOLDING YOUR HAND every inch of the way

- **It's focused on YOUR business** – pinpoints 5 KEY AREAS of your business with PROVEN STRATEGIES and RESULTS MONITORING to improve your profit performance

- **It provides complete accountability** – maintains momentum to make sure you TAKE ACTION when it is asked of you

Profit Pod features 5 key modules designed to take your business from its current state through to increased profitability in just 12 months. The modules are:

1. Reality check
2. Preparation

3. Launch pad
4. Acceleration
5. Outcome

Let's look at each of them one by one:

Stage 1 – REALITY CHECK

Reality check looks objectively at your business to pinpoint which areas need attention. It highlights issues in four distinct areas including Business Health, Brand and Marketing Impact, Financial Awareness and Sales Efficiency, providing comprehensive feedback and recommendations for action. This suite of reality check indicator tests is really simple to complete, typically taking less than 5 minutes per indicator, but at the same time providing valuable insights to pinpoint the areas of concern within your business that need focused attention. The reports you will receive here contain comprehensive feedback, strategies, recommendations and key actions related to your particular business.

Stage 2 – PREPARATION

This stage is all about getting your financial house in order. We provide you with many checklists, intuitive on-screen tools and other guidance on what you need to do to get your business ready for profit acceleration. The Preparation stage is a seven step process where you will also learn about the key finance and profit fundamentals that will be pivotal for your future success.

Stage 3 – LAUNCH PAD

Launch Pad uses the information gained in the Preparation Stage to create your personalised Profit Acceleration Plan using the unique Profit Acceleration Formula, designed to guarantee at least a 20% increase in profit over a 12-month period. You'll get tried-and-tested profit improvement strategy suggestions combined with the monthly milestones needed to make sure you achieve them. The odds are that you will have never created a plan like this before!

Stage 4 – ACCELERATION

Acceleration puts your Launch Pad plan into action using achievable monthly milestones and reminders that will keep you on track, monitor your progress and maintain momentum. The results that you need to enter each month will be monitored on an intuitive and easy-to-use traffic light based performance dashboard, which includes a unique Profit Navigator feature to make sure you stay completely on track. You will be held totally accountable for your results.

Stage 5 – OUTCOME

This stage comes at the end of the 12-month period where we look at your final results and compare them to your 'Launch Pad' plan and where you started from. The celebrations happen here because if you have done everything asked of you according to the plan, then your profit will be 20% higher – or more!

And remember, the best bit is, getting started is ABSOLUTELY FREE! Visit **www.profitpod.com** today.

Lightning Source UK Ltd.
Milton Keynes UK
UKOW05f1027171216
290228UK00001B/49/P

9 781684 188635